Multifunctional Cement-Based Sensors for Intelligent Infrastructure

Multifunctional Cement-Based Sensors for Intelligent Infrastructure: Design, Fabrication and Application covers the development and use of cement-based sensors for monitoring structural health, durability, and environmental conditions in concrete infrastructure.

Monitoring the performance and condition of bridges, buildings, and roads improves safety and longevity while preventing failures and reducing maintenance costs. Cement-based sensors offer low cost, ease of installation, and compatibility with existing building materials, and can also provide real-time monitoring data to detect and diagnose potential issues before they become major problems. This book sets out the principles of the sensing mechanisms, fabrication techniques, and performance evaluation along with several case studies. It also provides a glimpse into a future where concrete structures will not only stand as pillars of strength but also become an indispensable part of smart cities as the core of automation.

The book suits researchers, engineers, and practitioners involved in design, construction, and maintenance of concrete buildings and infrastructure.

Wengui Li is a Scientia Associate Professor in the School of Civil and Environmental Engineering and the group leader of *Intelligent Concrete and Infrastructure Materials* in the Centre for Infrastructure Engineering and Safety (CIES) at The University of New South Wales (UNSW Sydney), Australia. He is the recipient of Australian Research Council (ARC) Future Fellow and ARC DECRA Fellow.

Wenkui Dong earned his PhD from the University of Technology Sydney, Australia. Currently, he works as Postdoctoral Research Fellow at the Institute of Construction Materials at Technische Universität Dresden, Germany.

Surendra P. Shah is a Presidential Distinguished Professor at the University of Texas at Arlington, Walter P. Murphy Professor (emeritus) at Northwestern University, and a member of the National Academy of Engineering, USA.

Multifunctional Cement-Based Sensors for Intelligent Infrastructure

Design, Fabrication and Application

Wengui Li, Wenkui Dong and Surendra P. Shah

CRC Press is an imprint of the
Taylor & Francis Group, an **informa** business

Cover image: Wengui Li, Wenkui Dong and Surendra P. Shah

First edition published 2025
by CRC Press
2385 NW Executive Center Drive, Suite 320, Boca Raton FL 33431

and by CRC Press
4 Park Square, Milton Park, Abingdon, Oxon, OX14 4RN

CRC Press is an imprint of Taylor & Francis Group, LLC

ISBN: 978-1-032-66284-8 (hbk)
ISBN: 978-1-032-66365-4 (pbk)
ISBN: 978-1-032-66368-5 (ebk)

DOI: 10.1201/9781032663685

Typeset in Sabon
by KnowledgeWorks Global Ltd.

Contents

Preface

Concrete, the bedrock of contemporary civil infrastructures, serves as the cornerstone upon which societies construct their aspirations. As we enter an era demanding sustainability, resilience, and innovation, the application of concrete materials is undergoing extensive and profound development opportunities. This book delves into the domain of multifunctional cement-based sensors (CBSs)—a pioneering approach that transforms how we perceive and interact with concrete infrastructures. The integration of advanced sensing technologies within concrete emerges as a transformative solution in response to the growing need for infrastructure efficiency and the automation of future smart cities.

The concept of CBSs originated in the early 1990s, traversing a developmental journey from initial popularity to relative obscurity and then regaining prominence in current research. The inspiration is rooted in a shared commitment to addressing the challenges posed by aging infrastructure, environmental considerations, and the ever-evolving expectations of the built environment. CBSs typically consist of common construction materials combined with well-dispersed conductive fillers. Due to the formation of conductive networks within the composite, these sensors acquire characteristics such as piezoelectricity, piezoresistivity, electromagnetic shielding, etc. This trajectory highlights the significance of collaboration, exploration, and the relentless pursuit of the intricate relationship between materials science, sensor technology, and infrastructure engineering. It is not only an integration of interdisciplinary theories and designs but also an exploration of the domain of multifunctional sensors, investigating their design principles, fabrication techniques, and application in various concrete infrastructures.

This book is primarily based on the authors' previous research work in the field of CBSs, incorporating perspectives from pioneers in the field as well as recent publications and viewpoints. The entire book is divided into nine chapters. Chapter 1 introduces the historical development and definition of CBSs, their significance, and a comparison with conventional sensors. Chapter 2 mainly summarizes the essential materials for the production of CBSs, followed by Chapter 3 summarizing their fabrication, geometry, and layout. Chapter 4 systematically introduces the methods for signal reception and processing. Chapters 5 and 6 present the mechanical properties and durability of CBSs, mechanical self-sensing, and the impact of environmental actions on the self-sensing performances. Chapter 7 discusses CBSs with integrated multifunctional applications such as self-healing, self-heating, self-powering, energy harvesting, superhydrophobicity, etc. Chapter 8 reviews the potential application of CBSs in various civil infrastructures, ranging from pavements, buildings, bridges, and high-speed railways to oil wells and tunnels. Finally, Chapter 9 presents the challenges, prospects, and future developments of CBSs.

As you navigate through the pages of this book, we invite you to embark on a fascinating journey through the landscape where concrete materials meet sensor technology. It provides a glimpse into a future where concrete structures will not only stand as pillars of strength but also become an indispensable part of smart cities as the core of automation. We anticipate that this book will have a profound impact on stakeholders, including researchers, engineers, students, the construction industry, manufacturers, and suppliers, contributing to advancements in the design, fabrication, and application of CBSs for intelligent concrete infrastructure.

Wengui Li
The University of New South Wales, Sydney, Australia

Wenkui Dong
Technische Universität Dresden, Dresden, Germany

Surendra P. Shah
Northwestern University, Evanston, USA

Wengui Li

Chapter 1

Introduction to cement-based sensors

1.1 DEFINITION AND SIGNIFICANCE

Cement-based sensors (CBSs) pertain to a category of sensing devices wherein cementitious materials function as the predominant sensing medium. Along with the advantages of being compatible and economical, CBSs, especially those made with conductive fibers as a conductor, are often multifunctional. CBSs operate based on the material's piezoresistive response, thereby necessitating concrete with enhanced electrical conductivity. To decrease the electrical resistivity of cementitious composites, conductive fillers in various sizes from microscales to nanoscales have been employed; conductive fillers used to modify the cementitious composites include metallic materials of steel fibers, nickel powders, iron powders, and the carbon materials of carbon black (CB), carbon fiber (CF), carbon nanofiber (CNF), carbon nanotube (CNT), graphene nanoplate (GNP), graphite plate (GP), etc. Based on the piezoresistivity, CBSs can sense strain or stress changes in concrete, as well as monitoring of cracks, fatigue, temperature, humidity, and corrosion by simply measuring the altered concrete electrical resistivity. It is worth noting that an alternative approach involves employing CBSs relying on capacitance properties (rather than piezoresistivity), avoiding the need for highly conductive concrete [1]. In that case, when subjected to mechanical, environmental, or structural variations, such as strain, temperature fluctuations, or moisture content, the capacitance of the cementitious material is altered. Nevertheless, the latter variant remains relatively underexplored, prompting this book to primarily concentrate on CBSs capitalizing on the piezoresistive effect.

The incorporation of CBSs within concrete presents numerous significant advantages and merits. These sensors furnish invaluable observations pertaining to the conduct and state of concrete structures, thereby facilitating heightened safety, enhanced operational efficacy, and prolonged serviceability. The following are some of the key significances of using CBSs in concrete:

1. *Research and innovation:* The application of CBSs into concrete structures unlocks novel way for research and innovation. With the continuous progression of nanotechnology and sensor technology, researchers can delve into pioneering applications and optimize the functionality of these sensors, thereby bolstering their versatility across various domains.
2. *Early detection of structures:* CBSs exhibit the capability to discern minute alterations or irregularities within concrete structures that may elude visual detection. Their early detection proficiency concerning issues like microcracks or corrosion

DOI: 10.1201/9781032663685-1

enables engineers to proactively tackle these concerns prior to escalation, resulting in diminished repair expenses and reduced public disturbances.

3. *Structural health monitoring:* CBSs enable continuous and real-time monitoring of the structural health of concrete infrastructure, such as pavements, bridges, buildings, and tunnels. By providing data on the condition of concrete structures over time, CBSs facilitate predictive maintenance strategies. This approach maximizes the lifespan of infrastructure and yields substantial reductions in overall maintenance expenditures.
4. *Traffic and weight-in-motion detection:* CBSs can be utilized for monitoring road traffic conditions, including vehicle and pedestrian flow monitoring, velocity, and weight-in-motion detection. It enables prompt maintenance and repair interventions of pavement, thereby mitigating the risk of potential catastrophic failures.
5. *Environment-friendly monitoring:* CBSs offer a more sustainable and eco-friendly approach to structural health monitoring compared to traditional methods, which usually involve invasive inspections and the use of harmful chemicals. Sensors are produced from conductive fillers and cementitious materials, which are environment-friendly and readily available.
6. *Cost-effectiveness:* CBSs provide a cost-effective solution for long-term monitoring of concrete structures. The raw materials for some conductive fillers such as CB, steel fiber, and cementitious materials are relatively low cost. Besides, once embedded during construction, they can function for extended periods with minimal maintenance requirements, providing a cost-effective way to monitor infrastructure over its entire lifespan.
7. *Adaptability in multiple environments:* CBSs exhibit robustness and can be tailored to endure adverse environmental conditions. Compared to traditional sensors, the cementitious matrix has higher resistance to temperature, corrosion, and watery environments.
8. *Data-driven decision-making:* The data acquired through CBSs empowers engineers and decision-makers to make well-informed choices regarding maintenance, repair, or retrofitting strategies. This data-driven approach augments the overall efficiency and effectiveness of infrastructure management.

1.2 BACKGROUND AND DEVELOPMENT

The background of CBSs is closely related to the application of nanotechnology in the field of construction materials. Nanomaterials with their unique optical, electrical, and thermal properties have brought about a qualitative leap for the construction materials in these specific aspects. Take electrical properties for example. Electrical properties of cementitious composites have a large application foreground in multifunctional construction structures and materials, including but not limited to the self-heating and self-healing properties [2], crack and corrosion detection [3], temperature and humidity observation, electromagnetic interference (EMI) shielding [4], and the piezoresistivity-based stress/strain monitoring as the CBSs [5]. The self-heating property originates from the Joule effect that the electric energy of power-supplied conductive cementitious composites transfers into heat energy to increase the temperature. Because of increased temperature, the self-healing properties of cementitious composites can be strongly improved, especially for the conductive asphalt concrete [6]. Since the inner crack and corrosion can alter the electrical resistivity of cementitious composites, crack and corrosion detection can be achieved in the case of previous calibration and adjustment [7]. Similarly, the temperature and humidity alterations of cementitious composites can be detected by the electrical resistivity changes, and then the results of environmental temperature and humidity

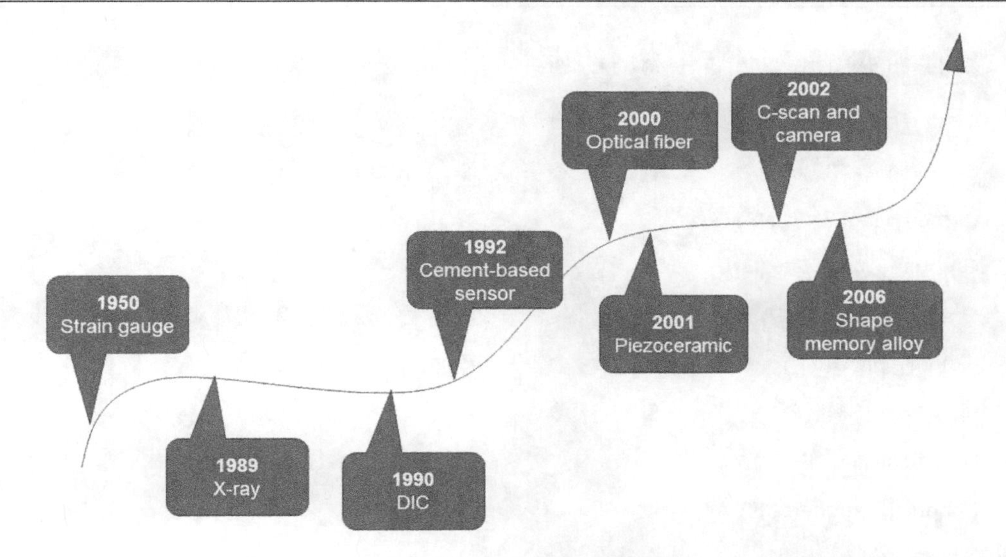

Figure 1.1 Number of publications on electrical and piezoresistive properties of conductive fillers–reinforced cementitious composites based on Scopus database up to 2023.

changes can be assessed [8]. Reflection and absorption losses of an electromagnetic wave are two main reasons responsible for EMI shielding, and better electrical conductivity can assure a rise in shielding effectiveness.

The investigations of electrical and piezoresistive properties of CBSs have especially attracted increasing attention. The piezoresistivity implies the correlation between electrical resistivity changes of cementitious composites and the imposed stress or deformation. Gauge factor represented by the fractional changes of resistivity under unit strain has been proposed to evaluate the sensing efficiency or piezoresistive sensitivity of conductive cementitious composites [9]. Recent studies have focused on incorporating different conductive fillers into CBSs to improve their sensitivity and performance. Fig. 1.1 shows the number of studies investigating the effects of carbon and metal materials on the electrical resistivity and piezoresistivity of cementitious composites based on the Scopus-indexed sources by 2020. It can be found that CNT is the most popular carbon-based filler in cementitious composites. Generally, the number of choosing fibrous materials (CNT, CF, steel fiber, and CNF) in CBSs is larger than sheet materials (GNP and GP) and far beyond the spherical CB, nickel powder, and slag. The reason might be the strongly improved electrical and mechanical properties and durability of cementitious composites filled with fibrous and sheet materials [10]. Moreover, researchers are increasingly enthusiastic about utilizing recently popular CNTs and graphene, among other new nanomaterials, to fabricate CBSs.

1.3 COMPARISON WITH TRADITIONAL SENSING TECHNOLOGIES

Successful adoption of a new technology for field concrete structure health monitoring (SHM) greatly depends upon the integration of the sensor function, accuracy, installation procedure, application conditions, and even costs. Fig. 1.2 lists the application timeline of different sensing techniques on SHM. In the early age of SHM, strain gauges are often developed into concrete structures for strain and deflection measurements. The application

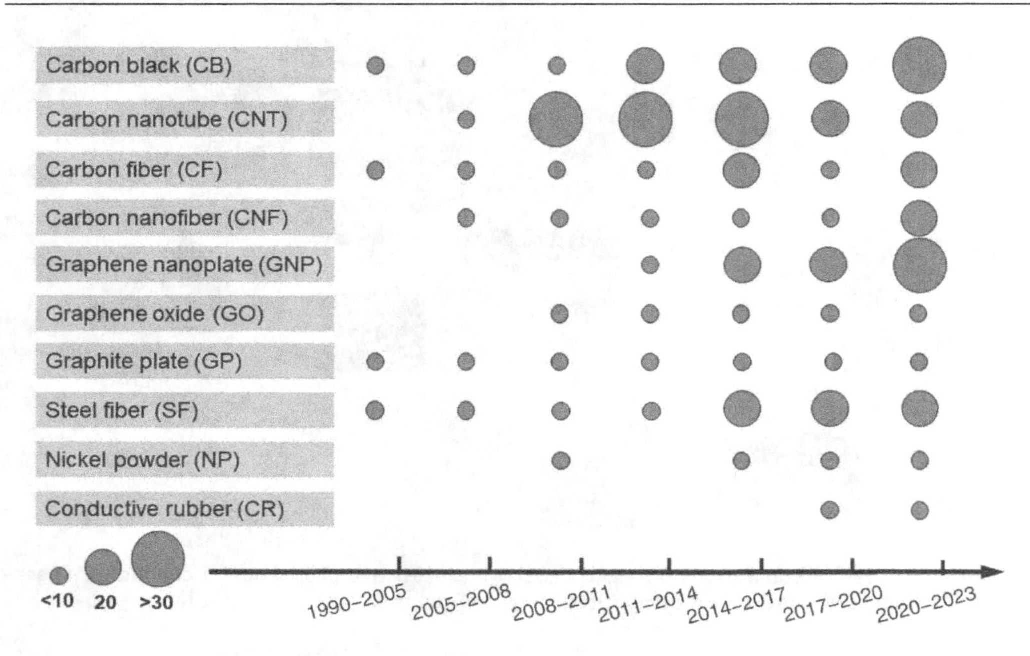

Figure 1.2 The application history of sensing techniques on SHM.

of strain gauges in construction site was testified by Backer et al. [11] with an adequate accuracy of stress measurements in comparison to the calculated stresses under a tensile force. Kovačič et al. [12] studied the signal collection and data processing of strain gauges for bridge loading tests in a practical project, and they obtained comparable results to those measured in a laboratory. However, even without economic considerations, the use of strain gauges is always limited due to the complexity of assembly, low sensitivity, thermal instability, and poor durability of the gauges. In contrast, optical fiber sensors have excellent performance, in terms of sensitivity and accuracy, which can be used to monitor microcracks in concrete through subtle alterations of intensity, wavelength, frequency, phase, and polarization of the received light signal. Leung et al. [13] first proposed a method for applying optical fiber sensor to monitor crack generation in concrete, which suggested that low content of optical fiber could monitor a great deal of emerged cracks. Furthermore, Butler et al. [14] took advantage of embedded optical fiber to evaluate stresses in prestressed concrete beams at an early age, through which the prestress losses could be detected accurately. Even with powerful sensitivity and accuracy, optical fiber sensors are still obstructed by their high cost and complicated installation in construction. In addition, the sensors can be easily damaged by either sharp components in matrixes (e.g., coarse aggregate) and/or improper operations during the embedding procedure.

Following optical fiber sensors are the applications of piezoelectric ceramics and shape memory alloy for the strain and stress observation of concrete structures. Song et al. [15] embedded ten piezoelectric ceramic patches in different cross sections of the concrete bent-cap for SHM, and they developed a damage index to evaluate the structure safety, which showed a close correlation with the observed damages. Su et al. [16] diagnosed the damage to hydraulic concrete structures by establishing a piezoelectric ceramic module, which was found especially effective for monitoring the hidden dynamic damages of hydraulic structures. Even with the foregoing advantages, it cannot be denied that these non-intrinsic embedded sensors typically have a difficulty of cohesion with their concrete substrate, thus having detrimental effects on mechanical properties and durability of the concrete elements. Another SHM technique is the nondestructive scanning method, including acoustic emission (AE), X-ray, or CT-scan. While detecting only surface cracks and deteriorations, this group of methods often requires a significant post-testing data process, which is time consuming. In addition, the accuracy of crack detection largely depends upon the resolution of scanning, which needs to be further studied, and, meanwhile, their costs are also often too high to be widely applied commercially. Digital image correlation (DIC) is a noncontact optical technique used for measuring deformation, displacement, and strain on a surface, which is an effective tool for the determination of deformation and fracture behavior of concrete structures through the examination of surface displacements [17]. In view of the above-mentioned sensing techniques, CBSs on account of the piezoresistive characteristics have been developed by some researchers to automatically monitor the deformation and cracks of concrete structures. Compared to the conventional sensors, CBSs are fully intrinsic and coupling well with the concrete components. They are seamlessly incorporated into the concrete matrix during the construction process. This integration ensures a more robust and long-lasting sensor system, as the sensor becomes an intrinsic component of the structure itself. Besides, the CBSs are much easier to install and manufacture with much less fabricating costs as well as have a good compatibility with cement matrices, better durability, and sensitivity when compared with other existing SHM techniques for concrete. To summarize the different sensing techniques, Table 1.1 provides a comparison of these different sensing techniques for SHM, including sensor installation methods, primary applications, and limitations [18].

Table 1.1 Comparison of different sensing techniques used in SHM

Sensing sensor and techniques	*Installation method*	*Features and applications*	*Limitations*	*References*
Strain gauge	Attached on the surface	Wide application; low costs around AUD6.0 each	Low sensitivity; worse durability; complex assembly; non-intrinsic	Neild et al. [11] and Backer et al. [19]
Optical fiber	Embedded	Accurate; high sensitivity; nonconductive; good durability	Dozens of AUD per meter; vulnerable; non-intrinsic	Leung et al. [13] and Butler et al. [14]
DIC	Noncontact	Wide application; nondestructive	Thousands of AUD each; extra analysis; non–real time	Golewski [20]
Piezoceramic	Embedded	Wide application; accurate; high sensitivity; good durability	Dozens of AUD each; non-intrinsic	Su et al. [16]
Shape memory alloy	Embedded	Accurate; high sensitivity; good damping property	Hundreds of AUD per meter; non-intrinsic	Mu et al. [21]
X-ray or CT-scan	Noncontact	Wide application; nondestructive	Thousands of AUD each; extra analysis; non–real-time; radiative	Suzuki et al. [22]
Camera	Noncontact	Convenient; easy operation	Hundreds of AUD each; vulnerable	Aghlara and Tahir [23]
CBSs	Embedded	Sensitivity; accurate; intrinsic; costs as low as cement; good durability	Need external voltage	Chen and Chung [24]

1.4 SUMMARY

CBSs represent a revolutionary advancement in the realm of smart construction materials and structural health monitoring. These sensors use cementitious materials with added nanotechnology to provide them with sensing capabilities. They are embedded directly within concrete during construction, thus becoming an integral part of the structure. By responding to various physical and chemical changes in their environment, they provide crucial data on the condition and behavior of concrete structures, leading to early identification of issues and the adoption of predictive maintenance protocols. Their distributed sensing capabilities and adaptability to harsh environments further enhance their utility in various applications, such as bridge monitoring, building safety assessment, and environmental studies. Overall, the advancement of CBSs has revolutionized our comprehension and administration of concrete infrastructure. By harnessing nanotechnology and exploiting the inherent properties of cementitious materials, these sensors offer a pathway toward safer, more resilient, and sustainable structures for the future.

REFERENCES

1. D. Chung, Y. Wang, Capacitance-based stress self-sensing in cement paste without requiring any admixture, Cement and Concrete Composites 94 (2018) 255–263.
2. H. Yang, H. Cui, W. Tang, Z. Li, N. Han, F. Xing, A critical review on research progress of graphene/cement based composites, Composites Part A: Applied Science and Manufacturing 102 (2017) 273–296.
3. H. Xiao, H. Li, J. Ou, Strain sensing properties of cement-based sensors embedded at various stress zones in a bending concrete beam, Sensors and Actuators A: Physical 167(2) (2011) 581–587.
4. D. Micheli, R. Pastore, A. Vricella, R.B. Morles, M. Marchetti, A. Delfini, F. Moglie, V.M. Primiani, Electromagnetic characterization and shielding effectiveness of concrete composite reinforced with carbon nanotubes in the mobile phones frequency band, Materials Science and Engineering: B 188 (2014) 119–129.
5. T. Schumacher, E.T. Thostenson, Development of structural carbon nanotube–based sensing composites for concrete structures, Journal of Intelligent Material Systems and Structures 25(11) (2014) 1331–1339.
6. Z. Wang, Q. Dai, X. Yang, Integrated computational–experimental approach for evaluating recovered fracture strength after induction healing of asphalt concrete beam samples, Construction and Building Materials 106 (2016) 700–710.
7. M.-J. Lim, H.K. Lee, I.-W. Nam, H.-K. Kim, Carbon nanotube/cement composites for crack monitoring of concrete structures, Composite Structures 180 (2017) 741–750.
8. W. Dong, W. Li, N. Lu, F. Qu, K. Vessalas, D. Sheng, Piezoresistive behaviours of cement-based sensor with carbon black subjected to various temperature and water content, Composites Part B: Engineering 178 (2019) 107488.
9. W. Li, W. Dong, Y. Guo, K. Wang, S.P. Shah, Advances in multifunctional cementitious composites with conductive carbon nanomaterials for smart infrastructure, Cement and Concrete Composites 128 (2022) 104454.
10. W. Dong, W. Li, X. Zhu, D. Sheng, S.P. Shah, Multifunctional cementitious composites with integrated self-sensing and hydrophobic capacities toward smart structural health monitoring, Cement and Concrete Composites 118 (2021) 103962.
11. H. De Backer, W. De Corte, P. Van Bogaert, A case study on strain gauge measurements on large post-tensioned concrete beams of a railway support structure, Insight-Non-Destructive Testing and Condition Monitoring 45(12) (2003) 822–826.
12. B. Kovačič, A. Štrukelj, N. Vatin, Processing of signals produced by strain gauges in testing measurements of the bridges, Procedia Engineering 117 (2015) 795–801.

13. C.K. Leung, N. Elvin, N. Olson, T.F. Morse, Y.-F. He, A novel distributed optical crack sensor for concrete structures, Engineering Fracture Mechanics 65(2-3) (2000) 133–148.
14. L.J. Butler, N. Gibbons, P. He, C. Middleton, M.Z. Elshafie, Evaluating the early-age behaviour of full-scale prestressed concrete beams using distributed and discrete fibre optic sensors, Construction and Building Materials 126 (2016) 894–912.
15. G. Song, H. Gu, Y. Mo, T. Hsu, H. Dhonde, Concrete structural health monitoring using embedded piezoceramic transducers, Smart Materials and Structures 16(4) (2007) 959.
16. H. Su, N. Zhang, H. Li, Concrete piezoceramic smart module pairs-based damage diagnosis of hydraulic structure, Composite Structures 183 (2018) 582–593.
17. D. Corr, M. Accardi, L. Graham-Brady, S. Shah, Digital image correlation analysis of interfacial debonding properties and fracture behavior in concrete, Engineering Fracture Mechanics 74(1) (2007) 109–121.
18. W. Dong, W. Li, Z. Tao, K. Wang, Piezoresistive properties of cement-based sensors: Review and perspective, Construction and Building Materials 203 (2019) 146–163.
19. S. Neild, M. Williams, P. McFadden, Development of a vibrating wire strain gauge for measuring small strains in concrete beams, Strain 41(1) (2005) 3–9.
20. G.L. Golewski, Measurement of fracture mechanics parameters of concrete containing fly ash thanks to use of Digital Image Correlation (DIC) method, Measurement 135 (2019) 96–105.
21. T. Mu, L. Liu, X. Lan, Y. Liu, J. Leng, Shape memory polymers for composites, Composites Science and Technology 160 (2018) 169–198.
22. T. Suzuki, T. Shiotani, M. Ohtsu, Evaluation of cracking damage in freeze-thawed concrete using acoustic emission and X-ray CT image, Construction and Building Materials 136 (2017) 619–626.
23. R. Aghlara, M.M. Tahir, Measurement of strain on concrete using an ordinary digital camera, Measurement 126 (2018) 398–404.
24. P.-W. Chen, D. Chung, Concrete as a new strain/stress sensor, Composites Part B: Engineering 27(1) (1996) 11–23.

Chapter 2

Materials for cement-based sensors

2.1 INTRODUCTION

CBSs are composite materials formed by combining a cementitious matrix with conductive fillers. The electrical resistivity of conventional cementitious composites is too high to be captured through the ordinary digital multimeter. To reduce the electrical resistivity of cementitious composites, conductive fillers in various sizes from micro- to nanoscales have been employed to modify the cementitious composites, including the metallic materials of steel fibers, nickel powders, iron powders, as well as the carbon materials of nano-carbon black (NCB), carbon fiber (CF), carbon nanofiber (CNF), carbon nanotube (CNT), graphene nanoplate (GNP) and graphite plate (GP), etc. [1]. Moreover, recycled metallic wastes and conductive rubber products were also considered sustainable conductive fillers of conductive cementitious composites. The effects of conductive fillers on the electrical conductivity and self-sensing properties of cementitious composites are complex because of their different shapes, sizes, aspect ratios, electrical conductivity, piezoresistivity, and dispersion difficulties in the cement matrix.

Supplementary cementitious materials (SCMs) are pozzolanic materials that are added to cement during the manufacturing process or used as a partial replacement of cement in concrete mixtures. Properly designed concrete mixtures with SCMs can result in more sustainable, durable, and high-performance CBSs. Therefore, it is essential to consider the specific type and dosage of SCMs used in CBSs, as the effects of SCMs on cement and concrete can exhibit variability depending on factors such as the specific type of SCM utilized, the type of cement employed, the water-to-binder (w/b) ratio in the mixture, and the prevailing environmental conditions.

This chapter offers an elaborate introduction on the constituent materials of CBSs, including matrix cementitious materials, SCMs, and an assortment of conductive fillers ranging from macroscales to nanoscales. Additionally, it explores the consequential impact of their combinations on the electrical conductivity and piezoresistive characteristics exhibited by CBSs.

2.2 MATRIX CEMENTITIOUS MATERIALS

2.2.1 Cement paste, mortar, and concrete

The cementitious materials of cement paste, mortar, and concrete are widely applied as the matrix materials for the production of CBSs. Normally the electrical conductivity of

 DOI: 10.1201/9781032663685-2

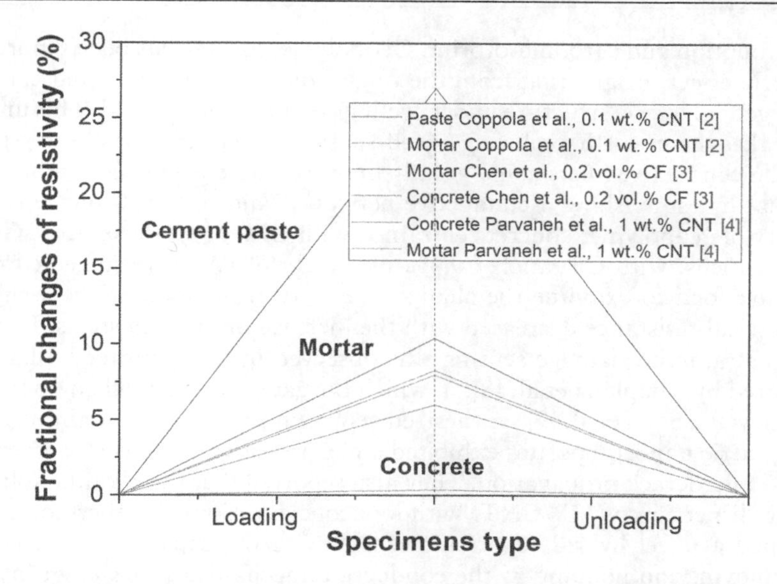

Figure 2.1 Fractional changes of resistivity in different cement matrices during loading and unloading.

cement paste is the highest, followed by cement mortar, and concrete has the lowest electrical conductivity due to the insulating coarse aggregates affecting the generation of conductive pathways. Cement mortar has an electrical resistivity of approximately 10^6–10^7 Ω·cm. Coppola et al. [2] conducted several experiments on CNT-reinforced cement paste and mortar, and compared their fractional changes of resistivity (FCR) under loadings. Chen and Chung [3] investigated CF on the electrical response of mortar and concrete, and compared their different fractional changes in resistivity. Parvaneh and Khiabani [4] proposed that the FCR for piezoresistivity-based concrete are still satisfactory enough as a strain sensor with an amount of CNT larger than 2% by the weight of cement. As depicted in Fig. 2.1, the FCR from these studies during the process of loading and unloading are presented and compared. Obviously, even mixed with different conductors and contents, the cement paste's FCR are times larger than mortar and dozens of times larger than that of concrete. This was also confirmed by García-Macías et al. [5] whose tests indicated that the resistivity changes of paste are dozens of times higher than those of mortar and concrete. Due to the existence of coarse aggregate, piezoresistive concrete sensors generally have high electrical resistivity (up to ~10^9 Ω·cm). Its application is limited but it still has attracted some research attempts. In summary, with the existence of aggregates, the sensitivity of CBSs is weakened, owing to the imperfect crack controlling capacity of conductive fibers under the interaction of aggregates. Meanwhile, aggregates tend to separate conductive passages by increasing the contact probability with the conductive phase, rather than boosting the contacts among conductors.

2.2.2 Alkali-activated materials (AAMs)

The utilization of AAMs and geopolymers in lieu of conventional cement presents numerous advantages, including substantial environmental and social benefits due to the opportunity for the utilization of waste streams, and the sustainability characteristics of low energy consumption and carbon footprint. Geopolymer mortar has been reported to have a piezoelectric effect, originating from the migration of mobile hydrated cations in the pores of the geopolymeric structure under cycling compression load [6]. This unique effect in AAMs makes them worth exploring as self-sensing materials. It has been reported that incorporating conductive fillers could greatly improve the self-sensing capability of structural material by establishing a conductive network. Vaidya and Allouche [7] explored the feasibility of geopolymer concrete entrained with CFs for strain sensing. Geopolymer concrete specimens, with CF ratio of 0.4% by weight of FA, were subjected to bending or compressive load to examine the change in electrical resistance. The results showed that the electrical resistance decreased with the increase in bending stress. Excellent performance in strain and damage sensing was observed in CF-reinforced alkali-activated pastes prepared by Vilaplana et al. [8], in which the gauge factor reached up to 661 under compressive cycles. Saafi et al. [9] synthesized geopolymer mortars containing MWCNTs. The geopolymeric nanocomposites exhibited a piezoresistive response together with high sensitivity to microcrack propagation. They also reported that the alkaline solution could facilitate the dispersion of MWCNTs with low concentrations. Furthermore, Saafi et al. [10] developed a novel hybrid superionic sensor with the capability to sense temperature, by employing ion hopping as the conduction mechanism. As shown in Fig. 2.2, a

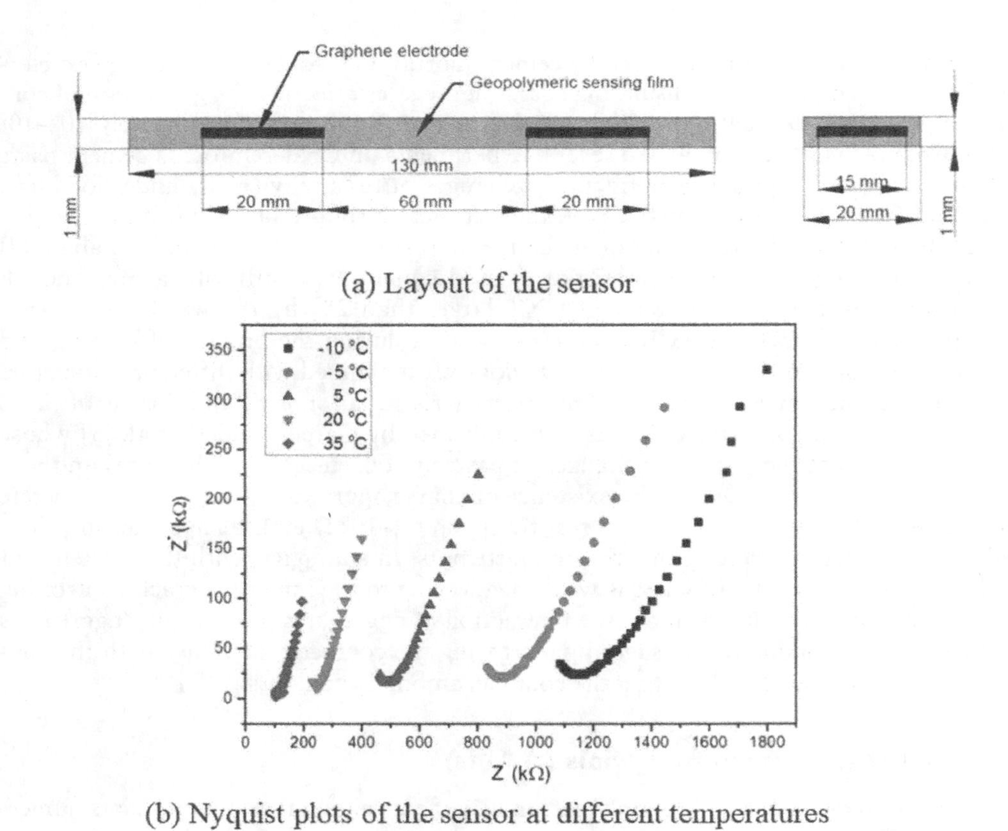

Figure 2.2 Graphene/geopolymer cement as a superionic conductor.

supersonic film made entirely from FA-based geopolymer and two embedded graphene electrodes constituted this sensor. It presents the Nyquist plots (Z′ and Z″ represent the real and imaginary components of impedance, respectively) for the sensor at different temperatures from 10°C to 35°C, indicating that the sensor can measure temperature with high sensitivity, as high as 21.5 kR/°C.

Table 2.1 presents the properties of self-sensing AAMs from the published studies, including the application of graphene and graphite. It was observed that the fractional change in electrical resistance of FA geopolymeric composites containing graphene exhibited linear response with high sensitivity to tensile and axial compressive stress [6]. It was suggested that the high strain sensitivity was due to the larger inter-contact area between the 2D graphene sheets which leads to a larger contact resistance change. As shown in Fig. 2.3, the strain measured with this developed material agreed well with the strain measured with the foil strain gauge [11]. When comparing the result with that of other existing self-sensing cementitious materials, it could be found that the CBSs produced by AAMs exhibit comparable or even higher sensitivity.

Table 2.1 Alkali-activated materials as matrix to produce CBSs

Group	*Conductive filler*	*Test parameter*	*Self-sensing parameters*	*Sensitivity*		*Reference*
				Guage factor	*Temperature sensing*	
Fly ash–based AAMs paste	GNP	Electrical resistance	Compressive strain; tensile strain	43.87 (under compression); 20.7 (under tension)	_	Saafi et al. [11]
Paste	GNP	Impedance and dielectric properties	Compressive strain; tensile strain; temperature	35 (under compression); 358 (under tension)	21.5 kΩ/°C (from −10°C to 35°C)	Saafi et al. [10]
Mortar	CNT	Impedance	Flexural strain; damage	15 (under flexure)	_	Saafi et al. [9]
Concrete	CF	Electrical resistance	Flexural strain; compressive strain	458 (under flexure)	_	Vaidya and Allouche [7]
Slag-based AAMs paste	CF	Electrical resistance	Compressive strain; damage	661 (under compression)	_	Vilaplana et al. [8]
Metakaolin paste	CNT	Electrical resistance	Compressive strain; flexural strain	663.3 (under compression); 724.6 (under flexure)	_	Bi et al. [12]
Other-based AAMs Potassium inorganic polymer	CNT/GNP	DC conductivity	Temperature	_	_	MacKenzie and Bolton [13]

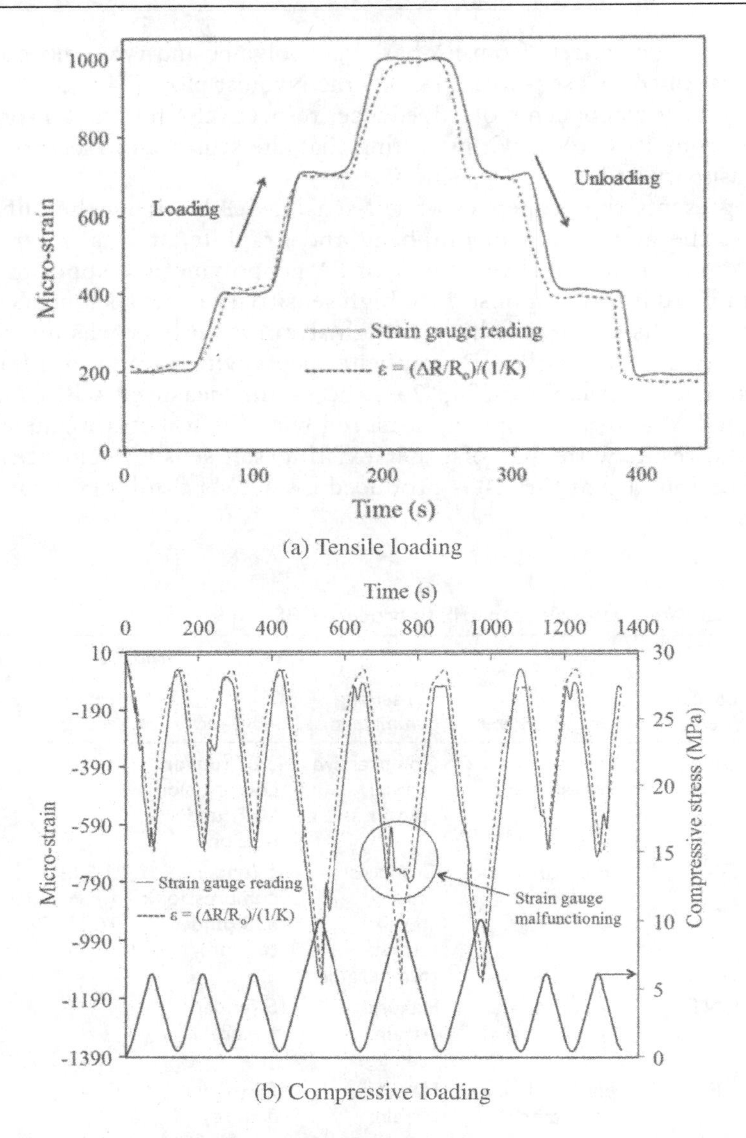

Figure 2.3 Response of graphene-modified geopolymeric composite under tension and compression cyclic stress.

2.2.3 Water-to-binder ratio

The w/b ratio influences both the slump flow and viscosity of cementitious materials, especially resulting in different dispersion efficiency of conductive fillers. Han et al. [14] investigated the piezoresistive sensitivity of composite with w/b ratios of 0.6 and 0.5, respectively, and concluded that the former is more sensitive to the electrical resistance changes under compression. Contrarily, Kim et al. [15] believed that lower water content benefits both fiber dispersion and piezoresistivity. As the w/b ratio decreased, the piezoresistive sensitivity of CNT-reinforced cement composite was improved and the

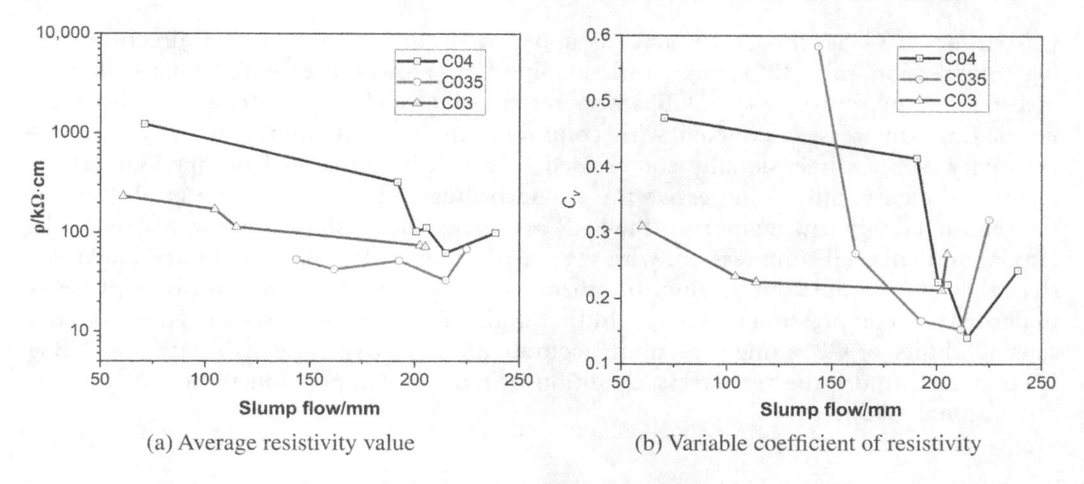

(a) Average resistivity value

(b) Variable coefficient of resistivity

Figure 2.4 Electrical properties of CNFs-filled cement paste with different slump flows and w/b ratios of 0.3, 0.35, and 0.4.

piezoresistivity has a close connection with rheological property; therefore, the controversy from the above studies may have come from the difference in composite rheology. By investigating the cement paste with the w/b ratios of 0.3, 0.35, and 0.4, Wang et al. [16] found that the slump flow of around 200 mm is most favorable for enhancing the dispersion of CNFs. As shown in Fig. 2.4, the optimal fluidity within a specific range is critical for promoting the effective dispersion of fillers and the formation of continuous conductive pathways. Both excessively high fluidity, leading to segregation, and excessively low fluidity, resulting in poor workability, hinder the uniform distribution of fillers and the establishment of interconnected conductive networks. In other words, in comparison to the impacts by w/b ratio, the slump flow and rheology might be the root responsible for the distribution of conductors, electron movements, and generation of the conductive paths in CBSs.

2.2.4 Poisson ratio

In application, CBSs embedded in concrete will go through different extents of Poisson effect, which might affect the piezoresistive responses. Under compression, the axial deformations of the embedded sensor and the concrete are assumed to be identical, while the transverse deformation of these two materials may be different. Due to the difference in Poisson ratios, the sensor and the examined concrete could generate interaction force in the interface. For silica fume (SF)-added CBS, it has been formulated by Sun et al. [17] and Xiao et al. [18] that the stress σ_i in surfaces of these two materials can be expressed as follows:

$$\sigma_i = (\gamma_1 - \gamma_2)\sigma \tag{2.1}$$

where σ_i is the stress in the transverse direction; γ_1 is the Poisson ratio of concrete; γ_2 is the Poisson ratio of the embedded sensor; and σ is the stress in the loading direction.

Gauge factor means the electrical resistance changes per unit strain of sensors. The Poisson effect under loading can cause this effect. Pioneering work has proved that for

CF-reinforced CBSs, the gauge factor is upper to –390 in the transverse direction during compression and –59 during tensile loading [19]. However, existing experiments for investigating piezoresistivity of CBSs were restricted to uniaxial compression rather than multiaxial compression coupled with confining stress. Thus, undoubtedly, the piezoresistivity of one-dimensionally compressed CBS will be different from its counterpart in three-dimensionally compressive forces. According to the tests by Wu et al. [20] on CF-reinforced self-sensing mortar, the FCR and gauge factor showed great differences in biaxial and uniaxial compression, with the gauge factor under biaxial compression larger than that of uniaxial compression. In other words, the piezoresistive sensitivity of CBSs under biaxial compression is superior to that under uniaxial compression. Therefore, the sensing ability of CBSs might be more accurate and sensitive if the calibration of CBSs is conducted under the real stress condition of triaxial compression rather under one-dimensional loading.

2.3 SCMs

2.3.1 Silica fume

SCMs are commonly used for mechanical and durability improvement of the cementitious composite, including the partial substitution of SF to cement in particular. It was confirmed that the SF had the capacity to promote the cement hydration process and the generation of calcium silicate hydrate through a reaction with calcium hydroxide, which is also called the pozzolanic reaction [21]. Hence, better performances on corrosion resistance, mechanical strengths, and durability were given to the SF–cementitious composite. Moreover, studies found that the substitution of SF to cement was beneficial to the dispersion of nanomaterials and diminish pore size because of its small particles and filling effect [22]. In terms of the impact of SF on the electrical conductivity/resistivity of plain cementitious composites, it seems that the electrical resistivity moderately increased with the increase of SF in the early hydration process, while a more substantial resistivity increase was observed afterward due to the severe pozzolanic reaction [23]. The investigation was also conducted on the pore solutions of SF-blended cement paste, it discovered similar results of the increased cement hydration because of the existence of nucleation sites by SF for crystallization, and the reduced alkaline metal ions to increase the electrical resistivity of cement paste [24]. Interestingly, probably owing to the nucleation effect, SF in the early hydration stage had nearly no response to alkalis which even enhanced the concentrations of alkaline metal ions and hydroxyl ions, thus highly likely to temporarily decrease the electrical resistivity of cementitious composite in very early hydration process [25]. Nevertheless, contradictory conclusions were drawn by Kim et al. for the significantly higher electrical conductivity for the SF–cementitious composite, especially before the first 14 curing days. Fig. 2.5 illustrates the electrical resistance alteration with the existence of SF [3, 26–28]. It is clear that for all investigations, the SF-modified composite has lower electrical resistance compared to the plain and reinforced composites. In general, as additives in cementitious composite are normally nonconductive and small-scale, their functions mainly result from the filling effect and minimizing the size of porosity to improve the piezoresistivity of sensors.

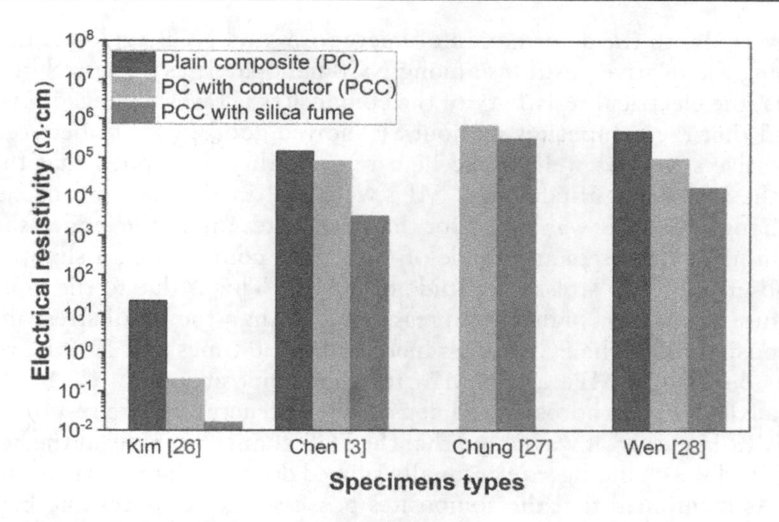

Figure 2.5 Effect of SF on the electrical resistivity of CBSs.

Fig. 2.6 illustrates the applied compressive stress and the FCR for the 2% NCB-filled cementitious composites with various contents of SF [29]. For the composites without SF, the FCR values were still consistent with the altered compressive stress, which decreased with the increase of stress and increased when the stress returned. Previous micro- and nanostructures of the NCB–cementitious composites without SF have shown their high porosity and severe NCB agglomerations. It seems that the applied stress is

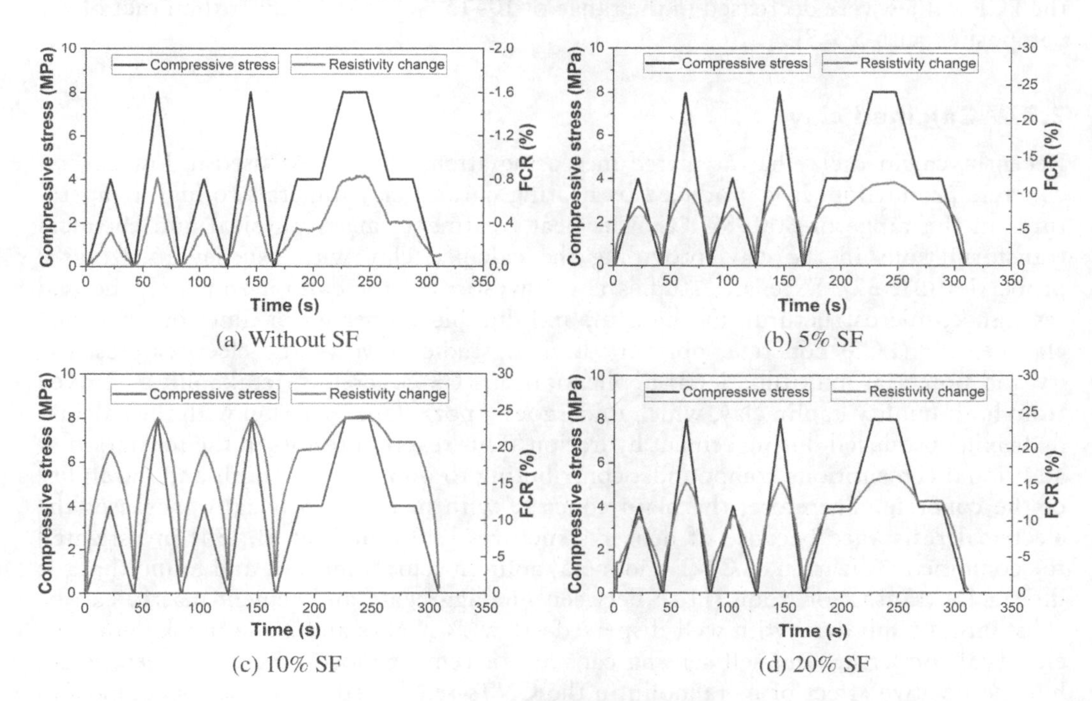

Figure 2.6 Piezoresistivity of CBSs with multiple contents of 0%, 5%, 10%, and 20% SF.

mainly converted into the deformation of micropores and NCB agglomerations, rather than changing the relative positions among NCB nanoparticles dispersed in the cement matrix; thus, the electrical resistivity of the composites is rarely changed. Even though, it was found that the composites without SF showed good linearity between the FCR and compressive stress, as well as excellent repeatability. It implied that the composites under the stress magnitudes of 8 MPa were still in the elastic range. Moreover, almost unchangeable FCR was found for the composites under the ladder-shaped loading maintaining at the stress amplitude of 4 MPa. In comparison, a slightly increased FCR was observed at the stress amplitude of 8 MPa. This is due to the more unstable microstructures caused by higher compressive stress than the small stress intensity for all the composites [30]. The FCR values increased by 20 times and 10-fold, respectively at the stress peaks of 4 MPa and 8 MPa for the composites with 5% SF. It could be deduced that the reduced porosity and denser microstructures were responsible for the increased FCR. However, it was found that the FCR changed rapidly at the beginning of the loading, and the changing rate gradually slowed down with the increase of compressive stress. As mentioned that the composites possessed fewer pores and better microstructures due to the SF filling effect, the small stress could lead to the altered positions of conductive NCB nanoparticles and result in the rapid resistivity changes. With the increase of compressive stress, the microstructures of composites became denser, and the porosity filled with SF started to be unstable, and that was why the FCR changes were reduced. In terms of the composites filled with 10% and 20% SF, similar FCR alterations could be found to that of the composites with 5% SF. The FCR values significantly increased and reached 20% and 25% for the composites with 10% SF at the stress amplitude of 4 and 8 MPa, respectively. However, for the SF dosage of 20%, the FCR values were decreased in the range of 10–15%, but still higher than that of the composites with 5% SF.

2.3.2 Calcined clay

Recently, calcined clay has attracted increasing attention as SCM used in low-carbon concrete production. It is produced by heating natural clay minerals to high temperatures in the range of 600–800°C. The heat treatment causes physical and chemical transformations in the clay, providing the calcined clay with excellent pozzolanic properties [31, 32]. Extensive studies have investigated the carbon emission, thermal resistance, microstructural, mechanical, and durable properties of limestone calcined clay cement (LC3) concrete, but very limited studies involve the electrical resistivity and potential multifunctional application of LC3 concrete. Metakaolin is derived from high-quality kaolin clay, which undergoes a pozzolanic reaction with the calcium hydroxide produced during cement hydration. This reaction results in the formation of additional cementitious compounds, contributing to improved strength and durability of the concrete. Therefore, the plain concrete with metakaolin usually increases the electrical resistance because of denser structures [33]. Singh et al. [34] investigated the combined utilization of CNT and metakaolin in cement mortar, and found that the interfacial transaction zoon (ITZ) between fine aggregate and cement matrix can be substantially improved with well-dispersed 0.1 wt.% CNTs and 20% metakaolin. The electrical properties and self-sensing capacity of cement mortar were not mentioned, but the positive effect of metakaolin in the CNTs-reinforced cementitious composites can be deduced.

2.3.3 Fly ash (FA)

Previous introductions in AAMs have mentioned the application of FA as precursor for the production of self-sensing geopolymer, hence, in this section we will only focus on the utilization of FA with cement in the CBSs. It has been reported that a high volume of FA (2.8 times of cement weight) in the fiber-reinforced cementitious composites will not greatly influence the piezoresistive performance of CBSs with CNF [35]. Deng and Li investigated the effect of FA content on the self-sensing performance of ECC mixture, and they found that with the FA-to-cement ratio elevated from 1.5 to 3.0, there was an observed augmentation in the number of cracks along the length of the specimen, concomitant with a reduction in crack width, thereby imparting advantageous effects on the durability and self-sensing behavior [36]. On that basis, the application of NCB with FA and cement further decreased the electrical resistivity of composites and increased the sensitivity to first cracking [37]. Konkanov et al. [38] substituted the cement by FA from 0, 5%, 10%, 15%, 20%, and 25% and investigated the effect of FA on the mechanical, electrical, and piezoresistive performances of cementitious materials. As listed in Table 2.2, the addition of FA does not result in a linear impact on sensitivity with both increases and decreases. Compared to the FA content, it seems like the enhanced workability and microstructures induced by fine and spherical FA particles are more critical to the self-sensing performance of CBSs.

2.3.4 Blast furnace slag

The incorporation of blast furnace slag in CBSs has been employed as a means of partial cement and fine aggregate replacement. For the cement replacement, the addition of slag in cement promotes the formation of hydration products, exhibiting improved performance compared to plain cement paste. However, when a high replacement level of GGBFS (30%) was used instead of cement, a noticeable retardation or slowdown of the hydration process was observed during the early-age curing [39]. Konkanov et al. [38] found that the addition of slag to OPC mortar resulted in a notable reduction of electrical resistivity, amounting to a threefold decrease by the end of the curing period. This effect can be attributed to the relatively elevated concentration of metal oxides present in these by-products [40]. On the contrary, Wang and Wang [16] found the reduced self-sensing capacity of the CNFs-reinforced cement paste containing 20% slag by weight of binder. The observed distinction can be attributed to the variation in conductive fillers utilized. In the former case, the absence of any conductive filler resulted in the predominant contributions of ion conduction and metal oxide conduction. Conversely, in the latter case, the incorporation of highly conductive CNFs led to the inhibition of self-sensing characteristics due to the presence of introduced slag. In addition, the steel slag powder has been used as mechanical filler of ultrahigh-performance fiber-reinforced concrete by Lee et al. [41], and they observed an increased tensile strength and satisfactory stress self-sensing performance with the incorporated amount of slag from 5% to 10% by weight of quartz powder.

Table 2.2 Stress sensitivity for the CBSs with different contents of FA

Specimen	*OPC*	*5% FA*	*10% FA*	*15% FA*	*20% FA*	*25% FA*
Stress sensitivity (MPa^{-1}) × 10^3	4.34	6.25	7.17	0.85	8.41	3.13

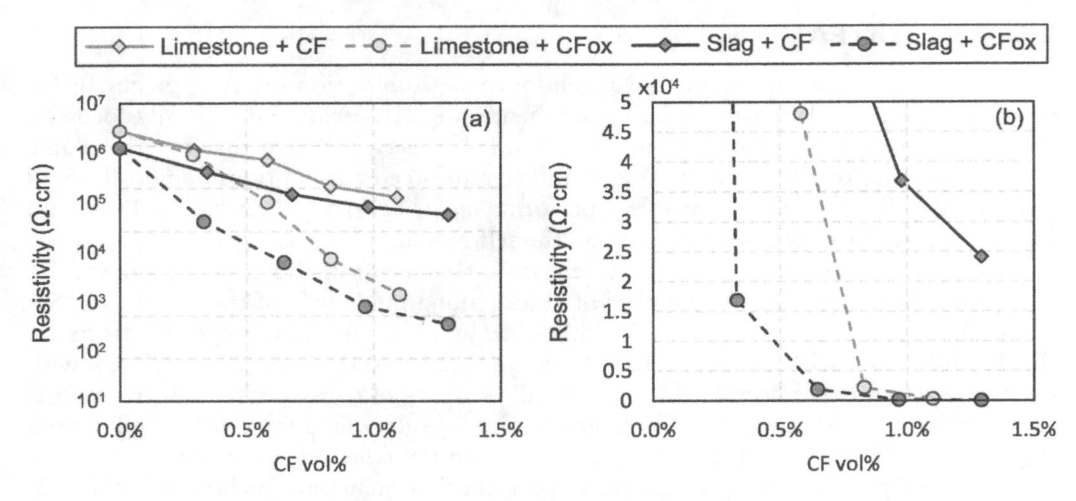

Figure 2.7 Electrical resistivity of equivalent dosage for the CF-filled cement composite with normal limestone aggregate or steel slag aggregate [42]. (a) Full scale. (b) Magnification of percolation threshold.

For aggregate replacement, copper slag and steel slag are by-products of copper/steel extraction and refining processes. Baeza et al. [42] incorporated electric arc furnace steel slags as fine aggregates in CF-reinforced cement composite, and they observed that the addition of steel slags resulted in enhanced compressive strength and electrical conductivity of the composites. As displayed in Fig. 2.7, this improvement led to the percolation threshold being achieved with lower CF contents. Similarly, it has been reported that 100% copper slag replacement of fine aggregate in cement mortar led to improved compressive strength and electrical conductivity [43]. Gulisano et al. [44] conducted an experiment where they incorporated slag aggregate along with graphene into the asphalt mixture. The resulting composite demonstrated remarkable self-sensing properties suitable for traffic monitoring and SHM systems.

2.3.5 Agricultural by-product

Rice husk ash (RHA), coconut husk ash (CHA), and sugarcane bagasse ash (SCBA) are agricultural by-products characterized by a substantial presence of amorphous silica and a porous microstructure. As partial replacements for cement, it is possible to reduce the cement content in concrete mixtures. This leads to a lower carbon footprint and energy consumption during cement production, as well as cost-savings. Gastaldini et al. [45] compared the electrical resistivity of concrete containing 10%, 20%, and 30% RHA to that of normal concrete, and found a higher resistivity with increasing RHA. They contributed the increasing resistivity as a result of the changes in microstructures of concrete. By synthesizing graphene derived from RHA, the electrically conductive cementitious materials can be developed with strengthened compressive strength and stress-sensing performance when incorporating RHA-derived graphene. The enhancements are even larger than the reinforcing effect by GNPs and MWCNTs [46, 47]. Gopalakrishnan and Kaveri [48] investigated the incorporation of 50% SCBA by weight of the binder along with graphene oxide (GO) into cement mortar. As a result of this combination, the electrical conductivity

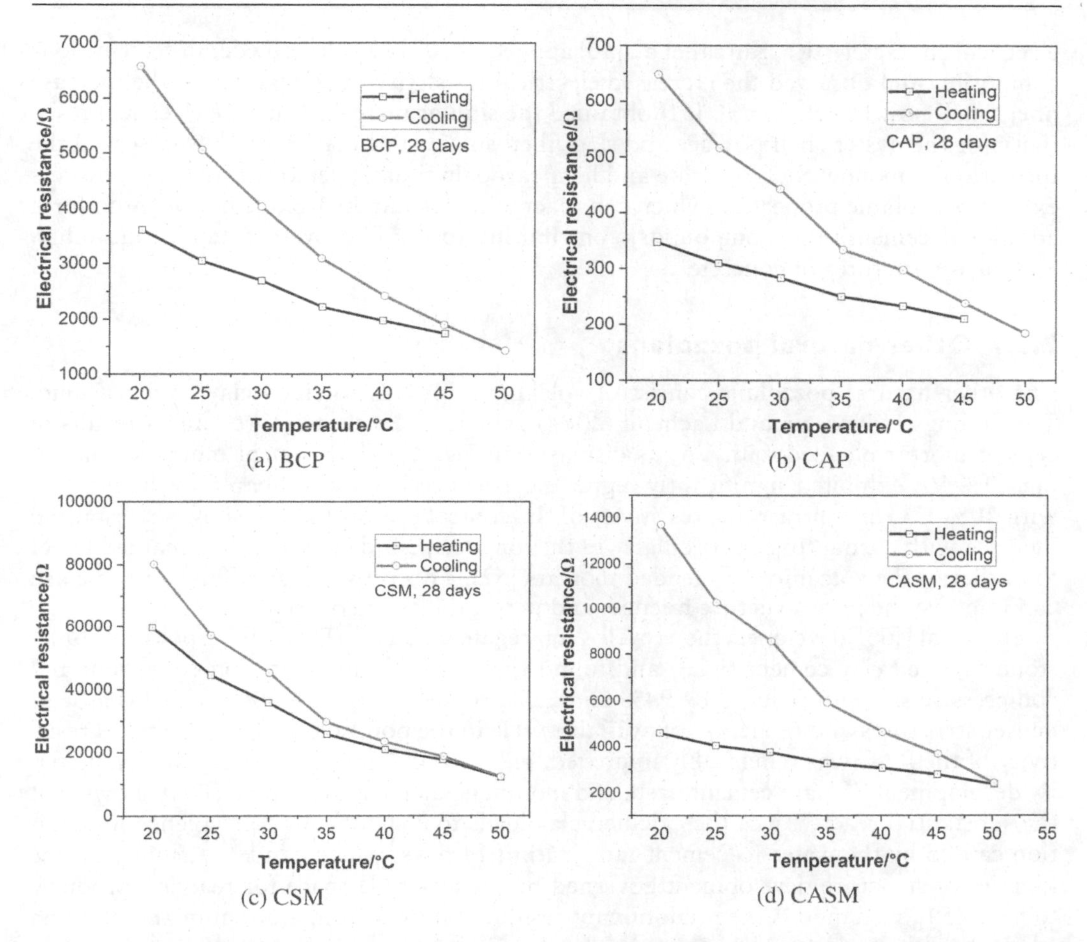

Figure 2.8 Electrical resistance changes as a function to temperature for the cement mortar containing/without CHA [49].

of the mortar exhibited a substantial enhancement, indicating its promising capability for self-sensing applications. Robert et al. [49] explored the effect of CHA as a cement replacement on the electrical characteristics of cement paste and mortar, and found that their conductivity could be clearly enhanced with inclusion of CHA. Fig. 2.8 shows the electrical resistance changes under treatment of heating and cooling processes for the cement composites with/without CHA, indicating that the composites are capable of intrinsic temperature monitoring automatically. In particular, the BCP, CAP, CSM, and CASM are the plain cement paste, plain cement mortar, CHA-modified cement paste, and CHA-modified cement mortar, respectively. However, the authors did not mention the reason for the decrease in electrical conductivity, but the denser structure and reduction in pores are very likely to lead to a decrease in conductivity after 28 days of curing.

2.3.6 Shell powder

The use of shell powder, such as oyster shell powder, seashell powder, and eggshell powder, as a partial cement replacement has attracted increasing attentions. In the aspect of electrical characteristics, Kumar et al. [50] applied 0–10% eggshell powder in cementitious composites and observed the increased electrical resistivity of concrete with the increase of eggshell powder. Han et al. [51] obtained the similar increased surface electrical resistivity for the oyster shell powder and slag-filled sustainable concrete. They observed the formation of monocarboaluminate and hemicarboaluminate, demonstrating the powders exhibit pozzolanic properties, which can react with calcium hydroxide in cement to form additional cementitious compounds, contributing to the improved strength, durability, and microstructures of concrete.

2.3.7 Other natural pozzolans

The other natural pozzolans consist of volcanic ash (VA), calcined clay, and red mud. The findings of Hossain and Lachemi (2004) presented the electrical resistivity results of cement mortar blended with VA, as displayed in Fig. 2.9. The cement mortars containing 40% VA exhibited significantly higher electrical resistivity compared to the mortars with 20% VA. In contrast, the resistivity of the control mortar did not show a substantial increase with curing time, particularly in the long term, and consistently remained lower than that of the volcanic ash-blended mortars [52]. This proves that adding volcanic ash can improve the pore structure fineness and permeability of concrete.

Guo et al. [53] developed the recycled aggregate concrete (RAC) incorporating limestone calcined clay cement (LC3) and found that the chloride migration coefficient and compressive strength reduced by 94% and 12%, respectively, with 50% LC3. Because of denser structures and reduction of available OH^- in the pore solution, the electrical resistivity of the RAC was remarkably improved. Fig. 2.10 illustrates the electrical conductivity development of plain cement paste and modified paste with FA and LC3. It shows that the substantial reactivity of the calcined clay leads to a significant convergence in variation caused by the water-to-cement ratio within 14 days, which can be attributed to the kinetics of structure development governed by the available space for reaction products to form [54]. Red mud is the predominant residue resulting from aluminum and alumina production through the Bayer process. The electrical resistivity of concrete containing red mud also showed the increased electrical resistivity and corrosion resistance because of the accelerated pozzolanic activity [55]. Moreover, the self-sensing capacity for the CBSs

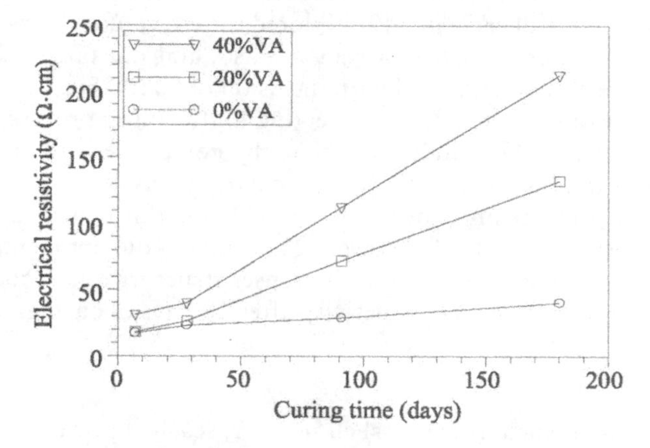

Figure 2.9 Electrical resistivity of cement mortar with curing time and VA content [52].

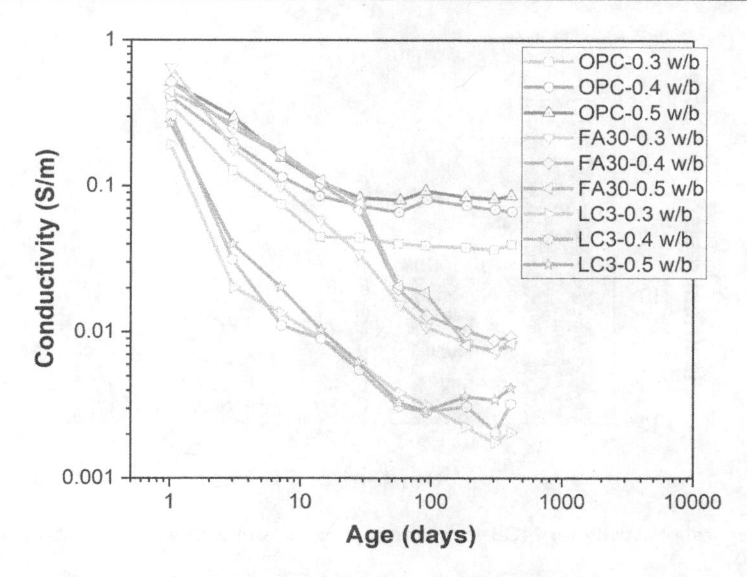

Figure 2.10 Electrical conductivity of plain cement paste, cement paste with 30% FA, and LC3 under w/b ratios of 0.3, 0.4, and 0.5 [54].

containing red mud has been partially investigated, while the piezoresistivity should be further improved with satisfied linearity, sensitivity, and repeatability [38]. Although the red mud contains some metal oxides such as iron oxide and aluminum oxide, the electrical conductivity is poor compared to the usually used conductive carbon fillers that might be responsible for the unremarkable piezoresistivity.

2.4 MACRO-REINFORCING MATERIALS

2.4.1 Polypropylene (PP) fiber

2.4.1.1 Introduction

Polypropylene (PP) fibers possessing excellent tensile strength are often used to improve the mechanical properties of cementitious composites. In this section, to improve the piezoresistive sensitivity of NCB-filled CBS as well as maintain its low cost, cheap PP fibers that have aspect ratio of 500 were applied in the NCB-filled cementitious composites. It is expected that the PP fibers enclosed with NCB can promote the generation of conductive passages and contact points in the CBSs, and thus improve the conductivity and piezoresistive sensitivity for structural health monitoring. Most importantly, this section provides a new method to enhance the electrical and piezoresistive properties of composite by adopting nonconductive fibers possessing high aspect ratio and tensile strength [56].

2.4.1.2 Electrical resistivity

Fig. 2.11 shows the electrical resistivity of the NCB-filled cementitious composites with different contents of PP fibers. It could be observed that the electrical resistivity

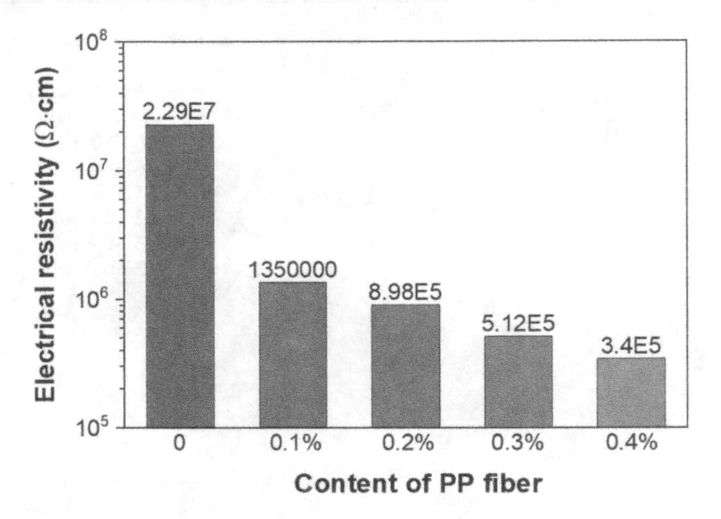

Figure 2.11 Electrical resistivity for NCB–cementitious composite with various contents of PP fiber.

dramatically decreased by more than one order of magnitude for the composites filled with 0.1 wt.% PP fibers. Afterward, with the increase of PP fibers, the electrical resistivity of the cementitious composite decreased at a slow rate and reached the value of 3.4×10^5 Ω·cm at the dosage of 0.4 wt.% PP fibers. With the presence of well-dispersed NCB nanoparticles, it shows that the PP fibers at a small concentration of 0.4 wt.% to the weight of binder have the capacity to decrease the electrical resistivity of the NCB cementitious composite by approximately two orders of magnitude.

To elucidate the phenomenon of decreased electrical resistivity, Fig. 2.12 shows the micromorphology of PP fibers with four different magnifications in the NCB-filled cementitious composite. Generally, it could be detected that the PP fibers were surrounded by the cement hydration products and NCB nanoparticles. The nonconductive hydration products made no differences on the electrical resistivity of PP fibers, while the enclosed NCB nanoparticles could generate the conductive passages through their contacts and greatly decrease the electrical resistivity of the PP fibers. Studies have demonstrated that the conductive fibrous fillers had better performance to enhance the electrical conductivity of the cementitious composite than the spherical fillers [57]. The high-aspect-ratio PP fibers with improved conductivity are just similar to the conductive fibrous fillers, which can stimulate the electrical conductivity of the whole cementitious composite by the inner connection among PP fibers. Second, in comparison to the density of NCB distribution in the PP fibers and cement matrix, it seems that the NCB nanoparticles had higher possibility to be absorbed on the surface of PP fibers, rather than spread in the cement matrix. It has been proposed that the NCB preferentially distributed in the epoxy with high polarity [58]. The NCB possesses the hydrophobic groups in the surface such as aryl group, ether group, and ester group [59], that is why the NCB nanoparticles have difficulties to disperse in the aqueous solution. However, it seems that the nonpolar PP fibers had much higher lipophilicity than cement matrix and aqueous solution, hence NCB nanoparticles tended to be absorbed by the surface of PP fibers. In other words, the addition of PP fibers led to the directional distribution of NCB nanoparticles in the cement matrix, especially along with the PP fibers, which was beneficial for the creation of conductive passages

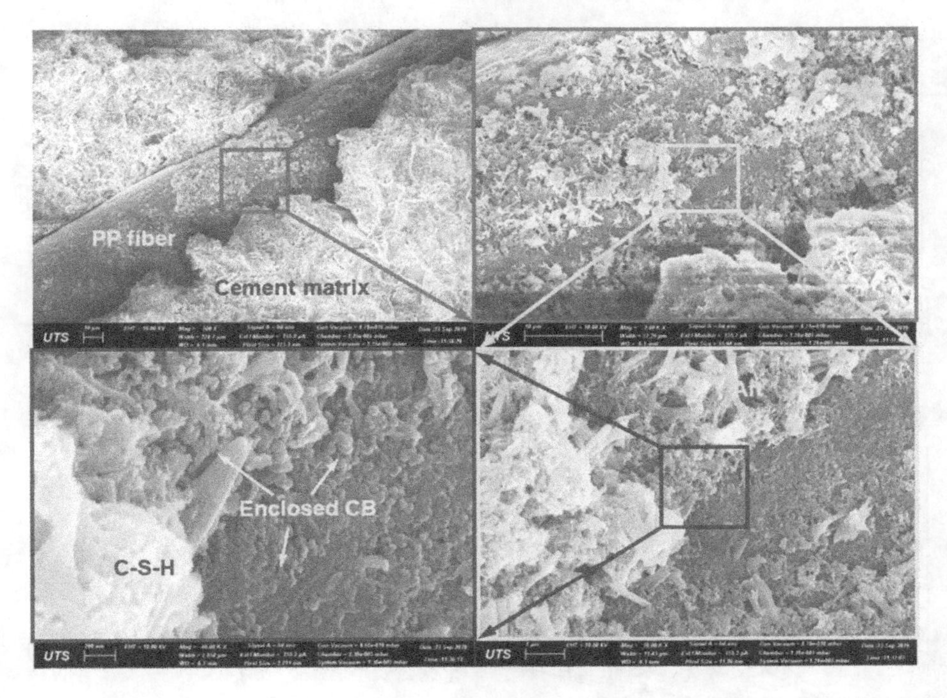

Figure 2.12 Microstructure of PP fibers in NCB–cement matrix and functional groups in NCB surface.

and decreased the electrical resistivity of the cementitious composite. However, the high concentration of NCB nanoparticles on the surface of PP fibers represented decreased NCB concentration in the cement matrix, which actually had a detrimental effect on the conductivity of cementitious composite. From the above experimental results, the conclusion can be drawn that the positive effect of NCB-enclosed PP fibers on the electrical conductivity of NCB-filled cementitious composite overwhelmed the negativity brought by lower NCB concentration in the cement matrix. More discussions on the reduced NCB concentration in cement matrix will be presented in the following section.

2.4.1.3 Self-sensing capacity

Fig. 2.13 illustrates the FCR as a function to compressive stress for the composites with various contents of PP fibers under various loading rates. For all composites, the FCR (absolute value, the same below) increased with the increment of compressive stress and decreased when the stress returned to zero and performed good piezoresistivity. Linear fittings of these results were attached, showing that the relationship between FCR changes and compressive strength was almost linear. Moreover, the overlapped or very close curves in the loading and unloading process indicated that the piezoresistivity of composite showed good repeatability. Therefore, it could be deduced that the NCB-filled cementitious composite both with/without PP fibers was eligible to be a stress sensor, regardless of the loading rates. Generally, the tendency of FCR changes was similar under various loading rates, which demonstrated that the loading rates almost possess no impact on the piezoresistivity. Since the loading rates in reality are always unpredictable

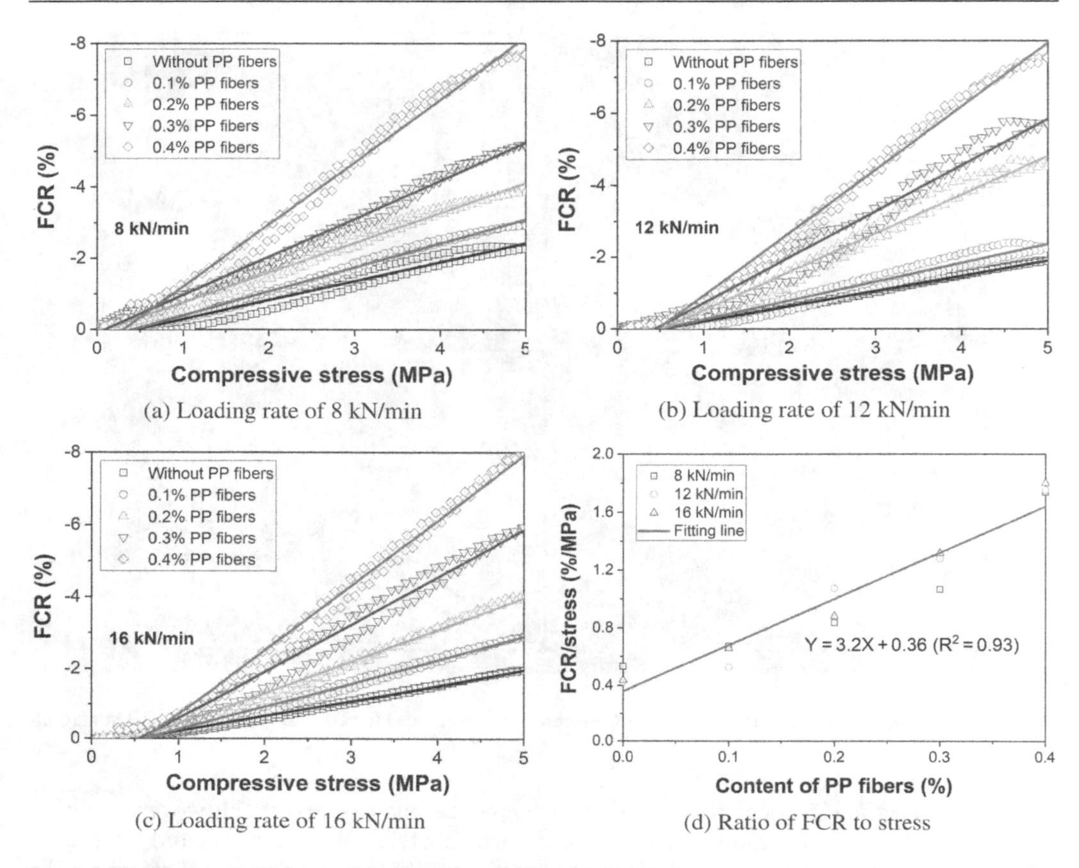

(a) Loading rate of 8 kN/min

(b) Loading rate of 12 kN/min

(c) Loading rate of 16 kN/min

(d) Ratio of FCR to stress

Figure 2.13 Fractional changes of resistivity as a function of compressive stress under various loading rates and ratio of FCR to stress with various PP contents.

and multiple, the results lay the foundation of applying the NCB/PP fibers-filled cementitious composite into real engineering project. In terms of the piezoresistive sensitivity, the composite without PP fibers reached the FCR peaks of approximately 2% at the stress magnitude of 5 MPa. However, it seems that the increasing rate of FCR increased with the increased content of PP fibers, and the largest FCR nearly 8% occurred for the composite filled with 0.4 wt.% PP fibers at the same stress magnitude. The results showed that the NCB cementitious composite filled with 0.4 wt.% PP fibers could improve the piezoresistive sensitivity by three times more than those without PP fibers. The SEM images have showed that the PP fibers were enclosed with conductive NCB nanoparticles, and the higher FCR mainly originated from the conductive PP fibers, which were more easily connected to form conductive passages given their high aspect ratio. Howbeit, the NCB nanoparticles were much more difficult to contact with each other due to the spherical properties, and that was why the FCR changes were lower for the NCB cementitious composite without PP fibers when subjected to identical compressive stress.

Fig. 2.13d illustrates the ratios of FCR to compressive stress as a function to the content of PP fibers. The ratio of FCR changes by unit stress can be a coefficient to evaluate the piezoresistive sensitivity. Regardless of the loading rates applied, the fitting curve showed that the piezoresistive sensitivity of NCB cementitious composite possessed linear relationship with the added dosage of PP fibers. Therefore, the fitting curve can be used to predict the compressive stress-sensing efficiency for the NCB cementitious composite filled with different contents of PP fibers. Generally, Eq. (2.2) to predict the compressive stress based on the statistical results in Fig. 2.13d is listed as follows:

$$\sigma_c = \frac{100\,\mathrm{FCR}}{3.2X + 0.36} \tag{2.2}$$

where σ_c is the compressive stress (MPa); FCR is the fractional changes of electrical resistivity; the constant 100 presents the magnified values of resistivity changes during fitting (i.e. 10% resistivity changes equal to 0.1); X is the content of PP fibers to the weight of binder in percentage (i.e., 0.1 wt.% PP fibers equals to 0.1). Based on the derived fitting formula, the compressive stress on the NCB-filled cementitious composite with/without PP fibers could be evaluated. In addition, the rough compressive stress could also be obtained for the composite filled with more than 0.4 wt.% PP fiber.

2.4.1.4 Mechanism discussion

Fig. 2.14 illustrates the distribution of NCB nanoparticles and PP fibers in the cementitious composite, as well as the contact points that lead to improved piezoresistivity. For the composite without PP fibers, the NCB nanoparticles have the capacity to connect with each other to form conductive points, which can be linked subjected to external forces and reduce the electrical resistivity. However, it is found that limited contact points can be observed due to the spherical properties of NCB nanoparticles, as shown in Fig. 2.14a. That is why the piezoresistive performance of NCB-filled CBSs is normally lower than the counterparts filled with fibrous conductors. On the one hand, with the addition of 0.1 wt.% PP fibers, the NCB concentration in cement matrix is decreased because a proportion of NCB nanoparticles are attached on the surface. As a result, it leads to the decreased contact points formed by NCB nanoparticles and somewhat weakens the piezoresistivity. On the other hand, the conductive PP fibers enclosed with

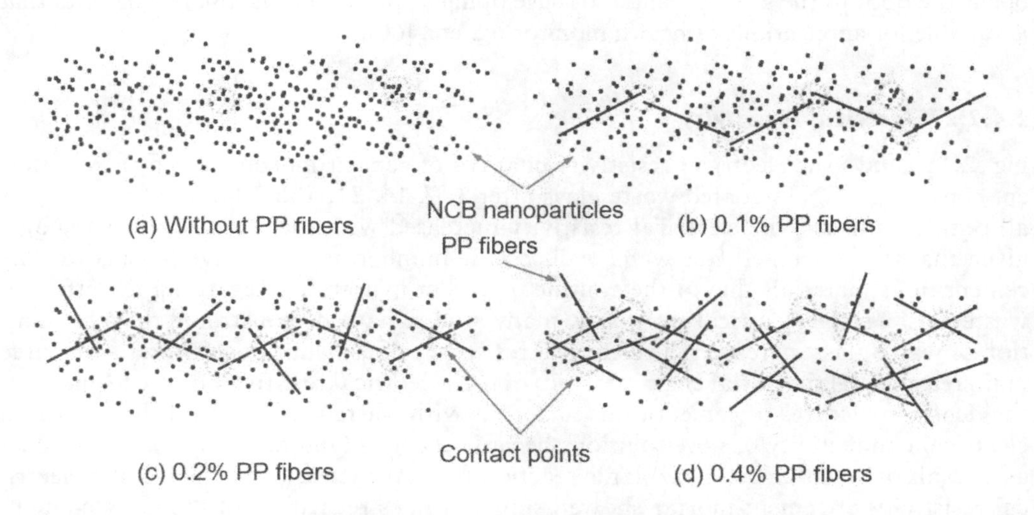

Figure 2.14 Schematic diagram of NCB, PP fibers, and potential contact points in cementitious composite.

NCB nanoparticles can significantly increase the contact points in the composite, because of their fibrous properties and high aspect ratio. The improved piezoresistivity indicates that the increased contact points by PP fibers overwhelm the decreased contact points, as plotted in Fig. 2.14b. In addition, with the further increase of PP fibers to 0.4 wt.%, more contact points by PP fibers can be observed, as shown in Fig. 2.14c and d. However, it is found that the contact points generated by NCB nanoparticles almost disappeared, due to the greatly decreased NCB concentration in the cement matrix. In summary, based on the piezoresistive results and the SEM images of PP fibers and cement matrix, the concentration of NCB in the cement matrix decreased with the increase of PP fibers. The improved piezoresistivity demonstrates that the contact points are easily formed by PP fibers, rather than NCB nanoparticles. Importantly, the number of contact points almost linearly increases with the increased PP fibers, which can be deduced from the linear altered ratios of FCR to stress with the increase of PP fibers.

2.4.2 Glass aggregate

2.4.2.1 Introduction

Many studies have widely explored the durability and mechanical properties of cementitious composites containing waste glass, none of them investigated the potential of using waste glass as conductive fillers. This book introduces the application of waste glass in the field of multifunctional cement and concrete construction with a new application of using waste glass as conductive fillers in cementitious composites. Given that the majority of waste glass is electrically isolated, a simple procedure of coating the waste glass cullet is carried out by immersing the alkali-washed waste glass into CNTs suspension followed by a drying process which evaporates the water molecular and leaves the conductive CNTs on the surface of waste glass cullet. Afterward, the waste glass cullet coated by CNTs is then applied in the cement mortar to replace the fine sand partially. The application of conductive waste glass in cementitious composites is determined to improve their electrical conductivity and achieve multifunctionality. The proposed coated waste glass opens the door to the great potential of developing a smart and sustainable material that is suitable for applications as health monitoring, etc. [60].

2.4.2.2 Electrical resistivity

Fig. 2.15 shows the electrical resistivity changes of cement mortar modified by different contents of CNTs-coated waste glass after 1, 7, 14, 21, and 28 days of curing. For all cement mortars, the electrical resistivity increased with the increase of curing age, given that the consumed free water reduced the number of conductive passages in the cementitious materials due to the continual cement hydration. Regarding the effect of waste glass on the electrical resistivity, many studies have demonstrated that the addition of waste glass increased the electrical resistivity of cementitious materials for three main reasons: (1) the initial characteristics of high electrical resistivity of waste glass, (2) the denser structures of cementitious materials with waste glass, and (3) the decreased electrical conductivity of pore solutions because the ions bind to the hydration product as a result of an enhanced pozzolanic reaction [61]. At the early curing age, the electrical resistivity of cement mortar showed similar values regardless of the glass content. This is due to the high water content in all mixtures at this stage. Afterward, with the

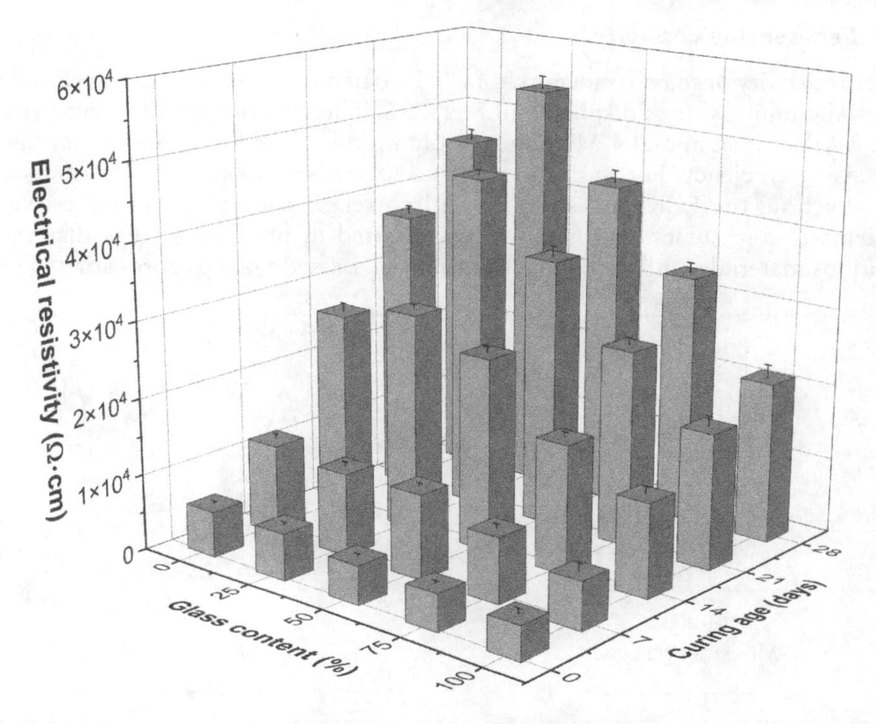

Figure 2.15 Electrical resistivity of modified cement mortar with CNT-coated waste glass at various glass contents and curing ages [60].

increased content of the waste glass, the electrical resistivity first showed an increasing tendency and then a decreasing tendency, with the critical content of 25% waste glass as fine aggregate. The above results have illustrated that the reduced water absorption was partially due to the enhanced microstructures and porosity. Similarly, the higher electrical resistivity of cement mortar with 25% waste glass than plain cement mortar was primarily contributed by the denser structures and pore refinement by waste glass particles [62]. Also, it implied that a low amount of CNT-coated waste glass failed to improve the electrical conductivity of cement mortar. In that circumstance, the majority of glass particles might be enclosed by nonconductive cement materials to block the conductive passages. However, with the growth of glass content, the connection among conductive CNT-coated glass particles became easier, which caused the later decreased electrical resistivity. On the one hand, the electrical resistivity of cement mortar increased with denser microstructures caused by glass particles. On the other hand, the electrical resistivity decreased with increase of waste glass because the attached CNTs improved the conductivity of glass particles and cementitious materials. The final electrical resistivity of cement mortar with CNTs-modified glass particles depends on the above-combined effects. The cement mortar modified with more than 25% CNT-coated waste glass possessed smaller electrical resistivity than that of plain cement mortar. In addition, considering the possibility of debonding CNTs coatings from glass particles during mixing, the escaped CNTs are another potential reason for the ameliorated electrical conductivity.

2.4.2.3 Self-sensing capacity

The piezoresistivity of cement mortar with CNT-coated waste glass under an elastic regime of uniaxial compression is displayed in Fig. 2.16. The selection of stress magnitudes of 1 MPa, 2 MPa, 3 MPa, and 4 MPa aims to obtain the gauge factor and assess the stress/strain sensing efficiency. For the plain cement mortar, similar to the performance under failure detection, the FCR values almost had no relationship to the compressive stress/strain. This is consistent with observations reported in previous studies that the dried cementitious materials without conductive fillers possessed poor piezoresistivity [29].

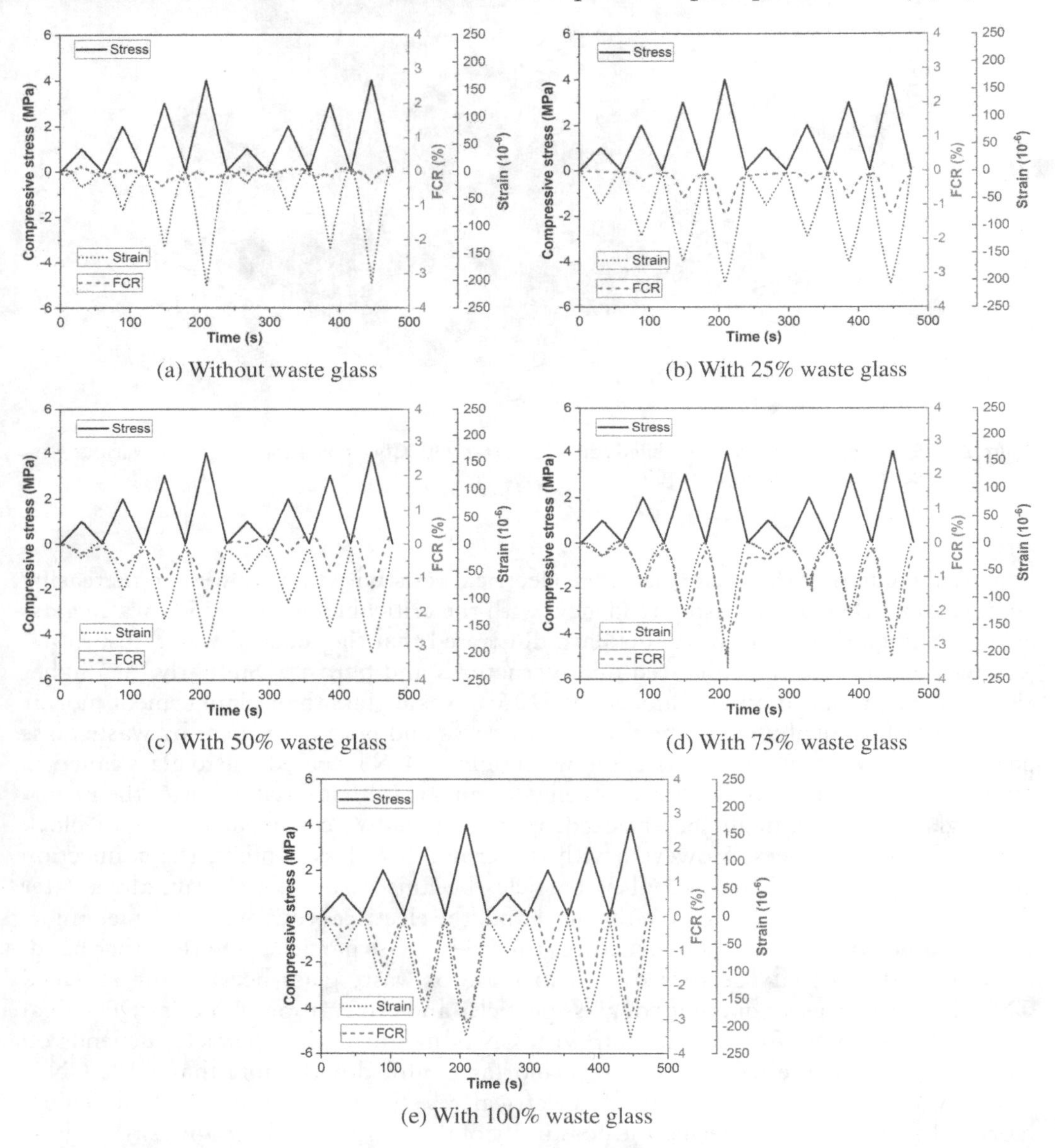

Figure 2.16 Piezoresistivity of modified cement mortar with varied contents of CNTs-coated waste glass under cyclic compression in elastic regime [60].

As shown in Fig. 2.16b–e, the piezoresistivity was significantly enhanced for the waste glass–modified cement mortar, with monotonously decreased and returned FCR values respectively in loading and unloading processes. It was found that the electrical resistivity changes of composites became larger with the increase of glass content. For the cement mortar with 25% waste glass, the FCR changes were nearly negligible under the stress magnitude of 1 MPa, while the counterparts filled with more waste glass showed excellent correspondence to the compressive stress. It implied that the cement mortar with insufficient amount of waste glass failed to detect small stress/strain. This phenomenon demonstrates that the number of conductive fillers is critical to forming conductive passages, and insufficient conductive fillers in the composites lead to the still discontinued conductive passages, especially under small stress/strain. All the cement mortars showed satisfactory piezoresistivity, when the stress magnitude exceeded 2 MPa. The piezoresistivity showed satisfactory repeatability under cyclic loading pattern. Still, the small fluctuations that slightly increased/decreased the FCR values emerged after the cyclic compression is finished. The damages of inner structures such as the generation of microcracks and the polarization due to the remained pore solutions and free ions are potential reasons responsible for the tiny growth of FCR. On the other hand, the permanent growth of conductive passages is responsible for the decreased FCR after compression. Overall, the small irreversible FCR changes implied local weaken structures (including both conductive and nonconductive phases) existing in the cement mortar. In comparison to the FCR changes under external stress/strain, the irreversible FCR was very limited but still should be eliminated before practical structural health monitoring.

To evaluate the stress/strain sensing performance, Table 2.3 lists the relationships among compressive strain and FCR values and their fitting curves. As previously mentioned that piezoresistivity might be affected by the magnitude of stress/strain, the piezoresistive performances of cement mortar respectively under stress ranges from 0 MPa to 4 MPa and from 1 MPa to 4 MPa were compared. Since the FCR of plain cement mortar had a poor relationship with compressive stress, their fitting curves and FCR values were not included. For the CNTs-coated glass-modified cement mortar, it was found that the gauge factor increased with the growth of waste glass. In addition, under two stress ranges, there was no significant difference in the gauge factor and linearity. The final gauge factor was obtained from the average values of the composites under two stress magnitudes. Overall, the cement mortar modified with CNT-coated waste glass showed excellent piezoresistivity, whose sensitivity was dozens of times higher than that of commercially available strain

Table 2.3 Gauge factor of cement mortar with multiple contents of waste glass in the linear and repeatable portion

	Stress from 0 to 4 MPa			*Stress from 1 to 4 MPa*			
Glass content	*Fitted curves*	*Degree of linearity (R^2)*	*Gauge factor*	*Fitted curves*	*Degree of linearity (R^2)*	*Gauge factor*	*Average gauge factor*
0%	–	–	–	–	–	–	–
25%	$Y = 52.1X + 0.11$	0.819	52.1	$Y = 53.9X + 0.14$	0.902	53.9	53.0
50%	$Y = 70.3X - 0.08$	0.954	70.3	$Y = 71.7X - 0.06$	0.948	71.7	71.0
75%	$Y = 144.9X - 0.21$	0.912	144.9	$Y = 137.7X - 0.29$	0.900	137.7	141.3
100%	$Y = 149.2X + 0.25$	0.960	149.2	$Y = 152.5X + 0.29$	0.961	152.5	150.9

gauge. The gauge factor reached approximately 53.0, 71.0, 141.3, and 150.9, respectively, for the composites with 25%, 50%, 75%, and 100% waste glass.

2.4.2.4 Mechanism discussion

To reveal the underlying mechanism, Fig. 2.17a–g present the microstructures of cement mortar modified with CNTs-coated waste glass, which mainly shows the surface morphology of the waste glass, CNTs and its boundaries to the cement matrix. Because of the considerable size differences between glass particles (mm) and CNTs (nm), images at different magnifications are chosen to clarify the microstructures. As shown in Fig. 2.17a and b, the cement matrix shows a good cohesion to the waste glass particle. On the surface of waste glass, the attached CNTs could be found twisted with cement hydration products in Fig. 2.17f, indicating that the coated CNTs possibly enhanced the cohesion of cement matrix to waste glass. Moreover, for the regions near the boundary in Fig. 2.17d–e, the cement matrix was mixed with CNTs which not only played a bridging effect to strengthen the cement matrix but also enhanced the formation of conductive passages to enhance the electrical conductivity and piezoresistivity. In addition to the relatively well-dispersed CNTs, Fig. 2.17g displays the CNT agglomerations in the boundaries. The aforementioned SEM images of the CNT-coated waste glass have demonstrated the possibility of CNTs agglomerations on the surface of waste glass. During the specimen preparation, it seemed that a proportion of CNTs agglomerations entered into the cement matrix. This part of CNTs also played an important role in the electrical conductivity enhancement and contributed to the lowest electrical resistivity for the cement mortar with 100% waste glass. However, since the CNTs agglomerations possess poor cohesion to cement matrix and lead to high porosity [63], the mechanical strengths (which were not involved here) can be weakened with the increased number of CNT agglomerations.

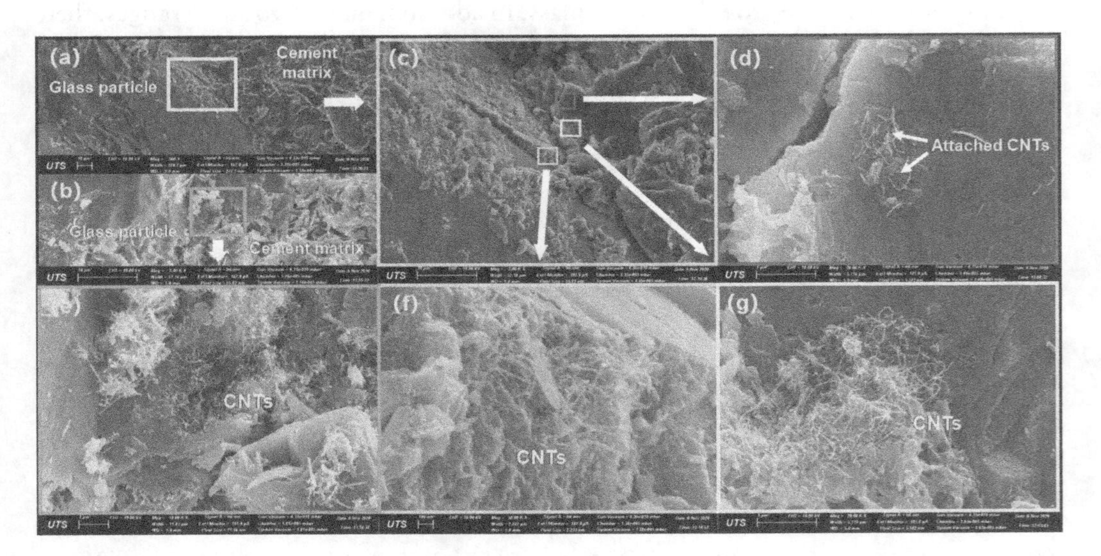

Figure 2.17 Microstructural morphology of cement mortar with CNT-coated waste glass [60].

2.4.3 Conductive rubber fiber (CRF)

2.4.3.1 Introduction

Waste rubber has been widely studied as an additive for production of cementitious materials to improve specific properties, such as sound absorption, thermal insulation, and damping property. The use of waste rubber is relatively inexpensive and can reduce the negative environmental impacts induced by the million tons of nonbiodegradable rubber wastes produced annually [64]. Nonetheless, it is well-known that untreated rubber is usually nonconductive and cannot be directly used to produce conductive cementitious sensors. Recently, with increasing demands on electromagnetic shielding materials, conductive rubber products incorporating carbon black, aluminum/silver, or nickel are becoming more prevalent in modern industries of aerospace, precision instruments, and automobiles. A specialized extrusion machine can be used to combine the nonconductive silicone and the conductive particles (mainly carbon black, Al/Ag, and Ni/C) firmly. The fully integrated cementitious composites become a perfect conductive rubber composites with electrical resistivity as low as 10^{-3} to 10^{-1} Ω·cm. With the inspiration from the use of recycled rubber for the concrete manufacturing for environment-friendliness and sustainability, the use of conductive rubber in cementitious composite is environment-friendly and nontoxic to both workforces and investigators in real project and experimental research.

2.4.3.2 Electrical resistivity

Different contents of conductive rubber from 20 to 140 fibers (0.32–2.24 vol.%) and w/b ratios of 0.34, 0.38, and 0.42 were attempted to explore the conductivity interference by amount of rubber and w/b ratios [65]. Fig. 2.18 shows the electrical resistivity

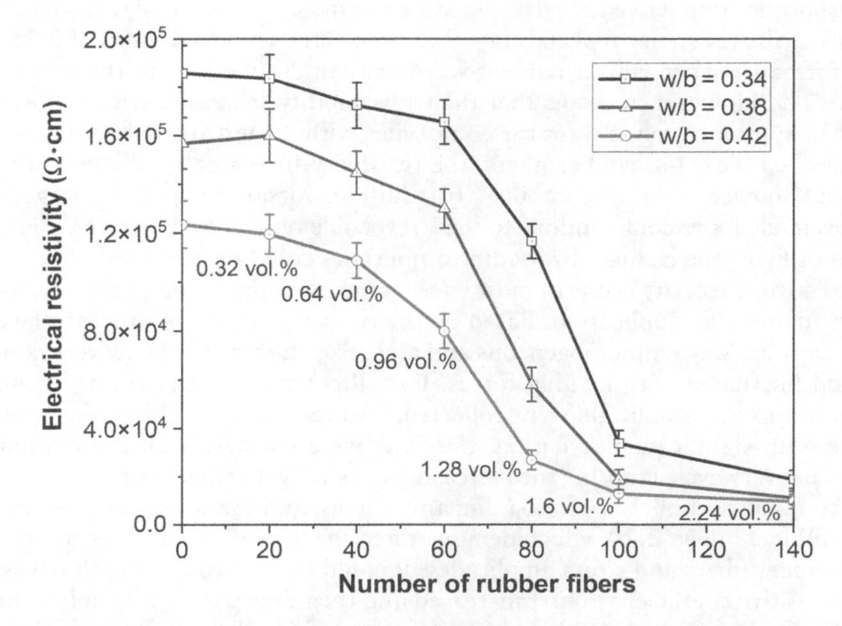

Figure 2.18 Electrical resistivity of the rubber–cement composite with different contents of rubber fibers at w/b ratios of 0.34, 0.38, and 0.42 [65].

of composites reinforced by different amounts of rubber fibers in three different w/b ratios: 0.34, 0.38, and 0.42. Obviously, conductivity percolation could be observed for all rubber–cement composites with the increase of rubber content. It could be seen that for composites at the w/b ratios of 0.34 and 0.38, their resistivity was greatly decreased with rubber contents from 60 fibers to 100 fibers (0.96–1.6 vol.%). However, for the composites at w/b ratio of 0.42, the resistivity decreased more gently with rubber content. This may be due to the reduced electrical resistivity by more water content, which diminished the positive effects of conductive rubber fibers on the electrical conductivity. Also, the tendency could be observed that the higher the w/b ratio, the lower the resistivity of the composite, demonstrating the influences on resistivity of the water content in cementitious composite.

2.4.3.3 Self-sensing capacity

Fractional changes of resistivity of rubber–cement composites at the w/b ratio of 0.34 are illustrated in Fig. 2.19, where sub-images represent the composites with different rubber contents from 20 fibers to 140 fibers. Generally, the resistivity decreased upon loading and increased upon unloading, despite that it was greatly fluctuated for the composite filled with 140 rubber fibers. Specifically, the composites embedded with less than 140 rubber fibers showed excellent repeatability on resistivity under cyclic stress below 10 MPa, while a continual increase on resistivity was observed for the composite with 140 rubber fibers. This phenomenon could be simply explained by the formed conductive rubber networks in the composite. As the percolation threshold was approximately 60–100 rubber fibers (0.96–1.6 vol.%), it could be assumed that the connected conductive passages have been established for the composite with 140 rubber fibers. Under the circumstances, a small disturbance on the composite might detach the neighboring rubber fibers and interrupt the conductive networks to cause the increased resistivity.

In terms of the resistivity repeatability, the composites at the w/b ratio of 0.34 showed good performance when rubber content was less than 80 fibers, with the deviation narrowed to ≤1%. However, it seems that the irreversibility increased with rubber content and reached approximately 2% for the composites with 80 and 100 rubber fibers. For the composite filled with 140 rubber fibers, the resistivity increased gradually with loading process and showed poor repeatability. In addition, for all composites, larger loading amplitudes made more contributions to the irreversible resistivity. However, compared to the values of FCR, the changes by loading amplitudes could be neglected.

Piezoresistive sensitivity is one of most vital factors to evaluate the application potential of CBS in engineering application. Based on the rubber installation process, the rubber–cement composite was nonhomogeneous and anisotropic; thus, the FCR were nonlinearly altered and fluctuated with loading forces. For all composites, the average values from these cycles for each amplitude were collected, and for the 140 rubber fibers-reinforced composite with significant fluctuation, the FCR were selected from the minimum and maximum in every cycle ignoring the increasing resistivity. Furthermore, the relationships between conductive rubber fibers, loading amplitudes, and the stress-sensitive coefficient (Fi) are exhibited in Fig. 2.20, which demonstrated the dependence of sensitive coefficient Fi on the rubber fibers and stress amplitudes. It could be seen that for each stress magnitude, the sensitive coefficient Fi first increased and then decreased with rubber content. It was the composite containing 80 rubber fibers that achieved the best sensitivity, with Fi of 1.725, 1.858, 1.750, and 1.420 at the amplitudes of 4, 6, 8, and 10 MPa, respectively. The composites reinforced by 60 or 100 rubber fibers possessed secondary sensitivity,

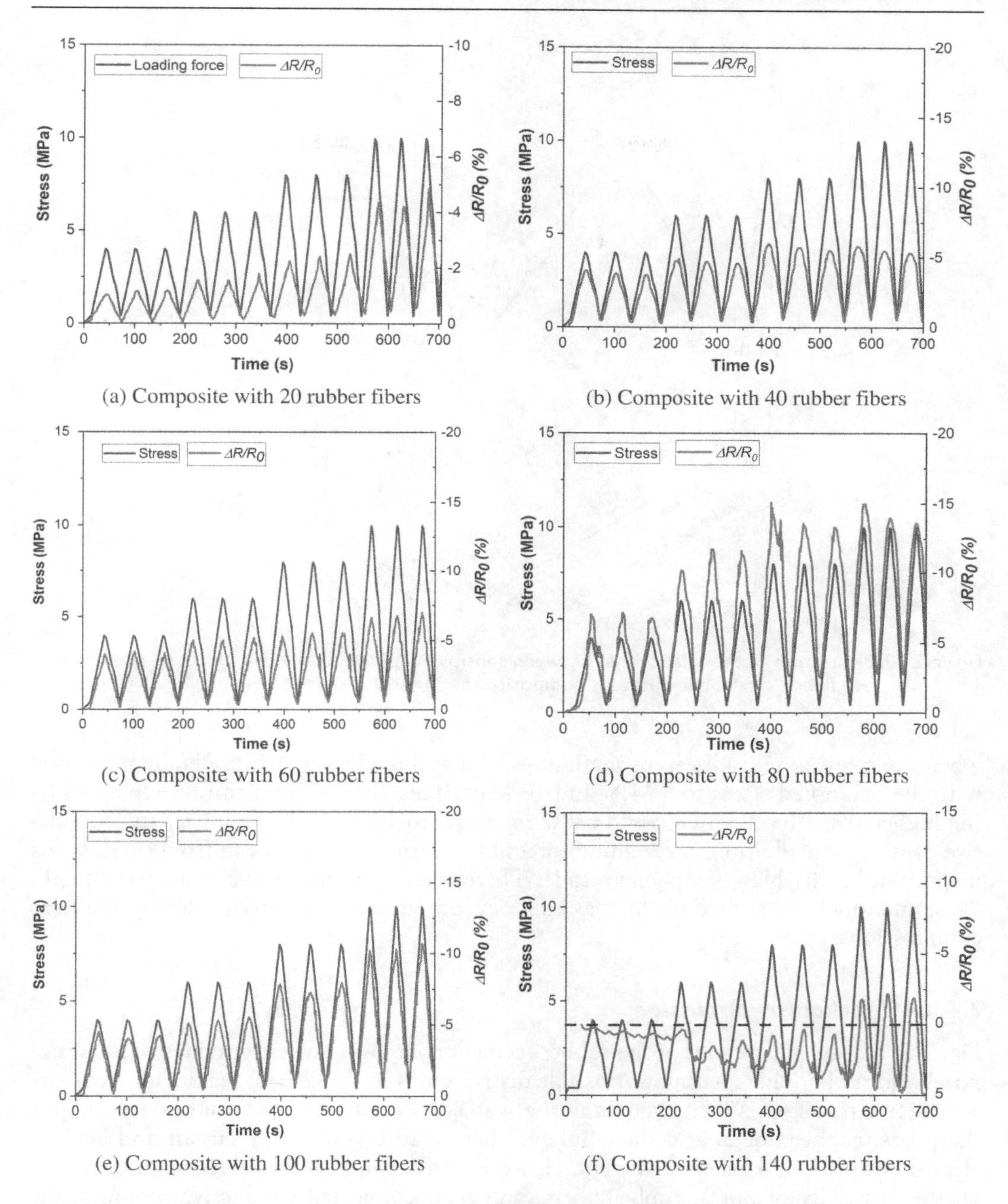

Figure 2.19 Fractional changes of resistivity with different rubber contents and stress magnitudes for the cementitious composites at w/b ratio of 0.34 [65].

and the worse sensitivity was emerged for the composite with 20 or 140 rubber fibers. Therefore, it could be concluded that the rubber–cement composites in the range of percolation threshold were more likely to change its resistivity under external forces, while the composites became insensitive to forces if outside the ranges. In terms of the influences of stress amplitudes on sensitivity, it was observed that the sensitive coefficient Fi was higher

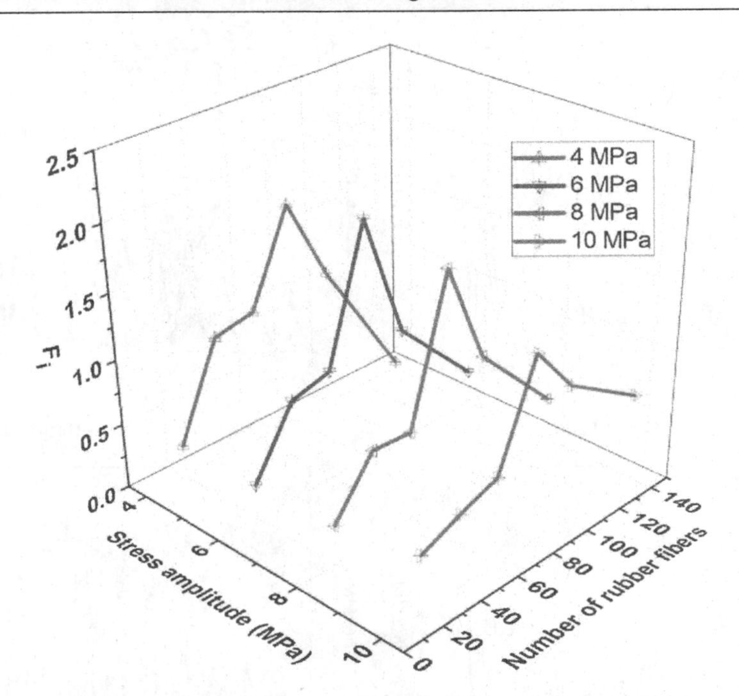

Figure 2.20 Illustration of the relationship between sensitive coefficient, stress magnitude, and the rubber fibers for the cementitious composite at w/b ratio of 0.34 [65].

when the stress amplitudes were small at 4 MPa and 6 MPa, which gradually decreased with the magnified stress to 8 MPa and 10 MPa. This phenomenon can be explained by the higher stress level which could cause more microcracks and disconnect the conductive passages. In addition, the loading forces are continually increasing from small stress magnitude to the high stress magnitude. Therefore, the previous cyclic loadings might cause microdefects and result in the reduction on electrical sensitivity during the next compression.

2.4.3.4 Mechanism discussion

Fig. 2.21 is a schematic plot of the rubber–cement composite, where conductive phases of pores and rubber fibers with four possible distributions in the cement matrix might create conductive passages. When the composite was connected to an external power supply, the pores, rubber fibers, and their linkages helped to complete the circuit and achieve electron transition to generate current. However, on account of the profound discrepancy on the resistivity of pores, rubber fibers, and their connected conductive passages, the conductivity amelioration brought by these conductive phases was in a huge difference. It caused the complicated piezoresistivity of rubber–cement composites because of the various degrees of piezoresistive expression by different conductive phases.

Pores containing solutions could improve the conductivity of cementitious composite because of the free movement of ions such as Ca^{2+}, SO_4^{2-}, OH^-, and so on. These include two conductive possibilities through pores, with one by already connected pores running through the composite and another by the neighbor pores which might be linked with

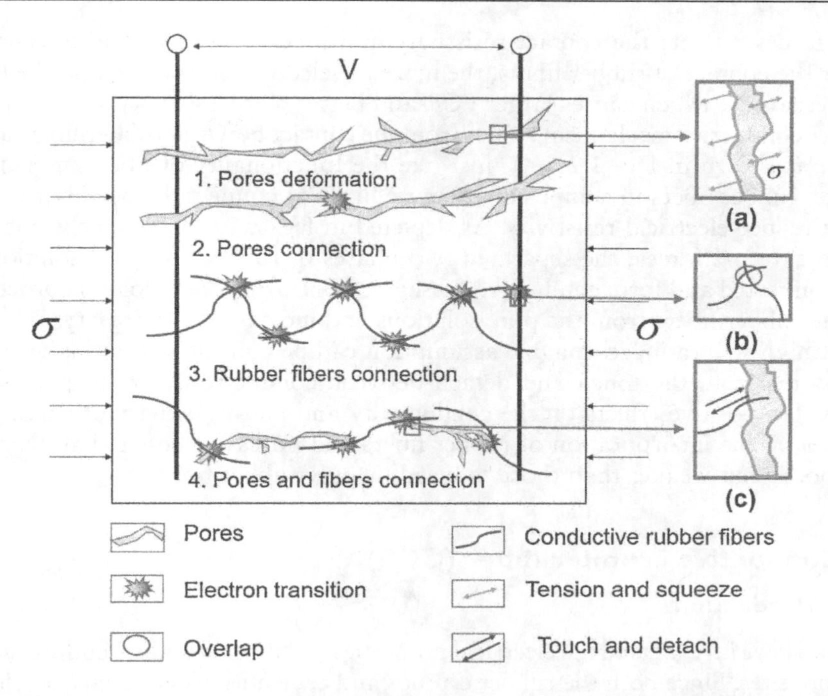

Figure 2.21 Schematic illustration for the electrical conductivity and piezoresistivity mechanism of rubber–cement composites [65].

each other under external forces, as shown in Fig. 2.21 (1 and 2). It can be concluded that the former has a direct relationship with both electrical conductivity and piezoresistivity of cementitious composite, while the latter is more related to the piezoresistivity only, until the separated pores are connected irreversibly with external forces. As illustrated in Fig. 2.21 (a), pore solutions can be squeezed in compression to lead higher concentration and better conductivity, and vice versa in tension. Hence, the FCR alter with the solution concentrations and generate the piezoresistivity. For the pores-connected passages with gaps, the conductivity amelioration and piezoresistivity might be more obvious if the gaps are eliminated under external forces. On the other hand, better conductivity and piezoresistivity could be observed for the composites with high w/b ratios because of the greater number of pores. However, the impacts of pores and pore solutions are very limited on the piezoresistivity, because of the nonconductive and compact structures of cement hydration products.

The conductivity and piezoresistivity induced by conductive rubber fibers are responsible for the conductive and piezoresistive properties of rubber–cement composite, and as mentioned, the electrical resistivity could be greatly reduced with increase of rubber fibers. As depicted in Fig. 2.21 (3), the separated rubber fibers will overlap or connect with each other to form conductive paths under external forces and alter electrical resistivity. Because of the excellent electrical conductivity of rubber fibers, a small-scale connection of rubber fibers can produce a significant fractional change of resistivity in composite. Moreover, not only the overlapped or connected rubber fibers, but also the tightness degree between two rubber fibers, which is determined by initial microstructures and external forces, affects the contact resistivity and piezoresistivity. It is considered that the closer the contact of rubber fibers, the more the electron transition, and the lower the contact resistivity, which causes larger FCR and better piezoresistivity.

Another conductive mechanism comes from the contact between rubber fibers and pores solutions, as shown in Fig. 2.21 (4). Just like the functionality of fibers in cementitious composite, rubber fibers in composite are more likely to connect the neighbor pores solutions and reduce electrical resistivity. As depicted in Fig. 2.21 (c), when the composite is subject to external forces, the separated two phases of rubber and pore solutions might become connected and form conductive passages. Contrarily, the unloading process might detach the rubber fibers from the pore solutions and increase the resistivity. For the composites at high w/b ratio, reasonable assumption can be concluded that higher resistivity changes come from the touch and detach between conductive fibers to pore solutions. Generally, the positive effects on the conductivity and piezoresistivity of rubber–cement composites by the incorporation of rubber fibers and pores are better than that brought by only pores, but weaker than those induced by only rubber fibers.

2.4.4 Conductive crumb rubber (CCR)

2.4.4.1 Introduction

Few studies have investigated the electrical properties of rubber crumbs–reinforced cementitious composites. Since both the rubber crumbs and cementitious materials are electrically isolated, the rubberized cementitious composites often illustrated higher electrical resistivity than the plain ones [66]. Recently, with the development of composite materials technology, special rubber products with excellent electrical conductivity were manufactured for modern industries such as electromagnetic shielding, microelectronic assembly, and automobile. To make full use of these conductive rubber scrapes, as mentioned in last section, attempts have been made on the improved electrical and piezoresistive properties of cementitious composites filled with conductive rubber fibers [67]. The results demonstrated the potential of conductive rubber products to manufacture conductive cementitious composites. This section reveals the application of conductive rubber crumbs (CRC) to improve the electrical conductivity and piezoresistivity of cementitious composites.

2.4.4.2 Electrical resistivity

Fig. 2.22 displays the electrical resistivity for the rubberized mortar filled with various contents of CRC at w/b ratios of 0.40, 0.42, and 0.45 with/without drying treatment [68]. For the composites without drying treatment, it could be seen that the electrical resistivity of the mortar decreased with the increase of w/b ratios. The lower resistivity for the composites at w/b ratios of 0.45 was mainly due to the higher water content in the cement matrix. In terms of the effect of CRC, the electrical resistivity of the rubberized mortar slightly varied when the added CRC reached 10%, where the decreased resistivity for the composites at w/b ratios of 0.40 and 0.45 and the increased resistivity for the composites at w/b ratio of 0.42 could be observed. The discrepancy represented that the low replacement ratio of fine sand by CRC nearly made no differences on the electrical resistivity of the rubberized cement mortar, in which the rubber crumbs were completely separated and enclosed by cement matrix. Afterward, with the increase of CRC content, the electrical resistivity of mortar monotonously decreased by more than one order of

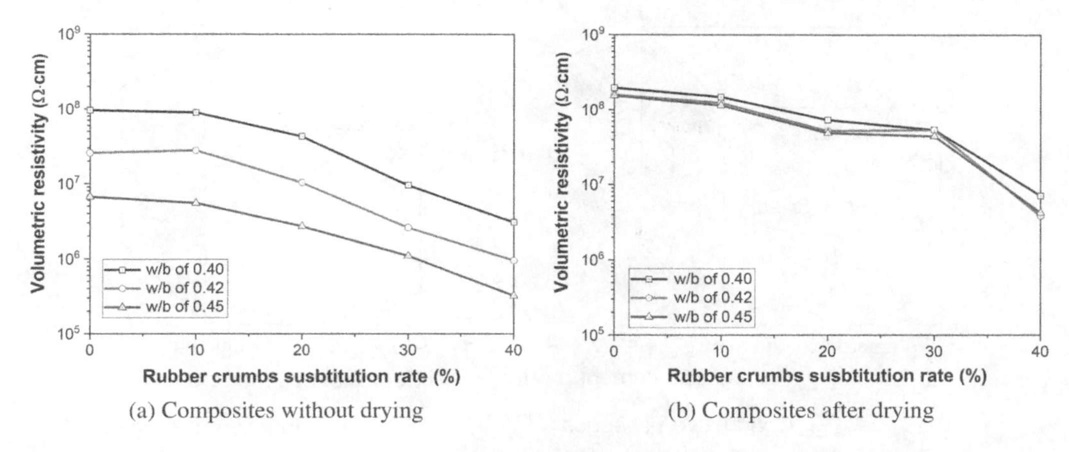

Figure 2.22 Electrical resistivity for the rubberized mortar with various contents of CRC at different w/b ratios [68].

magnitude when the CRC replacement ratio reached 40%. In terms of the dried composites in Fig. 2.22b, all the mortar showed dramatic increases on the electrical resistivity and the discrepancy between composites at various w/b ratios greatly reduced after the drying treatment. It was clear that the resistivity reduction mainly originated from the decreased water content and pore solutions in the composites. Therefore, since the mortar with a higher w/b ratio of 0.45 contained larger water content, the drying treatment could decrease the water content to the utmost, and that was why the resistivity increase reached the maximum compared to the counterparts at lower w/b ratios. Moreover, it was observed that the mortar with 30% rubber crumbs substitution rate illustrated much faster resistivity reduction than that with 20% rubber crumbs substitution rate. This was due to the fact that more rubber crumbs could bring more air bubbles and pores in the mixture, which were filled with conductive solutions before drying and exhibited better conductivity. However, the pore solutions were considerably decreased after the specimens were dried and displayed even higher resistivity than the mortar with lower rubber content. Still, it was seen that the rubberized cement mortar after drying gently decreased with the increase of rubber crumbs, especially when the substitution rate reached 40%. Different from the cementitious composites reinforced with commonly used conductive nanomaterials, it was seen that there never existed sudden and swift electrical resistivity reduction with the rising content of CRC from 10% to 40% for the composites no matter drying or not. In other words, it means that the conductive passages penetrating the whole specimens were never generated with the increase of CRC. Even though, the separated conductive passages might be elongated and partially connected by means of the contact CRC distributed in the cement matrix, and thus the electrical resistivity of the rubberized mortar could be reduced. The rubberized mortar filled with 40% CRC had a larger number of connected rubber crumbs, which further demonstrated the fact that the connected rubber crumbs should be responsible for the decreased electrical resistivity.

Fig. 2.23 represents the effects of conductive rubber crumbs to increase the electrical conductivity of the modified cement mortar, incorporating the above-mentioned cross-sectional morphology of rubberized cement mortar on the macro- and microscales. Different from the initially existed conductive passages filled by conductive solutions, the

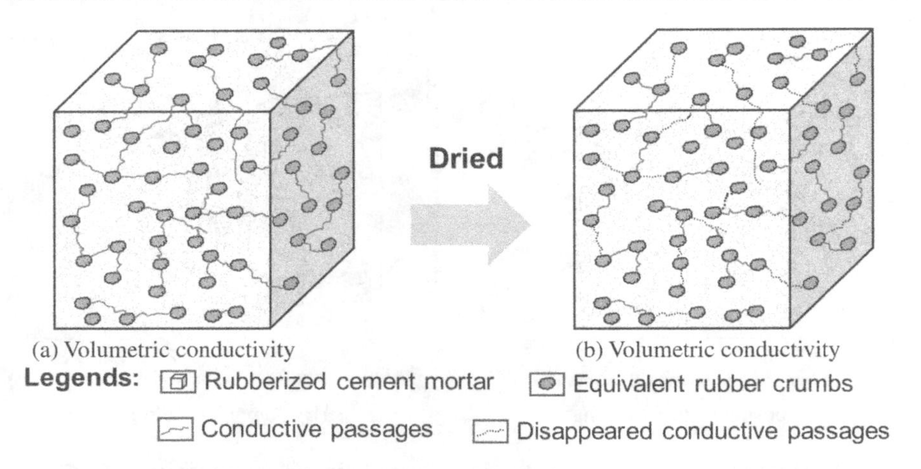

Figure 2.23 Schematic diagram of conductive passages in rubberized cement mortar before and after drying treatments [68].

added conductive rubber crumbs in the mortar could work as another solid electron and ions carrier and promote the free movements of conductive ions. Moreover, the introduction of rubber crumbs in the mixing process could bring additional air bubbles and increase the porosity, which were normally filled with conductive solutions and elongate the conductive passages. Another characterization of rubberized cement mortar was due to the different thermal and physical properties between rubber crumbs and cement matrix. Hence, the microcracks had a higher possibility to appear between two rubber crumbs. It means that the pore solutions–filled cracks could be easily connected with the assistance of conductive rubber crumbs, which directly leads to the electrical resistivity reduction. The more conductive the rubber crumbs in the cement mortar, the more the numbers of connected microcracks and the longer the conductive passages the composites possessed. In terms of the effect of water content, it could be deduced that the initial conductive passages disappeared owing to the decreased pore solutions after drying treatment. Different from the cementitious composites filled with conductive nanoparticles, whose electrical conductivity probably improved with the drying process due to the decreasing surrounded water content to reduce the contact resistance between nanoparticles [69], the lost water content and disappeared conductive passages connecting nearby rubber crumbs caused the worse conductivity for the rubberized cement mortar. It can be demonstrated that the enhanced conductivity of cementitious composites was not simply due to the connected rubber crumbs, but also due to the assistance of conductive pore solutions.

2.4.4.3 Self-sensing capacity

Fig. 2.24 plots the electrical resistivity changes of CRC-filled cement mortar at the w/b ratio of 0.40 while subjected to uniaxial cyclic compression. Generally, the FCR decreased with the applied load, and increased to the initial value during the unload stage. It was observed that all the CRC-filled cement mortar showed repeatable electrical resistivity after six cycles of compression, demonstrating that the rubberized cement mortar had excellent piezoresistive repeatability. In addition, the cement mortar filled with 40% CRC was provided with the highest FCR, by the average values reaching approximately 2.45%.

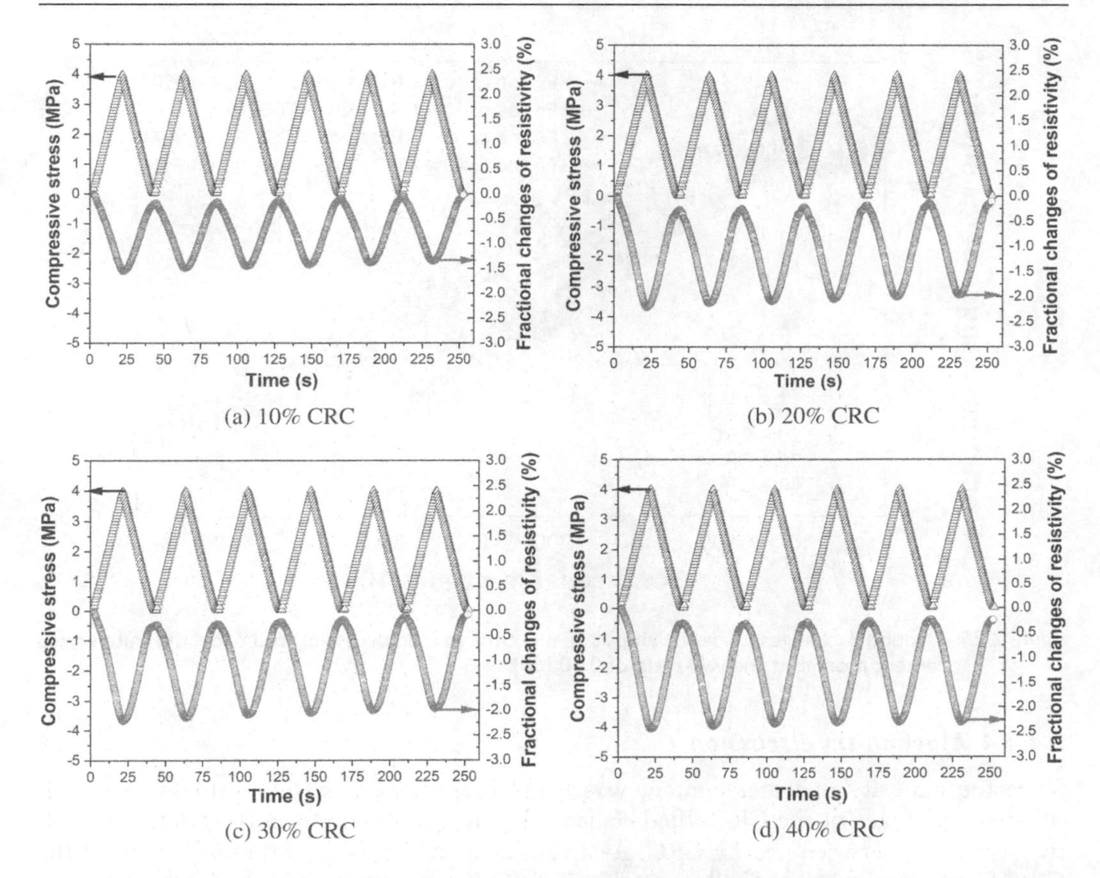

Figure 2.24 The FCR for the rubberized cement mortar at the w/b ratio of 0.40 under cyclic compression [68].

However, for the counterpart filled with 10% CRC, the FCR at the stress peak were only 1.48%. It can be deduced that the cement mortar filled with higher content of CRC had higher tendency to alter their electrical resistivity in the same stress environment. For the cement mortars filled with 20% and 30% CRC, whose FCR were very similar, reached approximately 2.07% and 2.14%, respectively.

Fractional changes of resistivity as a function to the compressive strain for the cement mortar filled with different contents of CRC are plotted, as shown in Fig. 2.25. It was seen that the compressive strain of cement mortar increased with the increase of CRC, illustrating the better deformability for the cement mortar filled with more CRC. This is consistent with the previous studies on the rubberized cement mortar, due to the lower elastic modulus of rubber crumbs [70]. Moreover, the gauge factors were calculated from the fitting lines, with the values of 72.2, 77.6, 62.7, and 57.6, respectively, for the cement mortar filled with 10%, 20%, 30%, and 40% CRC. It was found that the cement mortar filled with 20% CRC possessed the best piezoresistivity. In comparison to the commercial strain gauge which has the gauge factor of 2, the 20% CRC-filled cement mortar are given nearly 39 times higher sensitivity to monitor the compressive deformations. In particular, the reason for the decreased sensitivity for the cement mortar filled with more than 20% CRC is mainly due to the considerable increases of compressive strain.

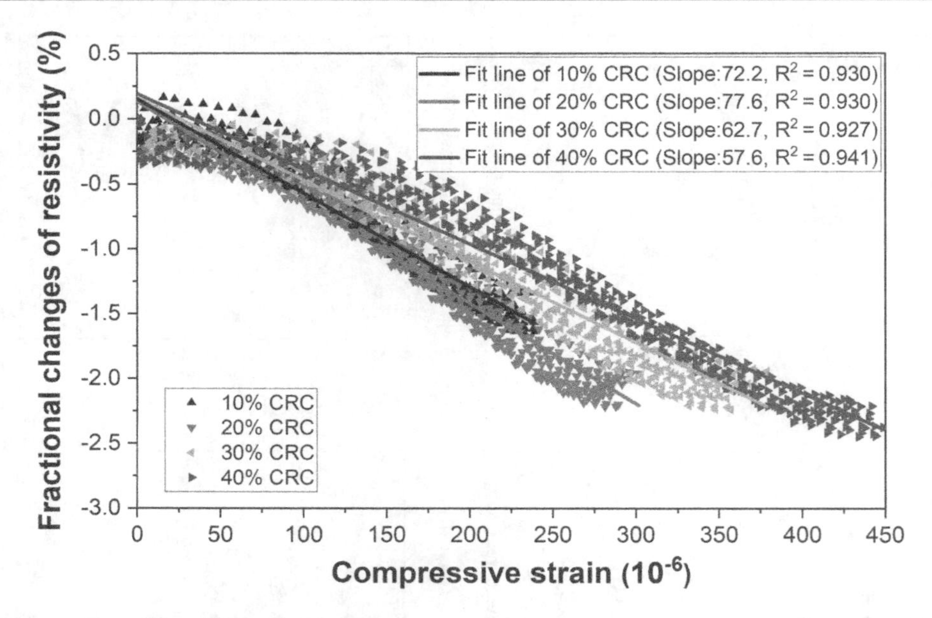

Figure 2.25 Fractional changes of resistivity as a function to compressive strain for the rubberized cement mortar at the w/b ratio of 0.40 [68].

2.4.4.4 Mechanism discussion

Since the majority of water content was dried before piezoresistive test, the electrical resistivity changes of the CRC-filled cement mortar are mainly from the electrical resistivity changes between nearby CRC. As a result, the microstructural morphology of the CRC-filled cement mortar is illustrated in Fig. 2.26, where the relative positions between nearby CRC and their relationship to the electrical resistivity changes can be explained. Generally, the relative positions of CRC in the cement mortar can be divided into (1) the state of complete isolation; (2) the closely neighboring state; and (3) the complete contact state.

The morphology of the complete isolation between CRC can be found from Fig. 2.26a with different magnifications in sub-images from 1 to 3. It was observed that the cement matrix totally intrudes into the space between CRC-1 and CRC-2. In those circumstances, since the CRC particles are never in contact with each other in the cement mortar, the electrical resistivity changes under uniaxial compression are mainly induced by the deformation of cement matrix. Based on the literatures, the dense microstructures of the cement matrix under cyclic compression do cause electrical resistivity changes. However, the factional changes of resistivity for the cement matrix itself is very low, even the cement mortar never dried before the piezoresistive test [1]. Therefore, for the 10% CRC-filled cement mortar with a large proportion of the electrical resistivity changes originating from the compressed cement matrix, lowest FCR are monitored in comparison to the cement mortar with higher contents of CRC, because most CRC particles are isolated in the mortar.

Another relative position between CRCs of neighboring state is illustrated in Fig. 2.26b, with different magnifications in the magnified images. It shows that the CRC-3 is completely isolated from the CRC-4 and CRC-5 by cement matrix, similar to the relationship

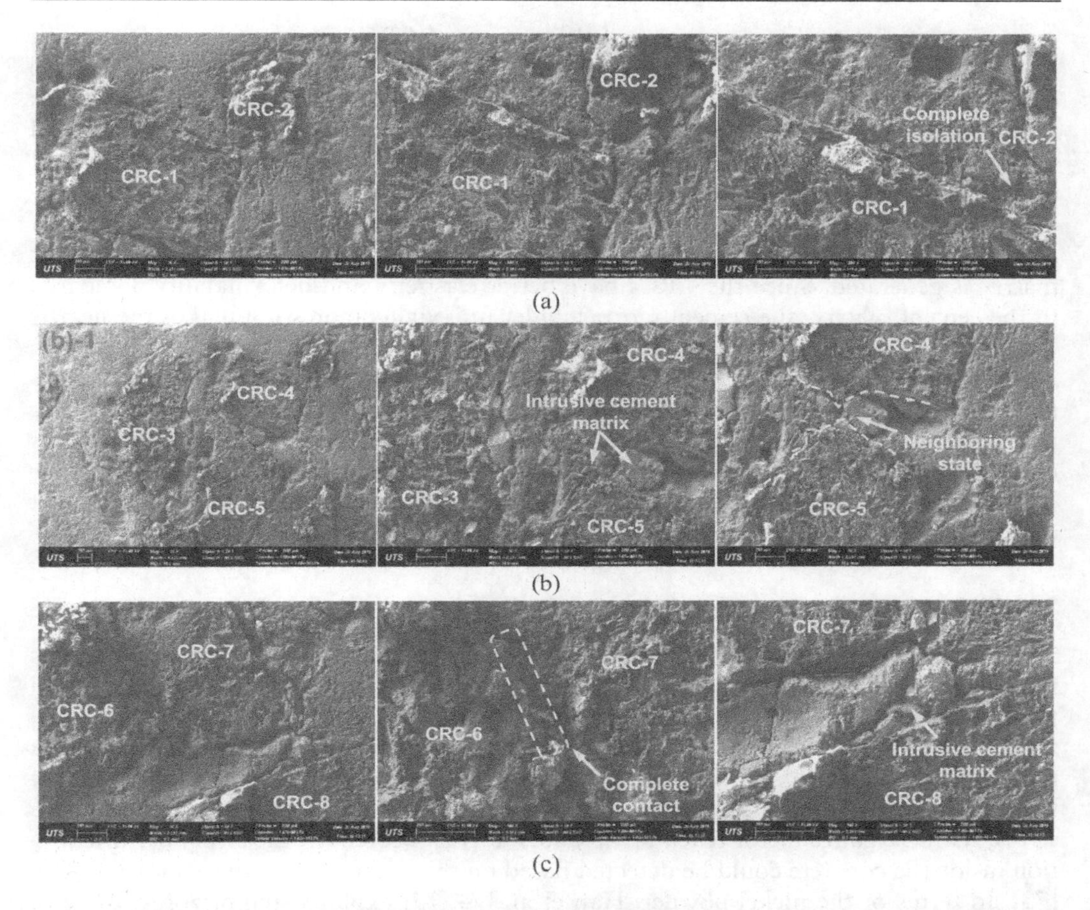

Figure 2.26 Microstructural morphology of CRC-filled cement mortar [68]. (a) Complete isolation. (b) Neighboring state. (c) Complete contact.

between CRC-1 and CRC-2. However, for the CRC-4 and CRC-5, their locations are much closer and nearly get contact. Since the distance between these rubber crumbs is really close, the cement matrix becomes difficult to intrude into the gaps during the casting procedures, resulting in micropores or loose microstructures in this area. Hence, the intrusive cement matrix with loose microstructures is easily damaged after being subjected to uniaxial compression, as shown in Fig. 2.26b (3). In such a situation, the neighboring CRCs have higher possibility to connect with each other under compression and cause the electrical resistivity reduction. In this section, the CRC in the neighboring state is gradually increased with the increase of CRC. That is why the FCR of CRC-filled cement mortar increased with the increased CRC content. On the other hand, due to the loose microstructures of cement matrix in the gaps between neighboring CRCs, permanent electrical resistivity reduction might occur because of the damaged nonconductive cement matrix and the connection between neighboring CRCs. That is why several cement mortars permanently decreased the electrical resistivity after the uniaxial compression.

As for the complete contact state of CRC in the cement mortar, the CRC-6 and CRC-7 have excellent connection with each other, as shown in Fig. 2.26c. Larger magnification on the interfacial transaction zones of the connected CRC indicates that only a small amount of cement matrix can intrude into the boundary region without the presence of fine aggregate, as shown in Fig. 2.26c (2). The self-sensing cementitious composites with the absence of aggregate normally are provided with higher piezoresistivity and self-sensing ability, hence the complete contact state of CRC with only a thin film of cement paste in the boundaries can significantly alter the electrical resistivity under uniaxial compression. In addition, for the cement mortar with excessive content of CRC, it can be predicted that there is a thoroughly connected CRC without the intruded cement matrix is generated. Since the CRCs have better elasticity and deformability compared to the cement matrix, the cement mortar under uniaxial compression makes the nearby CRC compress with each other and it also alters the contact resistivity between CRCs. Also, the compressed CRCs themselves can slightly lower the electrical resistivity due to the closer conductive fillers, and thus contribute to the FCR for the cement mortar under uniaxial compression. Overall, the piezoresistivity of dried CRC-filled cement mortar is mainly brought by the aforementioned three conductive mechanisms. The predominance of electrical conductivity by CRC is increased in the cement mortar with the increase of the CRC content, which causes higher FCR under uniaxial compression.

2.4.5 Other macro-materials

Studies mentioned above have partially illustrated the conductive cementitious composites filled with various macro-conductive materials, while the effect of CF, steel fiber, or nickel powder has not been reviewed on the piezoresistivity. For instance, Wen and Chung [71] found that the steel fiber–filled cement paste had better piezoresistive sensitivity than that reinforced with CF, especially under tensile strain. This has been further elucidated that the steel fiber–filled cement mortar was provided with excellent sensitivity under tensile stress, with the gauge factor reaching up to 5,195 [72]. Moreover, the steel fiber distribution inside the concrete could be detected based on the electrical resistivity measurement [73]. In terms of the nickel powder, Han et al. [74, 75] explored the piezoresistivity of cementitious composites filled with nickel powder and found the gauge factor as high as 1,336.5 and 1,929.5. Furthermore, since the nickel powder could be directionally aligned in the magnetic field, the improved piezoresistivity could be observed for the cementitious composites because of the more generated conductive networks [76]. However, in comparison to steel fiber, it must be noted that nickel powder is one of the heavy metals that might cause diseases in humans and contaminate the environment.

2.5 NANOTECHNOLOGY IN ENHANCING SENSOR FUNCTIONALITY

2.5.1 0D nanomaterials

2.5.1.1 Introduction

There are few studies adopting the 0D fullerene in cementitious composites due to its high price and limited mechanical improvement [77]. On the contrary, NCB has been widely used as conductive fillers in polymeric and cementitious composites for the manufacturing of tires or electrical conductivity amelioration. It has been reported that NCB could reduce the compressive strength of cementitious composites with the increase of

NCB concentration [78], while studies also found that the small dosage of NCB could increase the mechanical properties owing to the filling effect of NCB nanoparticles [79]. By investigating the hydration heat and crystal phases of cementitious composites with NCB, the highest compressive strength reaching approximately 53 MPa was observed for the composites filled with 0.5 wt.% NCB. With increasing curing age, the positive effects of NCB are diminished, and the compressive strength of NCB cementitious composites is lower than that of plain cement mortar [80]. Another investigation also found monotonically decreased compressive strength of NCB cementitious composites with an increase of NCB content [81]. This phenomenon might be due to the sphericity of the NCB, thus lack of cohesion and frictional resistance to the cement matrix. Also, the nanoscale NCB possessing high surface energy could enclose the cement particles and entrap the water to affect the cement hydration and hardening process.

2.5.1.2 Percolation threshold

Fig. 2.27 compares the electrical resistivity of NCB-filled cementitious composites by different studies. The electrical resistivity of cementitious composites decreased with the addition of NCB. However, it seems that the resistivity reduction intensity has a solid relationship with the types and particle sizes of NCB. Dai et al. [78] found the percolation threshold for NCB cementitious composites ranging from 0.36 vol.% to 1.34 vol.% (0.7–2.5 wt.%), while a slightly lower NCB content range of 0.5 wt.% to 2.0 wt.% was achieved by Dong et al. [81] and much higher NCB content range of 7.22 vol.% to 11.39 vol.% was observed by Li et al. [82]. Based on their description of the raw materials, the particle sizes of NCB in the above investigations were 33, 20, and 120 nm, respectively. It could be deduced that the NCB nanoparticles with a smaller scale can more easily reach the percolation threshold. Moreover, the above studies showed a discrepancy in the specimen preparation by the usage of a dispersion agent. The well-dispersed NCB

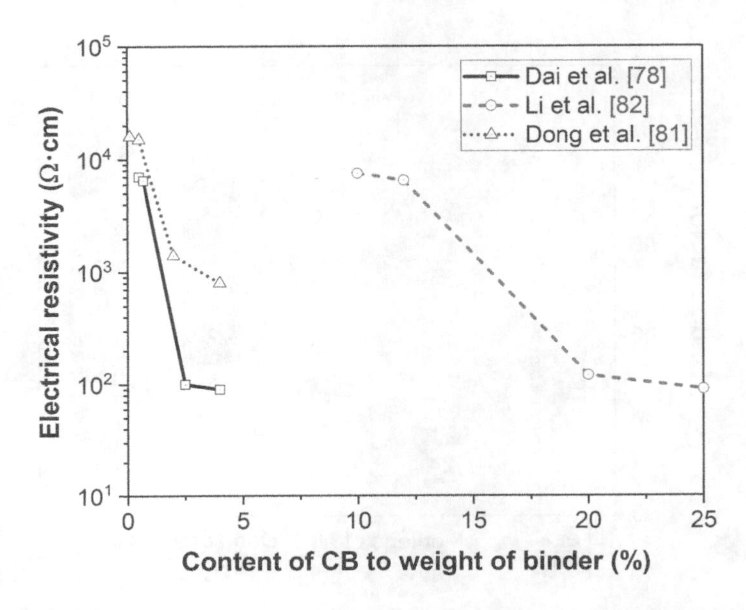

Figure 2.27 Percolation threshold of cementitious composites filled with NCB in different studies.

nanoparticles in the cementitious composites can stimulate the formation of the conductive passage and enhance electrical conductivity.

2.5.1.3 Self-sensing capacity

For the self-sensing or piezoresistivity of cementitious composites doped with NCB, the stable electrical output as a function of compressive stress/strain could be drawn, indicating that the NCB cementitious composites are capable of sensing stress and strain for structural health monitoring (SHM). As shown in Fig. 2.28, the piezoresistive sensitivity, which is also called as gauge factor, was proposed in various investigations on the NCB cementitious composites. The gauge factor ranging from 37.7 to 55.3 was obtained for the cementitious composites filled with 15.0–25 wt.% NCB by Li et al. [82]. Besides, they proposed a numerical model to evaluate the gauge factor under various strains, NCB concentrations, and temperatures [83]. Afterward, the lowest piezoresistive sensitivity was found by Monteiro et al. [84] with the gauge factor ranging from 24 to 30 for the composites filled with 10 wt.% and 7.0 wt.% NCB. Given the NCB nanoparticles had an identical particle size of 120 nm, the reduced gauge factor is considered owing to the introduction of fine aggregates, which might block the formation of conductive passages and weaken piezoresistivity. Later, they observed the slightly larger gauge factor 40–60 for the NCB cementitious composites when the NCB concentration decreased to 6.5 wt.% [85]. Because the aforementioned studies did not predisperse NCB nanoparticles before manufacturing the composites, the piezoresistive sensitivity for the NCB cementitious composites was obtained by Dong et al. [81] with the gauge factor ranging from 51 to 114, after the pretreatment of NCB dispersion by superplasticizer. Although super-high gauge factor larger than 300 was developed in the study, the composites were not only filled with NCB nanoparticles but also assisted by the macro-conductive rubber fibers. It is noted that higher gauge factor was observed by Li et al. [86] who applied NCB into

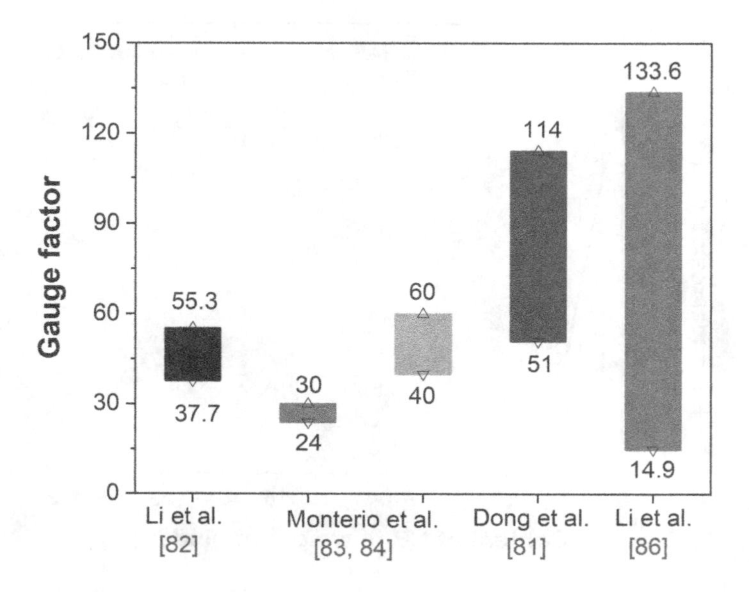

Figure 2.28 Various gauge factors for the NCB-filled cementitious composites under compression.

fibers-reinforced engineered cementitious composites (ECC). The results are similar to the previous investigation of using nonconductive PP fibers to improve the stress-sensing ability of NCB CBSs. The better sensitivity is mainly due to the newly formed conductive passages through the NCB-enclosed PP fibers. Therefore, the experimental results showed that the NCB cementitious composites only possessed moderate piezoresistive sensitivity.

2.5.2 1D nanomaterials

2.5.2.1 Introduction

CNF and CNT are the commonly used 1D conductive carbon materials in cementitious composites. In terms of CNF in cementitious composites, Chung [87] demonstrated that the CNF with a diameter of 100 nm was less effective to improve the electrical conductivity of cementitious composites, in comparison to the CF with 15-μm diameter. The investigator attributed the phenomenon to the more substantial contact resistance between conductive fillers to the cement matrix in a specific area per unit volume because of the small sizes of CNF. However, different results were obtained for the CF- and CNF-reinforced concrete, in which better electrical conductivity was found for the CNF-filled concrete [88]. The discrepancy also occurred on the mechanical properties of CNF-reinforced cementitious composites, since both significantly reduced and enhanced compressive strength of the composites were achieved. Because the CNF is easily self-agglomerated to form clusters, the worse electrical conductivity and reduced mechanical property are mainly due to the poor dispersion of CNF that causes microdefects in the cementitious composites. The improved mechanical strength is because of the CNF having the bridging effect to the cement matrix, as well as reducing the porosity in the composites. As for the piezoresistivity, the CBSs with CNFs and CFs were attached to reinforced concrete beams for strain and damaging monitoring. The results showed a higher gauge factor of 191.8 for the CNF-filled CBS, while only 49.5 for the CF-filled sensor in the compression [89].

Compared to the CF and CNF, studies are more focused on the electrical and piezoresistive properties of cementitious composites reinforced with CNT. By utilizing the electrical conductivity of CNT-reinforced cementitious composites, Kim et al. [90] investigated their self-heating efficiency under the input voltages of 3–20 V. In addition, Kim [91] proposed that the chloride penetration into cementitious composites could be monitored by the electrical conductivity alterations of embedded CNT–cement composites. They reported that the conductivity of composites was determined by the chloride ions movements and polarization effect. Based on the three impedance arcs/features in Nyquist plots, Wansom et al. [92] investigated the impedance spectroscopy of CNT-filled cementitious composites. They proposed that spectroscopy could distinguish the percolation from discontinuous CNT and characterize the agglomerations. The following sections introduce the electrical and piezoresistive performance of cementitious composites containing as-received and well-dispersed CNTs

2.5.2.2 CNT as-received

To characterize agglomerations, a SEM image in Fig. 2.29a, which possessed various sizes and shapes of CNT agglomerations, was separated into several images in Fig. 2.29c–f, where CNT agglomerations with the same diameter (*D*) were grouped in the same color [93]. The diameter (*D*) was determined by the averages distance of the lines passing through

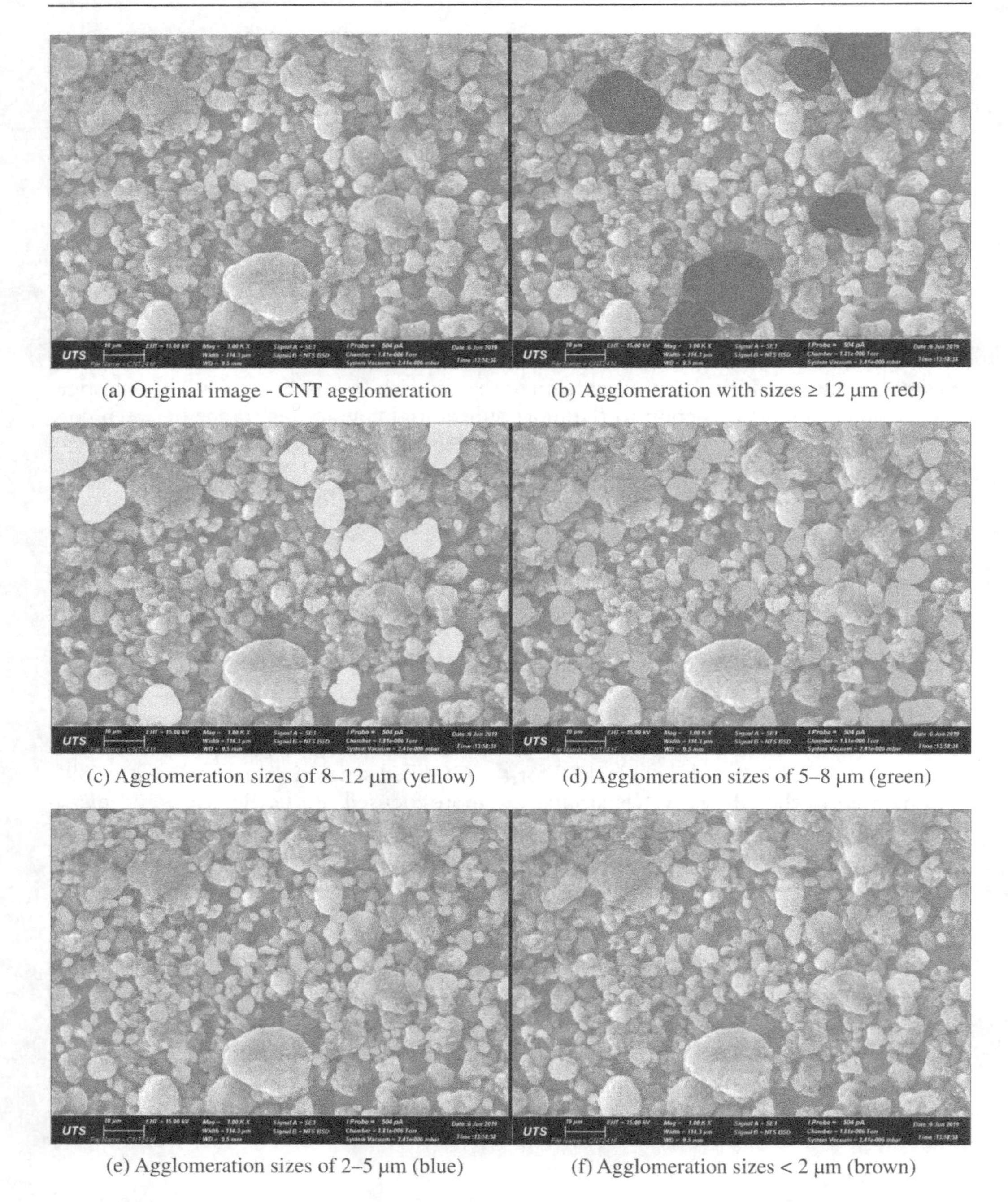

Figure 2.29 Classification of CNT agglomerations based on size distribution [93].

the centroid of agglomerations. Five groups were developed for agglomerations in the range of ≥12 μm, 8–12 μm, 5–8 μm, 2–5 μm, and ≤2 μm. The division based on the sizes not only facilitated the analysis but also was more convenient to distinguish the CNT agglomerations in different sizes, especially for the smaller one which might have been attached to the larger agglomerations. The shapes of CNT agglomerations were irregular

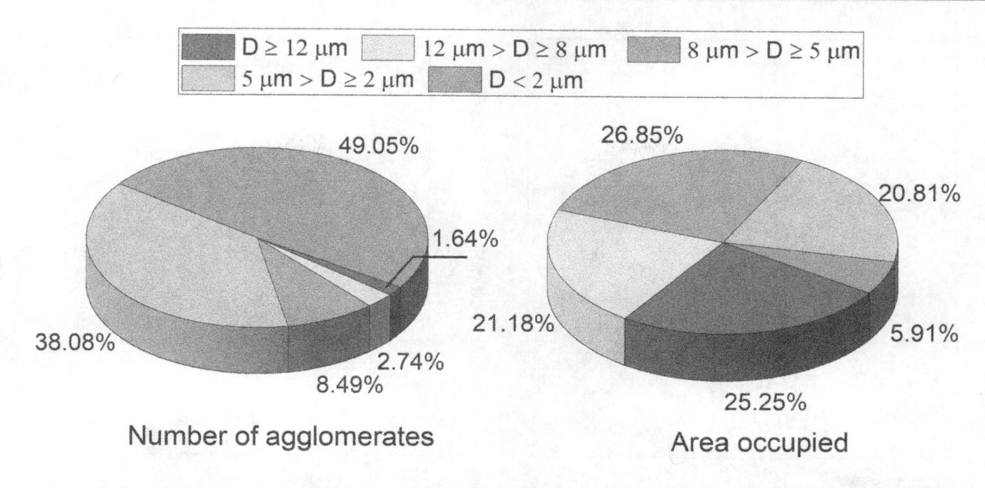

Figure 2.30 Size distribution of CNT agglomerations and the proportion of the occupied areas [93].

and varied with different sizes and roundness. It could be found that the number of larger sized agglomerations was lower than that of agglomerations in smaller sizes. Fig. 2.30 illustrates the diameter distribution of CNT agglomerations and the proportion of their occupied areas. It could be found that the CNT agglomerations in the diameter range of ≥12 μm possessed the smallest quantity (1.64%) but ranked second in the area occupied (25.25%). On the contrary, the smallest agglomerations with the diameter <2 μm accounted for the highest proportion in the aspect of numbers (49.05%) and the lowest ratio (5.91%) in the area occupied. The results implied that although a minority of CNT agglomerations were in larger sizes such as ≥12 μm, the volumetric proportion occupied by these agglomerations was in a large value. In addition, it could be observed that the majority of CNT agglomerations were in the sizes smaller than 8 μm (95.62%).

In terms of the self-sensing capacity, Fig. 2.31a–e shows the FCR of layer-distributed as-received CNT composite (LDCC) with different layers of CNT under the same cyclic compression. Despite the CNT powders with the absence of ultrasonic/surfactant dispersion, the piezoresistivity could still be observed for the LDCC. Previous investigations on the CNT CBSs mainly focus on the cementitious composites reinforced with dispersed CNT in aqueous solution, but none of them attempts to use the CNT powders or agglomerations for manufacturing piezoresistive cementitious composites. The main reason is that the influences of CNT on the mechanical properties of cementitious composites have been widely investigated, and the common senses have been established that the CNT agglomerations are detrimental to the mechanical properties of cementitious composites. In this section, the piezoresistivity could be achieved for the composites with only one layer of CNT powders (LDCC1), but meanwhile the satisfactory compressive strength was maintained well. Overall, the piezoresistive results indicated that the cementitious composites with even undispersed CNT could display the stress-sensing or piezoresistive property. It provides a new handy method without nanoparticle dispersion to manufacture the self-sensing CBSs, which could reduce the operational complexity and save energy in real construction sites.

However, different from the stable resistivity output of UDCC, the LDCC depicted various degrees of irreversibility in the FCR values under cyclic compression. This is due to the existence of weak CNT agglomerations and the induced porosity, which might be permanently deformed and alter the electrical conductivity of the composites. The

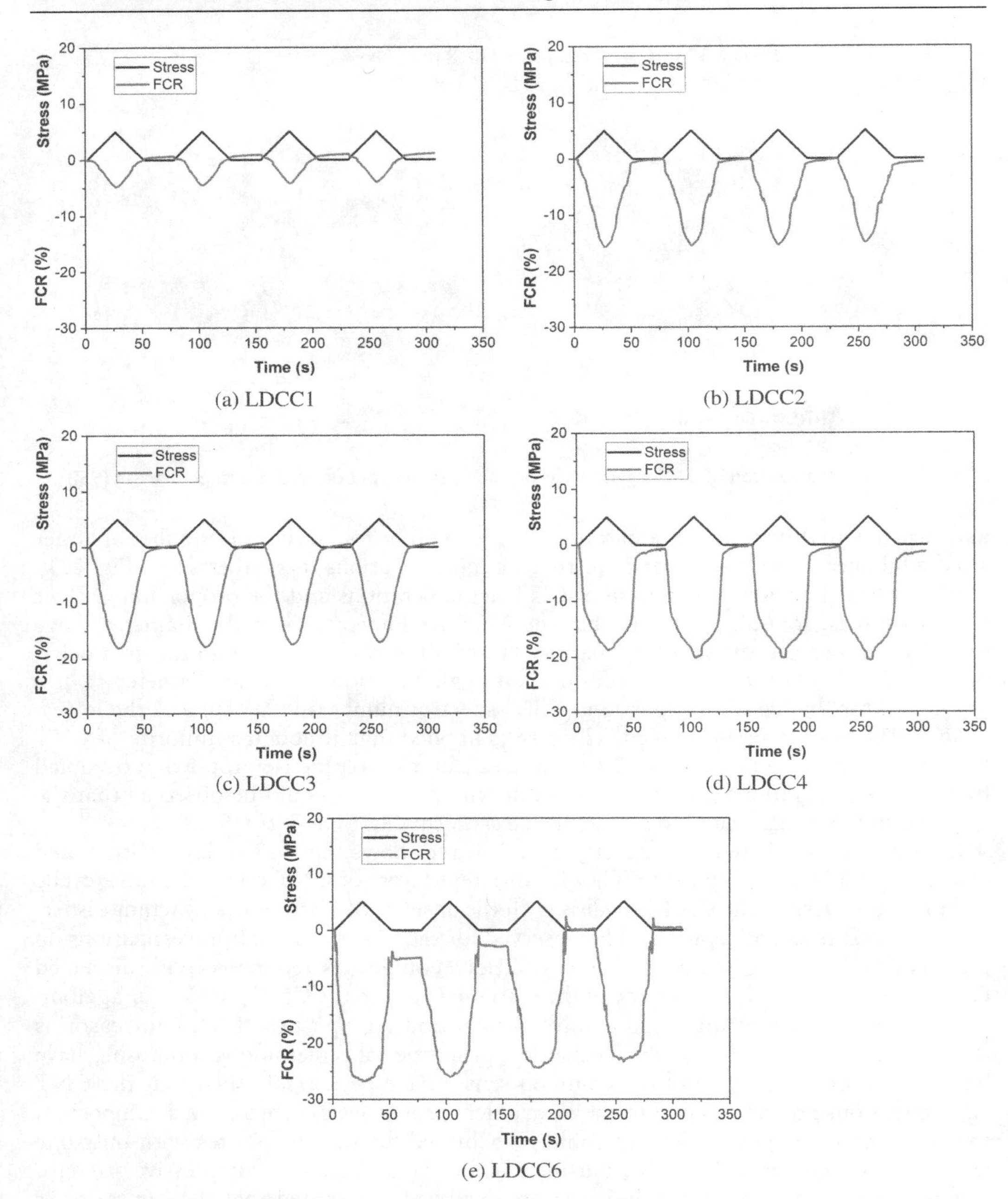

Figure 2.31 Electrical resistivity changes with the applied compressive stress for LDCC with different layers [93].

tendency could be seen that the irreversibility is increased with the number of CNT layers, especially for the LDCC4 with irreversible FCR of approximately 2% after cyclic compression. As for the LDCC6, significantly volatile FCR with the compressive stress could be seen, because of the excessive instable CNT agglomerations. It shows that the agglomerations decreased the resistivity in the first compression cycle, and gradually increased the resistivity in the following cycles. The changing rate attenuated with the compression implied that the irreversible deformations and linkages of CNT agglomerations were inclined to become stable when subjected to cyclic compression. In terms of their stress-sensing efficiency, the FCR varied with the contents and layers of CNT. The average values reached only 4.4% for the LDCC1. However, it could be seen that the FCR at the stress peaks increased with the CNT layers in the composites, with the values of 15.4%, 17.7%, 19.0%, and 23.2% for the LDCC2, LDCC3, LDCC4, and LDCC6, respectively [94].

2.5.2.3 Well-dispersed CNT

Fig. 2.32 illustrates the microstructures of composites at room temperature to assess the performance of well-dispersed CNT. Generally, intact microstructures could be detected on both the hydration products of C-S-H and the MWCNT. Just as many literatures mentioned reinforcements by MWCNT, the strengthen effect was mainly due to the MWCNT bridging and filling effect [94]. It was observed that C-S-H and MWCNT maintained excellent connections, which were responsible for the improved compressive strength and elastic modulus for the MWCNT-reinforced composites. In addition, the nucleation effect of MWCNT could be deduced from the SEM images, due to different thicknesses of MWCNT with attached hydration products observed.

To evaluate the piezoresistive sensitivity and repeatability, the cyclic compression having different stress magnitudes of 4 MPa, 6 MPa, 8 MPa, and 10 MPa were applied for piezoresistive test, with the loading rates of 0.33 kN/s, 0.5 kN/s, 0.67 kN/s, and 1.0 kN/s, respectively. For each stress magnitude, there existed three compression cycles. As shown in Fig. 2.33, a decreased electrical resistivity in the loading process and an increased resistivity in the unloading process could be observed for all composites. For the composites with 0.25% MWCNT, the FCR reached 6.2%, 9.4%, 12.1%, and 14.5%, respectively, at the stress magnitudes of 4 MPa, 6 MPa, 8 MPa, and 10 MPa, exhibiting acceptable linearity and excellent reversibility. In comparison to 0.25% MWCNT, smaller FCR and gentler resistivity jumps were observed for the 0.50% MWCNT-reinforced cementitious composites subjected to identical loadings and heat treatment. In particular, it was found that the FCR reached 6.8%, 8.8%, 10.8%, and 12.5%, respectively, at stress peaks of 4 MPa, 6 MPa, 8MPa, and 10 MPa for the composites without heat treatment. Although

Figure 2.32 Microstructure morphology of cementitious composites incorporating 0.25% MWCNT [93].

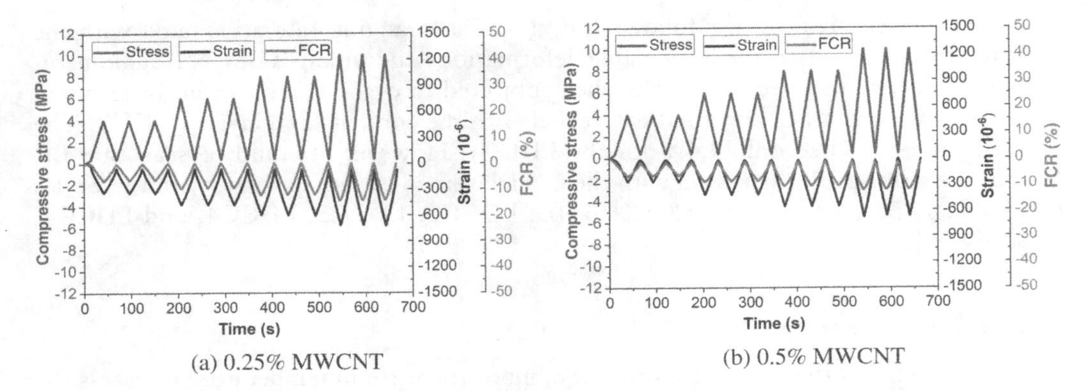

(a) 0.25% MWCNT (b) 0.5% MWCNT

Figure 2.33 Fractional changes of resistivity of MWCNT–cementitious composites related to compressive stress and strain [95].

the FCR in the first three cycles were relatively higher achieving 6.8% than 6.2%, the later increment of FCR for the composites with 0.25% MWCNT were considerably weakened for the composites with 0.50% MWCNT. Similar to composites with 0.25% MWCNT, the composites with 0.50% MWCNT still maintained stable piezoresistivity, and exhibited satisfactory repeatability [95].

2.5.3 2D nanomaterials

2.5.3.1 Introduction

The 2D graphene consists of a 2D single layer of sp^2-hybridized carbon atoms bonded in a hexagonal honeycomb lattice. GNP has an elastic modulus of 1.0 TPa, a tensile strength of 130 GPa, and electron conductivity of 10^7–10^8, which are also higher than that of CNTs and hence give GNP a wide range of potential applications in cementitious composites. In addition, the GNP used in cementitious composites usually had multiple layers. According to the different number of layers, GNP has a single layer or few layers (not more than ten layers), and the GNP having more than ten layers is considered as GNP nanoplates (GNPs). GNP and GNPs share highly similar properties due to their massive similarities in composition and structure. However, compared with GNP, the increasing number of layers leads to a decrease in the surface area of GNPs, which possibly has an adverse effect on the GNPs-based cementitious composites [96]. On the other hand, the cost of GNPs produced is much lower than that of the single-lager GNP, which makes GNPs more advantageous in cementitious composites when considering the production costs [97].

2.5.3.2 Electrical resistivity

In addition to the excellent durability and mechanical properties brought to the cementitious composites, improvements in electrical conductivity are still inspiring. Le et al. [98] found that the electrical conductivity of high-concentration GNP-reinforced mortar was not only insensitive to moisture content but also equivalent to the elastic compliance that made the damage self-sensing possible. Similarly, Liu et al. [99] investigated different failure patterns of GNP-filled cementitious composites under compression, and observed that the reaction of resistivity relates to the damage evolution processes. Also, they suggested that the concentration of GNP should be higher than 6.4% by weight of cement to ensure satisfactory conductivity and piezoresistivity [100]. Furthermore, investigators have studied the electrical conductivity of cementitious composites filled with GO and found even higher resistivity compared to the plain composites. However, Saafi et al. [11] investigated the electrical and piezoresistive behaviors of fly ash–based geopolymer filled with reduced GO and observed satisfactory conductivity and piezoresistivity. Generally, it can be deduced that the introduction of oxygen-containing groups on the graphene film decreases the electrical conductivity of GNP and deteriorates the conductivity improvement on the cementitious composites. The reduced GO regains the electrical conductivity that could be used for CBSs manufacturing.

Fig. 2.34a compares the electrical resistivity of GNP- and GP-filled cementitious composites with different concentrations. When the content of conductors is less than 1%,

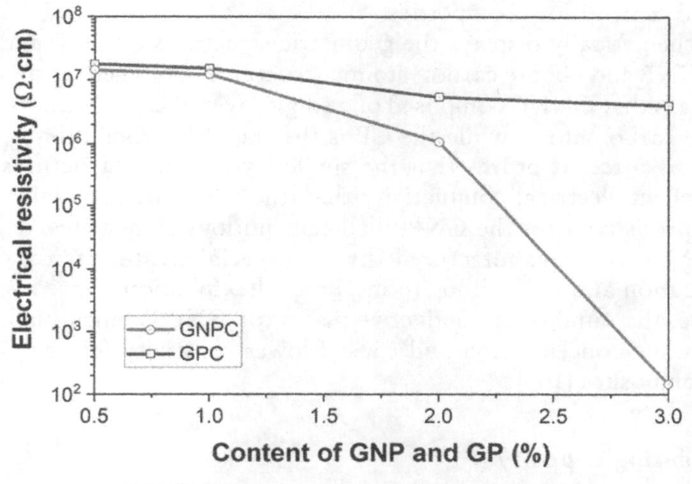

(a) Electrical resistivity of GNPC and GPC

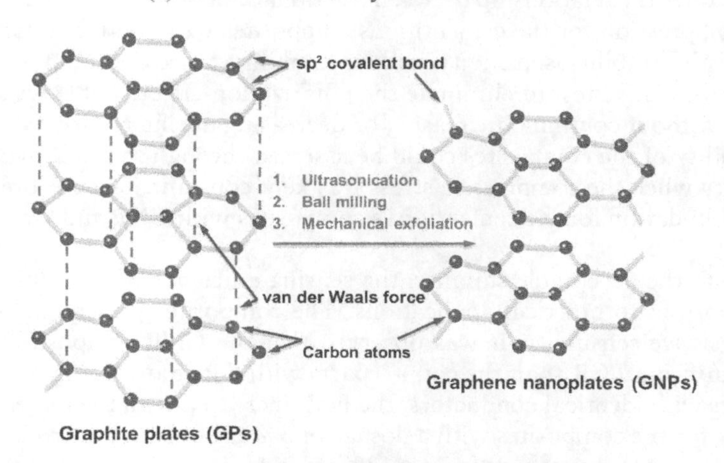

(b) Atomic structures between GNP and GP

Figure 2.34 Electrical resistivity and atomic structures of cementitious composites with different contents of GNP and GP [101].

the electrical resistivity of GNP- and GP-filled cementitious composites were very similar, and both were as high as 1×10^7 Ω·cm. It showed that the contents of GNP and GP from 0.5% to 1% failed to considerably reduce the electrical resistivity of the cementitious composites. Still, the electrical resistivity of GNPC was slightly lower than the counterparts filled with GP. Afterward, the electrical resistivity of GNP composites experienced a moderate reduction to approximately 1×10^6 Ω·cm as the content increased to 2%, while the GP composites only slightly decreased electrical resistivity to nearly 6×10^6 Ω·cm at the same concentration. As for the concentration of 3%, the electrical resistivity of GNP composites considerably decreased to only 1×10^2 Ω·cm. It implies that the percolation occurred for GNP composites as the content of GNP increased from 2% to 3%. However, the electrical resistivity of GP composites did not illustrate an apparent reduction, with the values still up to 4×10^6 Ω·cm. The number of conductive plates for GNP is much higher than that of GP under the same concentration and caused lower resistivity for the GNP-reinforced cementitious composites.

Fig. 2.34b schematically displays the geometric structures of GNP and GP in atomic size. Both the GNP and GP are carbon atomic structures combined by sp^2 covalent bond. Their difference is that GNP is composed of a single layer of carbon atoms with the thickness of a single carbon atom, while the GP is the stacked carbon atomic layers because of van der Waals force. It proves that the single-layer atomic structures provide GNP with more excellent electrical conductivity than the GP, which partially contributed to lower electrical resistivity for the GNP-filled cementitious composites. Moreover, it was seen that GNP could be manufactured through special treatments on the layered GP from ultrasonication and ball milling to mechanical exfoliation. Once overcome the van der Waals force, the number of conductive plates for GNP is much higher than that of GP under the same concentration and caused lower resistivity for the GNP-reinforced cementitious composites [101].

2.5.3.3 Self-sensing capacity

Fig. 2.35 illustrates the relationship between fractional changes of electrical resistivity and the applied compression for the cementitious composites with GNP and GP, to compare their piezoresistive stability, sensitivity, and repeatability. Since the composites were dried before the piezoresistive test to eliminate the polarization effect, stable resistivity output was observed without continual increasing or decreasing during resistivity measurement. Also, the stability of the composites could be assessed through the well-maintained electrical resistivity when the compressive stress was kept constant. The piezoresistive stability can lay a foundation for the cementitious composites with GNP and GP to become the stress sensor.

In addition to the acceptable stability, the sensing efficiency of the cementitious composites is important to practical applications. The composites presented completely different piezoresistive sensitivity. It was observed that the GNP composite was provided with more significant FCR than the counterparts with GP at any dosages. Moreover, for the composites with identical conductors, the FCR increased with the increase of GNP or GP. Especially for the composites with a dosage of 3% conductors, the resistivity changes were as great as approximately 30% and 13%. Combining the results of electrical resistivity, it was found that the better electrical conductivity for the cementitious composites benefited the piezoresistive sensitivity. This can be explained by the number of conductive passages and the contact points in the composites. The high concentration of conductors

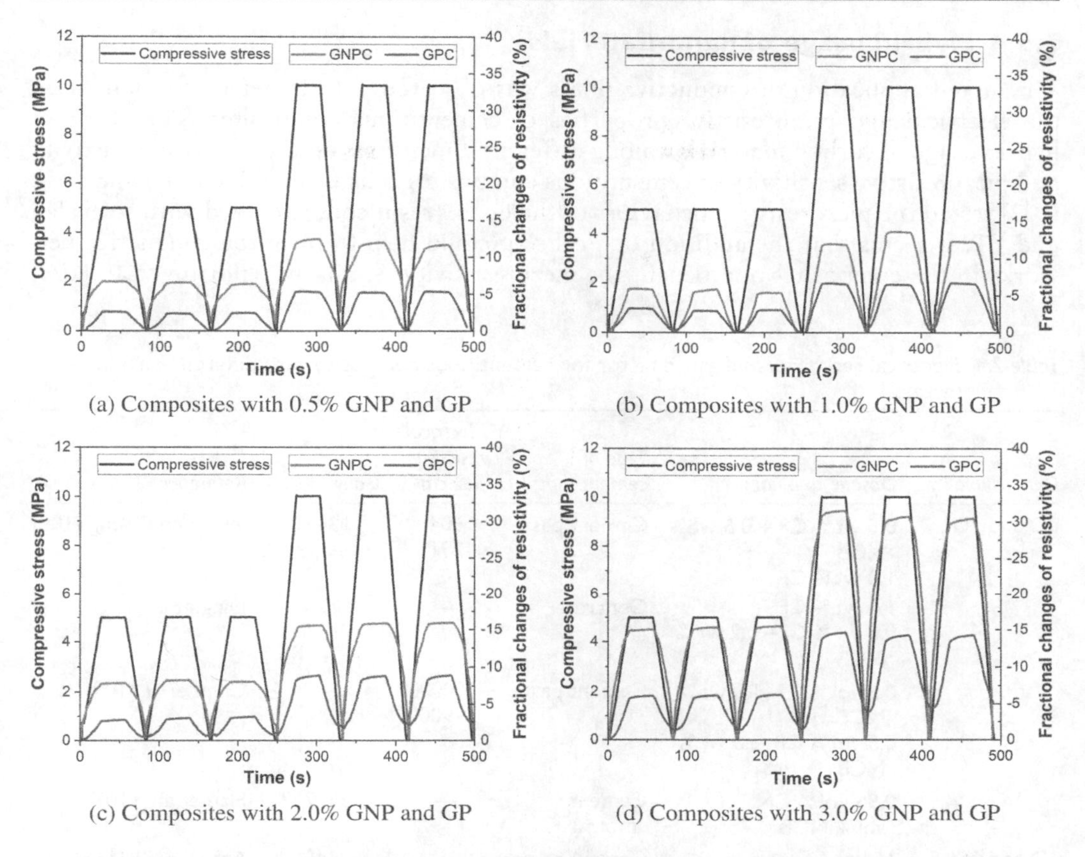

Figure 2.35 Fractional changes of resistivity for cementitious composites with GNP and GP under cyclic compression (GNPC or GPC) [101].

and lower electrical resistivity of composites represented the more generated conductive passages and more contact points. During the piezoresistive test, the more contact points between nearby conductors tended to link or separate with each other and caused the higher FCR.

In terms of the piezoresistive repeatability, repeatable electrical resistivity could be observed for all the GNP composites because of the small resistivity irreversibility after the cyclic compression. As for the GP composites, satisfactory repeatability was observed in the case of the small dosages at 0.5% and 1.0%. However, considerable irreversibility immerged when the concentration increased to 2.0% and 3.0%, with the irreversible resistivity reaching approximately 2.5% and 5.0%, respectively. This is consistent to the above-mentioned mechanical properties and the microstructures of the cementitious composites with GNP and GP. The GNP had an excellent connection to cement matrix and made denser microstructures of GNP–cement composites with higher compressive and flexural strength. In contrast, the GP tended to cause worse cohesion with the cement matrix due to the weak van der Waals force between sheets which might lead to slippage and the spalling of GP, and resulted in reduced mechanical properties. Therefore, the irreversible electrical resistivity could be easily generated for the composites with a high dosage of GP, owing to the lower compressive strength and the loose microstructures [101].

2.5.4 Hybrid usage of nanomaterials

The mixed application of conductive fillers was observed with better performances on the electrical and piezoresistive properties of cementitious composites. The effect of hybrid usage of carbon materials among different dimensions on the electrical resistivity and piezoresistive sensitivity of cementitious composites is listed in Table 2.4. Ding et al. [102] tested the piezoresistive behaviors of the epoxy resin concrete filled with steel slag and GP. It seems that the addition of steel slag could help to form the conductive networks in the composites and benefit the piezoresistivity. Similarly, Yıldırım et al. [103]

Table 2.4 Electrical resistivity and gauge factor for cementitious composite with different carbon materials

Dimensionality	*Dosage to binder*	*Type of cement matrix*	*Electrical resistivity (Ω·cm)*	*Gauge factor*	*Reference*
0D and 1D	0.5 wt.% CF + 0.5 wt.% NCB 1.0 wt.% CF	Cement paste	604 791	13–17 ~332	Wen and Chung [108]
	1.0 wt.% CF 0.8 wt.% CF+ 0.2 wt.% NCB	Concrete	—	25.6 13.5	Ding et al. [102]
	0.5 vol.% CF+1.0 wt.% NCB 0.5 vol.% CF+2.0 wt.% NCB	Cement paste	~ 5000000 ~ 900	—	Chen et al. [109]
	0.96 vol.% CNT+1.44 vol.% NCB	Cement mortar	—	625–727	Han et al. [110]
1D and 1D	15 vol. CF 15.0 vol. CF+1.0 vol. CNT	Cement paste	—	445 422	Azhari et al. [111]
	1.0 vol.% CNT 0.5 vol.% CNT+0.1 vol.% CF	Cement paste	—	167 160	Lee et al. [112]
	1.0 wt.% NCB+1.5 wt.% CNT+0.2 wt.% CF 1.0 wt.% NCB+1.5 wt.% CNT+0.4 wt.% CF	Cement paste	10,050 237	—	Hong et al. [113]
1D and 2D	1.0 wt.% CF 0.7 wt.% CF+2.5 wt.% GP 0.5 wt.% CF+5.0 wt.% GP	Concrete	~ 1120 ~ 990 ~ 970	—	Chen et al. [114]
	4.0 wt.% GNP + 0.66 wt.% CF	Cement mortar	—	295	Belli et al. [105]
0D, 1D, and 2D	12.7 vol.% GP 7.3 vol.% GP + 3.6 vol.% CF 7.3 vol.% NCB + 3.6 vol.% CF	Concrete	140,000 ~ 1200 ~ 1400	—	Wu et al. [106]

investigated the mechanical and shear damages of self-sensing ability of chopped CF and CNT-reinforced concrete, and found that the CF could significantly improve the flexural strengths and ductility of concrete. They found that the addition of CF could improve the sensing efficiency compared to the concrete beams only with CNT. By manufacturing the asphalt concrete filled with GP and CF, Liu and Wu [104] found that the specimens are more effective for compressive strain monitoring. Also, the combined use of GNP and CF in cementitious composites led to high conductivity and piezoresistivity [105]. On the other hand, the partial substitution of costly conductive fillers by cheap ones could dramatically reduce engineering expenditure. Wu et al. [106] compared the electrical conductivity of the asphalt concrete filled with single NCB, CF, GP, and their hybrids, and they found that the CF is best to improve the conductivity, followed by GP and NCB. Fortunately, the replacement of expensive CF by low-cost NCB and GP can reduce the engineering cost and maintain similar electrical conductivity. In addition, although the electrical conductivity and piezoresistivity were not involved, the better EMI shielding for the cementitious composites with GO-deposited CF than those only filled with CF indicated that the hybrid utilization of conductive carbon materials was an effective alternative [107].

2.5.5 Conductive mechanism

The electrical conductivity of the cementitious composite depends on whether or not conductive passages are created throughout the specimen. With the increase of fillers, the electrical percolation threshold was proposed to present the beginning of the formation of conductive passages, where the electrical conductivity sharply increased within a small change in the concentration of fillers. In other words, the percolation threshold illustrates the smallest dosage ranges of conductive fillers that significantly increase the electrical conductivity of cementitious composites. Before the formation of continual conductive passages, various connecting states of conductive fillers represent different conductivity of cementitious composites. With a small dosage, the conductive fillers are completely separated by the cement matrix and the composites show a high electrical resistivity. When the conductive fillers are at a moderate concentration, there exists the neighboring state of fillers and partial conductive paths might be established due to the tunnelling conductivity. Once the content of conductive filler exceeds the percolation threshold, the complete contact among fillers can be observed, and hereafter the electrical conductivity of composites experienced a significant enhancement.

However, it is quite different for the 0D, 1D, and 2D carbon materials to connect each other and form conductive passages, which makes their percolation threshold changeable in cementitious composites. As shown in Fig. 2.36a–c, the 1D carbon materials with a high aspect ratio perform better on declining the electrical resistivity of cementitious composites than those filled with 0D and 2D materials [115, 116]. More electrical contact points tend to be established among 1D fibrous fillers rather than the 0D spherical fillers. Additional usage of 2D nanoplate stimulates the formation of contact points, but the efficiency is relatively lower than that of 1D CNT under an identical volume fraction. The smaller size nanoparticles perform better than the micro- or macro-sized carbon materials in the cementitious composites. Still, the application of macro-fibers can enhance the electrical conductivity of cementitious composites. As illustrated in Fig. 2.36d and e, conductive NCB nanoparticles are connected to form conductive passages. Hence, the cementitious composite is provided with moderate

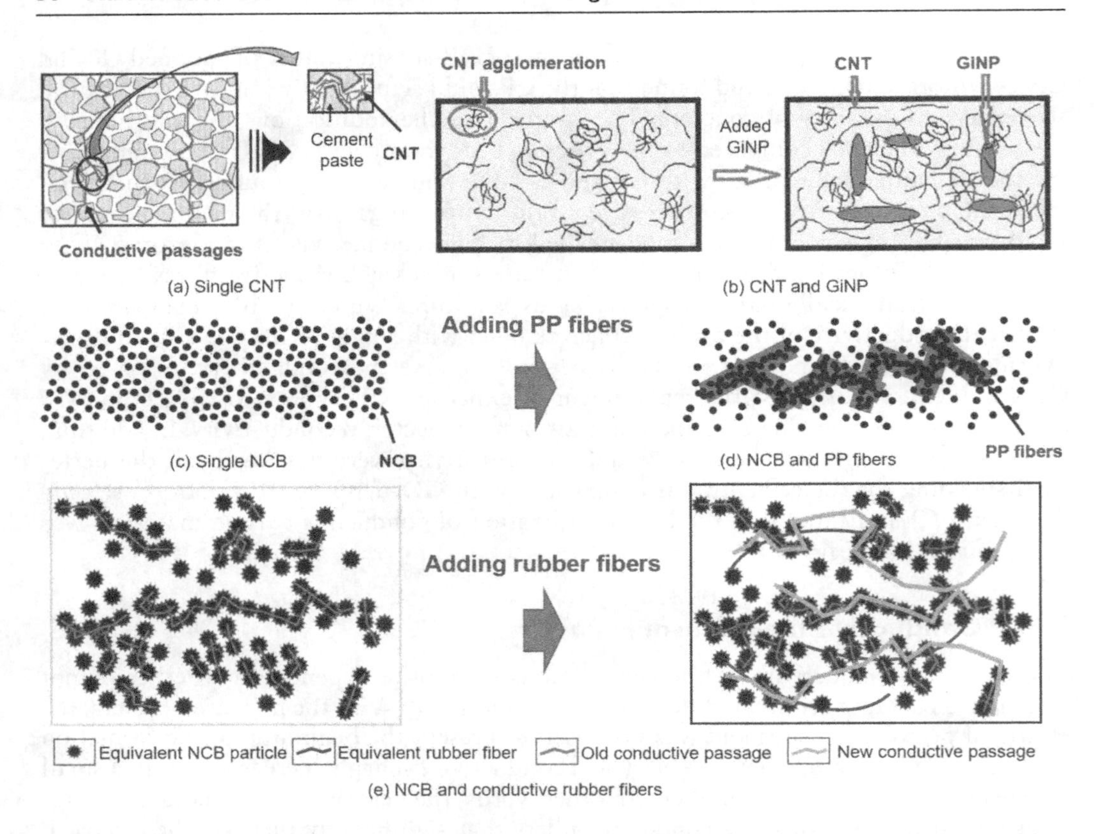

Figure 2.36 Conductive mechanism of the cementitious composites filled with various conductive fillers [97].

electrical conductivity and piezoresistivity. However, with the addition of macro-PP fibers and conductive rubber fibers, new conductive passages are established by the connection of macro-fibers and NCB to fibers. Also, the directional distribution of NCB nanoparticles on the surface of PP fibers enhances the formation of conductive passages because the PP fibers possess a higher aspect ratio than that of NCB. The appropriate concentration of macro-fibers increases the contact points and the conductive passages in cementitious composites, which are beneficial to the conductivity and piezoresistivity. Similarly, for the combined usage of NCB and CNT in cementitious composites, the CNT elongates the conductive passages and enhances the electrical conductivity [117].

2.6 SUMMARY AND CONCLUSIONS

This chapter provides a comprehensive overview of the essential components utilized in the development of CBSs. The raw materials for production of CBSs, including matrix cementitious materials, SCMs, conductive macro-reinforcing materials, and nano-carbon materials from NCB, CNT, CNF, GNP, and GP, have been summarized, in terms of their effect on improvement of conductivity and piezoresistivity of CBSs. By exploring the diverse materials and their synergistic effects, the chapter offers valuable insights into the design and fabrication of CBSs with enhanced sensing capabilities. The following are some key conclusions:

1. The existence of aggregate affects both electrical conductivity and piezoresistivity of CBSs, because the nonconductive aggregate hinders the formation of conductive pathways. The utilization of AAMs and geopolymer as matrix of CBSs may be a good option since the ionic conductivity is enhanced in the AAMs compared to that of cement matrix. The w/b ratio influences the slump flow and rheology of CBSs, as well as the dispersion efficiency of conductive fillers. Therefore, a proper w/b ratio is critical to the distribution of conductors, electron movements, and generation of the conductive paths in CBSs. The Poisson's ratio of CBSs is small, but its influence on the resistance and piezoresistive performance cannot be ignored. Under complex stress circumstances, the sensing ability of CBSs can be more accurate if the calibration of CBSs is conducted under the real stress condition of triaxial compression rather under one-dimensional loading.
2. The use of SCMs reduces the production cost of CBSs and enhances their sustainability, durability, mechanical strength, and even piezoresistive sensitivity of CBSs. Usually, the application of SF stimulates uniform distribution of conductive fillers and promotes cement hydration in CBSs. FA is able to enhance the workability and microstructures to influence the self-sensing performance of CBSs. Slags containing various metal oxides provide additional conductive particles in the composite, which often increase the electrical conductivity and bring piezoresistivity. The application of agricultural by-products, shell power, and other natural pozzolans in CBSs decreases the cement usages, reduces the cost for the sensors' production, and paves the way for future intelligent, low-carbon, and green building materials.
3. Usage of macro-reinforcing materials ranges from PP fiber, glass aggregate, CRF, CCR, steel fiber, and other metal fillers. The introduction of PP fibers benefits the orientation distribution of NCB and improves the electrical and self-sensing performance of CBSs. Glass aggregate often possesses poor electrical conductivity, but it can be used as conductive fillers in cement-based cements after special CNT coating treatment. The electrically conductive rubber products of CRF and CCR are rarely utilized in cementitious materials to enhance the conductivity and piezoresistivity. This chapter provides a novel approach to recycle CRF and CCR in cementitious materials for the manufacturing of CBSs for SHM. The metal fillers can efficiently increase the electrical conductivity and piezoresistivity of cementitious materials, but its high prices increase manufacturing costs of CBSs.
4. Nanotechnology has significantly improved the efficiency and reliability of CBSs for SHM. The 0D NCB are qualified additives to improve the electrical conductivity and piezoresistivity of cementitious composites. However, the conductivity promotion efficiency is lower than 1D and 2D carbon materials, since the spherical CB is more challenging to form conductive passages. It usually leads to the reduced fresh and mechanical properties of cementitious composites. The 1D CNF, CNT, and 2D GNP could significantly improve the electrical conductivity and piezoresistivity of cementitious composites as well as mechanical properties. The enhancement is either due to the bridging effect of fibers or the filling effect of carbon materials in the nanoscale. The 2D GP could improve the conductivity and piezoresistivity of cementitious composites, but at lower effectiveness, because the thicker plates and movable layers have cohesion problems to cement matrix.

REFERENCES

1. W. Dong, W. Li, Z. Tao, K. Wang, Piezoresistive properties of cement-based sensors: Review and perspective, Construction and Building Materials 203 (2019) 146–163.
2. L. Coppola, A. Buoso, F. Corazza, Electrical properties of carbon nanotubes cement composites for monitoring stress conditions in concrete structures, Applied Mechanics and Materials 82 (2011) 118–123.
3. P.-W. Chen, D. Chung, Concrete as a new strain/stress sensor, Composites Part B: Engineering 27(1) (1996) 11–23.
4. V. Parvaneh, S.H. Khiabani, Mechanical and piezoresistive properties of self-sensing smart concretes reinforced by carbon nanotubes, Mechanics of Advanced Materials and Structures 26(11) (2019) 993–1000.
5. E. García-Macías, A. D'Alessandro, R. Castro-Triguero, D. Perez-Mira, F. Ubertini, Micromechanics modeling of the uniaxial strain-sensing property of carbon nanotube cement-matrix composites for SHM applications, Composite Structures 163 (2017) 195–215.
6. Z. Tang, W. Li, Y. Hu, J.L. Zhou, V.W.Y. Tam, Review on designs and properties of multifunctional alkali-activated materials (AAMs), Construction and Building Materials 200 (2019) 474–489.
7. S. Vaidya, E.N. Allouche, Strain sensing of carbon fiber reinforced geopolymer concrete, Materials and Structures 44(8) (2011) 1467–1475.
8. J.L. Vilaplana, F.J. Baeza, O. Galao, E. Zornoza, P. Garcés, Self-sensing properties of alkali activated blast furnace slag (BFS) composites reinforced with carbon fibers, Materials 6(10) (2013) 4776–4786.
9. M. Saafi, K. Andrew, P.L. Tang, D. McGhon, S. Taylor, M. Rahman, S. Yang, X. Zhou, Multifunctional properties of carbon nanotube/fly ash geopolymeric nanocomposites, Construction and Building Materials 49 (2013) 46–55.
10. M. Saafi, G. Piukovics, J. Ye, Hybrid graphene/geopolymeric cement as a superionic conductor for structural health monitoring applications, Smart Materials and Structures 25(10) (2016) 105018.
11. M. Saafi, L. Tang, J. Fung, M. Rahman, F. Sillars, J. Liggat, X. Zhou, Graphene/fly ash geopolymeric composites as self-sensing structural materials, Smart Materials and Structures 23(6) (2014) 065006.
12. S. Bi, M. Liu, J. Shen, X.M. Hu, L. Zhang, Ultrahigh self-sensing performance of geopolymer nanocomposites via unique interface engineering, ACS Applied Materials & Interfaces 9(14) (2017) 12851–12858.
13. K.J.D. MacKenzie, M.J. Bolton, Electrical and mechanical properties of aluminosilicate inorganic polymer composites with carbon nanotubes, Journal of Materials Science 44(11) (2009) 2851–2857.
14. B. Han, X. Yu, E. Kwon, J. Ou, Effects of CNT concentration level and water/cement ratio on the piezoresistivity of CNT/cement composites, Journal of Composite Materials 46(1) (2012) 19–25.
15. H. Kim, I. Park, H.-K. Lee, Improved piezoresistive sensitivity and stability of CNT/cement mortar composites with low water–binder ratio, Composite Structures 116 (2014) 713–719.
16. H. Wang, X. Gao, R. Wang, The influence of rheological parameters of cement paste on the dispersion of carbon nanofibers and self-sensing performance, Construction and Building Materials 134 (2017) 673–683.
17. M.-Q. Sun, R.J.Y. Liew, M.-H. Zhang, W. Li, Development of cement-based strain sensor for health monitoring of ultra high strength concrete, Construction and Building Materials 65 (2014) 630–637.
18. H. Xiao, H. Li, J. Ou, Self-monitoring properties of concrete columns with embedded cement-based strain sensors, Journal of Intelligent Material Systems and Structures 22(2) (2011) 191–200.
19. S. Wen, D. Chung, Uniaxial tension in carbon fiber reinforced cement, sensed by electrical resistivity measurement in longitudinal and transverse directions, Cement and Concrete Research 30(8) (2000) 1289–1294.
20. B. Wu, X.-j. Huang, J.-z. Lu, Biaxial compression in carbon-fiber-reinforced mortar, sensed by electrical resistance measurement, Cement and Concrete Research 35(7) (2005) 1430–1434.

21. Z. Zhang, B. Zhang, P. Yan, Comparative study of effect of raw and densified silica fume in the paste, mortar and concrete, Construction and Building Materials 105 (2016) 82–93.
22. X. Li, A.H. Korayem, C. Li, Y. Liu, H. He, J.G. Sanjayan, W.H. Duan, Incorporation of graphene oxide and silica fume into cement paste: A study of dispersion and compressive strength, Construction and Building Materials 123 (2016) 327–335.
23. B.J. Christensen, T.O. Mason, H.M. Jennings, Influence of silica fume on the early hydration of Portland cements using impedance spectroscopy, Journal of the American Ceramic Society 75(4) (1992) 939–945.
24. J. Larbi, A. Fraay, J. Bijen, The chemistry of the pore fluid of silica fume-blended cement systems, Cement and Concrete Research 20(4) (1990) 506–516.
25. S. Diamond, Effects of microsilica (silica fume) on pore-solution chemistry of cement pastes, Journal of the American Ceramic Society 66(5) (1983) C-82–C-84.
26. G. Kim, B. Yang, K. Cho, E. Kim, H.-K. Lee, Influences of CNT dispersion and pore characteristics on the electrical performance of cementitious composites, Composite Structures 164 (2017) 32–42.
27. D.D. Chung, Dispersion of short fibers in cement, Journal of Materials in Civil Engineering 17(4) (2005) 379–383.
28. S. Wen, D.D.L. Chung, Carbon fiber-reinforced cement as a thermistor, Cement and Concrete Research 29(6) (1999) 961–965.
29. W. Dong, W. Li, Y. Guo, X. He, D. Sheng, Effects of silica fume on physicochemical properties and piezoresistivity of intelligent carbon black-cementitious composites, Construction and Building Materials 259 (2020) 120399.
30. W. Dong, W. Li, L. Shen, Z. Sun, D. Sheng, Piezoresistivity of smart carbon nanotubes (CNTs) reinforced cementitious composite under integrated cyclic compression and impact, Composite Structures 241 (2020) 112106.
31. K. Scrivener, F. Martirena, S. Bishnoi, S. Maity, Calcined clay limestone cements (LC3), Cement and Concrete Research 114 (2018) 49–56.
32. B.B. Sabir, S. Wild, J. Bai, Metakaolin and calcined clays as pozzolans for concrete: A review, Cement and Concrete Composites 23(6) (2001) 441–454.
33. N. Singh, S.P. Singh, Carbonation and electrical resistance of self compacting concrete made with recycled concrete aggregates and metakaolin, Construction and Building Materials 121 (2016) 400–409.
34. R.K. Singh, P.A. Danoglidis, S.P. Shah, M.S. Konsta-Gdoutos, On the interactions of carbon nanotubes with metakaolin cementitious binders: Effects on ITZ, modulus and toughness, Construction and Building Materials 408 (2023) 133605.
35. D.K. Hardy, M.F. Fadden, M.J. Khattak, A. Khattab, Development and characterization of self-sensing CNF HPFRCC, Materials and Structures 49(12) (2016) 5327–5342.
36. H. Deng, H. Li, Assessment of self-sensing capability of carbon black engineered cementitious composites, Construction and Building Materials 173 (2018) 1–9.
37. Y. Huang, H. Li, S. Qian, Self-sensing properties of engineered cementitious composites, Construction and Building Materials 174 (2018) 253–262.
38. M. Konkanov, T. Salem, P. Jiao, R. Niyazbekova, N. Lajnef, Environment-friendly, self-sensing concrete blended with byproduct wastes, Sensors 20(7) (2020) 1925.
39. B.M. Tariq, D. Muhmmad, A. Majed, E. Mahmoud, A. Ammar, Self-sensing cement composite for traffic monitoring in intelligent transport system, Magazine of Civil Engineering 5 (105) (2021) 10505.
40. S.-H. Hong, T.-F. Yuan, J.-S. Choi, Y.-S. Yoon, Effects of steelmaking slag and moisture on electrical properties of concrete, Materials 13(12) (2020) 2675.
41. S.-J. Lee, I. You, S. Kim, H.-O. Shin, D.-Y. Yoo, Self-sensing capacity of ultra-high-performance fiber-reinforced concrete containing conductive powders in tension, Cement and Concrete Composites 125 (2022) 104331.
42. F.J. Baeza, O. Galao, I.J. Vegas, M. Cano, P. Garcés, Influence of recycled slag aggregates on the conductivity and strain sensing capacity of carbon fiber reinforced cement mortars, Construction and Building Materials 184 (2018) 311–319.
43. T. Hemalatha, B. Sangoju, G. Muthuramalingam, A study on copper slag as fine aggregate in improving the electrical conductivity of cement mortar, Sādhanā 47(3) (2022) 141.

44. F. Gulisano, T. Buasiri, F.R.A. Apaza, A. Cwirzen, J. Gallego, Piezoresistive behavior of electric arc furnace slag and graphene nanoplatelets asphalt mixtures for self-sensing pavements, Automation in Construction 142 (2022) 104534.
45. A.L.G. Gastaldini, G.C. Isaia, T.F. Hoppe, F. Missau, A.P. Saciloto, Influence of the use of rice husk ash on the electrical resistivity of concrete: A technical and economic feasibility study, Construction and Building Materials 23(11) (2009) 3411–3419.
46. I. Rhee, Y.A. Kim, G.-O. Shin, J.H. Kim, H. Muramatsu, Compressive strength sensitivity of cement mortar using rice husk-derived graphene with a high specific surface area, Construction and Building Materials 96 (2015) 189–197.
47. I. Rhee, J.S. Lee, Y.A. Kim, J.H. Kim, J.H. Kim, Electrically conductive cement mortar: Incorporating rice husk-derived high-surface-area graphene, Construction and Building Materials 125 (2016) 632–642.
48. R. Gopalakrishnan, R. Kaveri, Using graphene oxide to improve the mechanical and electrical properties of fiber-reinforced high-volume sugarcane bagasse ash cement mortar, The European Physical Journal Plus 136 (2021) 1–15.
49. U.W. Robert, S.E. Etuk, S.A. Ekong, O.E. Agbasi, N.E. Ekpenyong, S.S. Akpan, E.A. Umana, Electrical characteristics of dry cement–based composites modified with coconut husk ash nanomaterial, Advances in Materials Science 22(2) (2022) 64–77.
50. P.C. Kumar, T. Shanthala, K. Aparna, S.V. Babu, Experimental Investigation on the Combined Effect of Fly Ash and Eggshell Powder as Partial Replacement of Cement, in: B. Kondraivendhan, C.D. Modhera, V. Matsagar (Eds.) Sustainable Building Materials and Construction, Springer Nature, Singapore, 2022, pp. 371–378.
51. Y. Han, R. Lin, X.-Y. Wang, Performance of sustainable concrete made from waste oyster shell powder and blast furnace slag, Journal of Building Engineering 47 (2022) 103918.
52. K. Hossain, M. Lachemi, Corrosion resistance and chloride diffusivity of volcanic ash blended cement mortar, Cement and Concrete Research 34(4) (2004) 695–702.
53. M. Guo, G. Gong, Y. Yue, F. Xing, Y. Zhou, B. Hu, Performance evaluation of recycled aggregate concrete incorporating limestone calcined clay cement (LC3), Journal of Cleaner Production 366 (2022) 132820.
54. Y. Dhandapani, M. Santhanam, Investigation on the microstructure-related characteristics to elucidate performance of composite cement with limestone–calcined clay combination, Cement and Concrete Research 129 (2020) 105959.
55. D.V. Ribeiro, J.A. Labrincha, M.R. Morelli, Effect of the addition of red mud on the corrosion parameters of reinforced concrete, Cement and Concrete Research 42(1) (2012) 124–133.
56. W. Dong, W. Li, K. Wang, Y. Guo, D. Sheng, S.P. Shah, Piezoresistivity enhancement of functional carbon black filled cement-based sensor using polypropylene fibre, Powder Technology 373 (2020) 184–194.
57. F.J. Baeza, O. Galao, E. Zornoza, P. Garcés, Effect of aspect ratio on strain sensing capacity of carbon fiber reinforced cement composites, Materials & Design 51 (2013) 1085–1094.
58. R. Tchoudakov, O. Breuer, M. Narkis, A. Siegmann, Conductive polymer blends with low carbon black loading: Polypropylene/polyamide, Polymer Engineering & Science 36(10) (1996) 1336–1346.
59. J. Donnet, Structure and reactivity of carbons: From carbon black to carbon composites, Carbon 20(4) (1982) 267–282.
60. W. Dong, Y. Guo, Z. Sun, Z. Tao, W. Li, Development of piezoresistive cement-based sensor using recycled waste glass cullets coated with carbon nanotubes, Journal of Cleaner Production 314 (2021) 127968.
61. M. Kamali, A. Ghahremaninezhad, An investigation into the hydration and microstructure of cement pastes modified with glass powders, Construction and Building Materials 112 (2016) 915–924.
62. J.-X. Lu, P. Shen, H. Zheng, B. Zhan, H.A. Ali, P. He, C.S. Poon, Synergetic recycling of waste glass and recycled aggregates in cement mortars: Physical, durability and microstructure performance, Cement and Concrete Composites 113 (2020) 103632.
63. S. Zhuang, Q. Wang, Inhibition mechanisms of steel slag on the early-age hydration of cement, Cement and Concrete Research 140 (2021) 106283.
64. R. Siddique, T.R. Naik, Properties of concrete containing scrap-tire rubber: An overview, Waste Management 24(6) (2004) 563–569.

65. W. Dong, W. Li, K. Wang, Z. Luo, D. Sheng, Self-sensing capabilities of cement-based sensor with layer-distributed conductive rubber fibres, Sensors and Actuators A: Physical 301 (2020) 111763.
66. W.H. Yung, L.C. Yung, L.H. Hua, A study of the durability properties of waste tire rubber applied to self-compacting concrete, Construction and Building Materials 41 (2013) 665–672.
67. W. Dong, W. Li, G. Long, Z. Tao, J. Li, K. Wang, Electrical resistivity and mechanical properties of cementitious composite incorporating conductive rubber fibres, Smart Materials and Structures 28(8) (2019) 085013.
68. W. Dong, W. Li, K. Wang, K. Vessalas, S. Zhang, Mechanical strength and self-sensing capacity of smart cementitious composite containing conductive rubber crumbs, Journal of Intelligent Material Systems and Structures 31(10) (2020) 1325–1340.
69. W. Dong, W. Li, N. Lu, F. Qu, K. Vessalas, D. Sheng, Piezoresistive behaviours of cement-based sensor with carbon black subjected to various temperature and water content, Composites Part B: Engineering 178 (2019) 107488.
70. N.-P. Pham, A. Toumi, A. Turatsinze, Rubber aggregate–cement matrix bond enhancement: Microstructural analysis, effect on transfer properties and on mechanical behaviours of the composite, Cement and Concrete Composites 94 (2018) 1–12.
71. S. Wen, D.D.L. Chung, A comparative study of steel- and carbon-fibre cement as piezoresistive strain sensors, Advances in Cement Research 15(3) (2003) 119–128.
72. E. Teomete, O.I. Kocyigit, Tensile strain sensitivity of steel fiber reinforced cement matrix composites tested by split tensile test, Construction and Building Materials 47 (2013) 962–968.
73. J. Lataste, M. Behloul, D. Breysse, Characterisation of fibres distribution in a steel fibre reinforced concrete with electrical resistivity measurements, NDT & E International 41(8) (2008) 638–647.
74. B. Han, Y. Yu, B. Han, J. Ou, Development of a wireless stress/strain measurement system integrated with pressure-sensitive nickel powder-filled cement-based sensors, Sensors and Actuators A: Physical 147(2) (2008) 536–543.
75. B. Han, B. Han, J. Ou, Experimental study on use of nickel powder-filled Portland cement-based composite for fabrication of piezoresistive sensors with high sensitivity, Sensors and Actuators A: Physical 149(1) (2009) 51–55.
76. H. Xiao, M. Liu, G. Wang, Anisotropic electrical and abrasion-sensing properties of cement-based composites containing aligned nickel powder, Cement and Concrete Composites 87 (2018) 130–136.
77. W. Sekkal, A. Zaoui, Novel properties of nano-engineered cementitious materials with fullerene buckyballs, Cement and Concrete Composites 118 (2021) 103960.
78. Y. Dai, M. Sun, C. Liu, Z. Li, Electromagnetic wave absorbing characteristics of carbon black cement-based composites, Cement and Concrete Composites 32(7) (2010) 508–513.
79. W. Dong, W. Li, L. Shen, S. Zhang, K. Vessalas, Integrated self-sensing and self-healing cementitious composite with microencapsulation of nano-carbon black and slaked lime, Materials Letters 282 (2021) 128834.
80. M.-G. Pârvan, G. Voicu, A.-I. Bădănoiu, Study of hydration and hardening processes of self-sensing cement-based materials with carbon black content, Journal of Thermal Analysis and Calorimetry 139 (2020) 807–815.
81. W. Dong, W. Li, L. Shen, D. Sheng, Piezoresistive behaviours of carbon black cement-based sensors with layer-distributed conductive rubber fibres, Materials & Design 182 (2019) 108012.
82. H. Li, H.-g. Xiao, J.-p. Ou, Effect of compressive strain on electrical resistivity of carbon black-filled cement-based composites, Cement and Concrete Composites 28(9) (2006) 824–828.
83. H. Xiao, H. Li, J. Ou, Modeling of piezoresistivity of carbon black filled cement-based composites under multi-axial strain, Sensors and Actuators A: Physical 160(1–2) (2010) 87–93.
84. A.O. Monteiro, P.B. Cachim, P.M. Costa, Self-sensing piezoresistive cement composite loaded with carbon black particles, Cement and Concrete Composites 81 (2017) 59–65.
85. A. Monteiro, A. Loredo, P. Costa, M. Oeser, P. Cachim, A pressure-sensitive carbon black cement composite for traffic monitoring, Construction and Building Materials 154 (2017) 1079–1086.

86. M. Li, V. Lin, J. Lynch, V.C. Li, Multifunctional Carbon Black Engineered Cementitious Composites for the Protection of Critical Infrastructure, in: G.J. Parra-Montesinos, H.W. Reinhardt, A.E. Naaman (Eds.), High Performance Fiber Reinforced Cement Composites 6: HPFRCC 6, Springer, Dordrecht, The Netherlands, 2012, pp. 99–106.
87. D.D.L. Chung, Electrically conductive cement-based materials, Advances in Cement Research 16(4) (2004) 167–176.
88. S. Erdem, S. Hanbay, M.A. Blankson, Self-sensing damage assessment and image-based surface crack quantification of carbon nanofibre reinforced concrete, Construction and Building Materials 134 (2017) 520–529.
89. F.J. Baeza, O. Galao, E. Zornoza, P. Garcés, Multifunctional cement composites strain and damage sensors applied on reinforced concrete (RC) structural elements, Materials 6(3) (2013) 841–855.
90. G. Kim, F. Naeem, H. Kim, H.-K. Lee, Heating and heat-dependent mechanical characteristics of CNT-embedded cementitious composites, Composite Structures 136 (2016) 162–170.
91. H.-K. Kim, Chloride penetration monitoring in reinforced concrete structure using carbon nanotube/cement composite, Construction and Building Materials 96 (2015) 29–36.
92. S. Wansom, N. Kidner, L. Woo, T. Mason, AC-impedance response of multi-walled carbon nanotube/cement composites, Cement and Concrete Composites 28(6) (2006) 509–519.
93. W. Dong, W. Li, Z. Luo, Y. Guo, K. Wang, Effect of layer-distributed carbon nanotube (CNT) on mechanical and piezoresistive performance of intelligent cement-based sensor, Nanotechnology 31(50) (2020) 505503.
94. L. Jianlin, L. Qiuyi, C. Shunjian, L. Lu, H. Dongshuai, Z. Chunwei, Piezoresistive properties of cement composites reinforced by functionalized carbon nanotubes using photo-assisted Fenton, Smart Materials and Structures 26(3) (2017) 035025.
95. W. Dong, W. Li, K. Wang, B. Han, D. Sheng, S.P. Shah, Investigation on physicochemical and piezoresistive properties of smart MWCNT/cementitious composite exposed to elevated temperatures, Cement and Concrete Composites 112 (2020) 103675.
96. W. Baomin, D. Shuang, Effect and mechanism of graphene nanoplatelets on hydration reaction, mechanical properties and microstructure of cement composites, Construction and Building Materials 228 (2019) 116720.
97. W. Li, F. Qu, W. Dong, G. Mishra, S.P. Shah, A comprehensive review on self-sensing graphene/cementitious composites: A pathway toward next-generation smart concrete, Construction and Building Materials 331 (2022) 127284.
98. J.-L. Le, H. Du, S.D. Pang, Use of 2D graphene nanoplatelets (GNP) in cement composites for structural health evaluation, Composites Part B: Engineering 67 (2014) 555–563.
99. Q. Liu, W. Wu, J. Xiao, Y. Tian, J. Chen, A. Singh, Correlation between damage evolution and resistivity reaction of concrete in-filled with graphene nanoplatelets, Construction and Building Materials 208 (2019) 482–491.
100. Q. Liu, Q. Xu, Q. Yu, R. Gao, T. Tong, Experimental investigation on mechanical and piezoresistive properties of cementitious materials containing graphene and graphene oxide nanoplatelets, Construction and Building Materials 127 (2016) 565–576.
101. W. Dong, W. Li, K. Wang, S.P. Shah, Physicochemical and piezoresistive properties of smart cementitious composites with graphene nanoplates and graphite plates, Construction and Building Materials 286 (2021) 122943.
102. Y. Ding, Y. Yang, R.-G. Liu, T. Xiao, J.-H. Tian, Study on pressure sensitivity of smart polymer concrete based on steel slag, Measurement 140 (2019) 14–21.
103. G. Yıldırım, M.H. Sarwary, A. Al-Dahawi, O. Öztürk, Ö. Anıl, M. Şahmaran, Piezoresistive behavior of CF-and CNT-based reinforced concrete beams subjected to static flexural loading: Shear failure investigation, Construction and Building Materials 168 (2018) 266–279.
104. X. Liu, S. Wu, Study on the graphite and carbon fiber modified asphalt concrete, Construction and Building Materials 25(4) (2011) 1807–1811.
105. A. Belli, A. Mobili, T. Bellezze, F. Tittarelli, P. Cachim, Evaluating the self-sensing ability of cement mortars manufactured with graphene nanoplatelets, virgin or recycled carbon fibers through piezoresistivity tests, Sustainability 10(11) (2018) 4013.
106. S. Wu, L. Mo, Z. Shui, Z. Chen, Investigation of the conductivity of asphalt concrete containing conductive fillers, Carbon 43(7) (2005) 1358–1363.

107. J. Chen, D. Zhao, H. Ge, J. Wang, Graphene oxide-deposited carbon fiber/cement composites for electromagnetic interference shielding application, Construction and Building Materials 84 (2015) 66–72.
108. S. Wen, D. Chung, Partial replacement of carbon fiber by carbon black in multifunctional cement–matrix composites, Carbon 45(3) (2007) 505–513.
109. B. Chen, B. Li, Y. Gao, T.-C. Ling, Z. Lu, Z. Li, Investigation on electrically conductive aggregates produced by incorporating carbon fiber and carbon black, Construction and Building Materials 144 (2017) 106–114.
110. B. Han, L. Zhang, S. Sun, X. Yu, X. Dong, T. Wu, J. Ou, Electrostatic self-assembled carbon nanotube/nano carbon black composite fillers reinforced cement-based materials with multi-functionality, Composites Part A: Applied Science and Manufacturing 79 (2015) 103–115.
111. F. Azhari, N. Banthia, Cement-based sensors with carbon fibers and carbon nanotubes for piezoresistive sensing, Cement and Concrete Composites 34(7) (2012) 866–873.
112. S.-J. Lee, I. You, G. Zi, D.-Y. Yoo, Experimental investigation of the piezoresistive properties of cement composites with hybrid carbon fibers and nanotubes, Sensors 17(11) (2017) 2516.
113. Y. Hong, Z. Li, G. Qiao, J. Ou, W. Cheng, Pressure sensitivity of multiscale carbon-admixtures–enhanced cement-based composites, Nanomaterials and Nanotechnology 8 (2018). doi: https://doi.org/10.1177/1847980418793529
114. M. Chen, P. Gao, F. Geng, L. Zhang, H. Liu, Mechanical and smart properties of carbon fiber and graphite conductive concrete for internal damage monitoring of structure, Construction and Building Materials 142 (2017) 320–327.
115. G. Kim, B. Yang, G. Ryu, H.-K. Lee, The electrically conductive carbon nanotube (CNT)/cement composites for accelerated curing and thermal cracking reduction, Composite Structures 158 (2016) 20–29.
116. Y. Perets, L. Aleksandrovych, M. Melnychenko, O. Lazarenko, L. Vovchenko, L. Matzui, The electrical properties of hybrid composites based on multiwall carbon nanotubes with graphite nanoplatelets, Nanoscale Research Letters 12(1) (2017) 406.
117. J.L. Luo, Z.D. Duan, T.J. Zhao, Q.Y. Li, Self-sensing property of cementitious nanocomposites hybrid with nanophase carbon nanotube and carbon black, Advanced Materials Research 143 (2011) 644–647.

Chapter 3

Fabrication, geometry, and layout of cement-based sensors

3.1 INTRODUCTION

The fabrication of cement-based sensors (CBSs) involves the incorporation of macro-, micro-, or nanoscale electrically conductive materials into the cement matrix during the mixing process. As mentioned in the previous chapter, these functional materials include carbon materials, metal materials, conductive polymers, sand, and glass particles with enclosing conductive particles, etc., depending on the desired sensing capabilities and applications. The homogeneous dispersion of these materials within the cement is critical to ensure the reliable and accurate performance of the sensors. Due to the existence of van der Waals force, nanomaterials can form agglomerates of nanoparticles in the aqueous solution, resulting in floating, precipitation, and uneven dispersion [1, 2]. The uneven dispersion can cause nanomaterials to form defects in CBSs and limit their benefits. To ensure that the addition of nanomaterials can fully improve the performance of CBSs, nanomaterials should be well dispersed first in water and the matrix. Taking 2D GNP and GO as an example, the dispersion procedure includes the following two methods: mechanical methods and surfactant modifications. Table 3.1 summarizes the dispersion methods for nanomaterials in cementitious composites. In addition, the methods to evaluate the quality of the dispersion were also compared in this section, ranging from ultrasonic treatment, surface modification, mineral addition, and hybrid mixing of carbon materials with different geometry, etc.

The layout of CBSs within concrete structures depends on the specific application and the parameter being monitored. For strain or stress sensing, the sensors are strategically positioned in regions of concrete structures expected to experience high load or deformation, such as critical structural elements or potential failure points. The sensing performance of small-scale CBSs differs between being embedded within the structure and being left unembedded because this involves the transmission of forces from the structure to the sensor during the loading process. In addition, when the sensor is embedded in the concrete structure, it experiences a more direct and intimate interaction with the surrounding concrete matrix. This close contact enhances the transfer of mechanical signals from the structure to the sensor, resulting in more accurate and sensitive measurements. However, when the sensor is fixed on the surface, there is an additional interface between the sensor and the concrete surface, which can lead to signal attenuation and reduced sensitivity. In this chapter, we have mainly discussed the performance of CBSs embedded in the different forms of concrete structures.

CBSs have shown great promise in structural health monitoring and smart infrastructure applications. However, when it comes to large-scale production and widespread

DOI: 10.1201/9781032663685-3

Table 3.1 Dispersion methods for graphene-based nanomaterials in cementitious composites

			Dispersion methods			
Matrix	*Nanomaterials*	*wt.% binder*	*Mechanical*	*Chemical*	*Test methods*	*References*
Paste	GOs	0.02, 0.04, 0.06	Ultrasonic	—	—	Li et al. [3]
Paste	CNT	0.1	Ultrasonic	Pluronic F-127	Microstructures Flexural strength	Parveen et al. [4]
Paste	GOs	0.01–0.04	Ultrasonic (5 min)	—	X-ray computed tomography and photoelectron spectroscopy	Li et al. [5]
Paste	GOs	0.02	Varied sonication times (0–60 min)	Four surfactants	UV-vis spectroscopy	Chuah et al. [6]
Paste	GOs	0.06	Ultrasound probe sonication (2 h)	Polycarboxylate-ether	—	Long et al. [1]
Paste	GNPs	0.1, 0.2	Stirrer and ultrasonicator (40 min)	Polycarboxylate superplasticizer	UV-vis spectroscopy	Bai et al. [7]
Paste	GNPs	1.0	BioLogics ultrasonic	Polycarboxylate superplasticizer	UV-vis spectrophotometer	Du et al. [8]
Paste	GNPs	0.01, 0.05	Stirrer and ultrasonicator (60 min)	Silane coupling agent	—	Guo et al. [9]
Paste	GOs	0.03–0.07	Magnetic stirrer and ultrasonicator (3 min)	Nitric-acid–treated nano-Fe_3O_4	—	Imanian Ghazanlou et al. [10]
Paste	GOs	0.45%	Ultrasonication (0–15 min)	Polycarboxylate-ether	UV-vis spectrophotometer	Jing et al. [11]
Paste	GOs, nano-silica	0.02–0.04%	Ultrasonication (3 h)	Alkaline environment ammonia solution and tetraethylorthosilicate	UV-vis spectroscopy	Lin et al. [12]
Paste	GNPs	0.01, 0.025, 0.05	Ultrasonic (30 min)	Different superficial active agents	UV-vis spectroscopy	Liu et al. [13]
Paste	GOs	0.03	Electromagnetically stirred and ultrasonic (5–15 min)	Polycarboxylate superplasticizer	UV-vis spectroscopy	Qin et al. [14]
Paste	GOs	0.01–0.08	Ultrasonic bath (8 h)	Sodium dodecyl sulfate	—	Şimşek et al. [15]
Mortar	GNPs	0.5–3.0	Ultrasonication dispersion (60 min)	Polycarboxylate-based superplasticizer	UV-vis spectroscopy	Dong et al. [16, 17]
Mortar	GOs	0.02, 0.04	Ultrasonic (30 min)	Various superplasticizers.	UV-Vis spectrometry	Yan et al. [18]
Mortar	GNPs	0.1–1.0	Magnetic stirrer, ultrasonicator (3 h)	Surfactant Tributyl phosphate (TBP)	UV-Vis spectrometry	Abedi et al. [19]
Concrete	GNPs	0.02–0.06%	Ultrasonication	A superplasticizer-added solution	—	Adhikary et al. [20]

application, several challenges must be addressed to ensure their successful integration into the construction industry. For example, ensuring uniform properties of CBSs poses challenges due to the anisotropic nature of the cement matrix and variations in the distribution of conductive fillers. The achievement of well-dispersed nanomaterials remains challenging, and their incorporation often involves complex and expensive manufacturing processes. The workability of cementitious composites may be reduced as the amount of added carbon nanomaterials increases, adversely affecting practical construction processes. The fluctuations in temperature and humidity can impact the electrical resistivity of cementitious composites, which, in turn, may influence the correlation with the targeted index of CBSs. The development of an accurate data collection system is essential to collect data and evaluate the sensing efficiency of multifunctional CBSs. The toxicity of nanomaterials for the production of CBSs is still unknown, which should be urgently conducted in further investigations.

3.2 MANUFACTURING PROCESSES: DISPERSION, MIXING, CASTING, AND CURING

3.2.1 Dispersion of conductive filler

Uniform dispersion, referring to the even distribution of sensing elements throughout the entire cement matrix, is crucial when incorporating nanomaterials into cement-based materials because it directly influences the sensing performance and reliability of the CBSs. The hybrid dispersion method first developed by Konsta-Gdoutos et al. showed that a well-dispersed small amount of CNT can profoundly influence the mechanical property and piezoresistivity of CNT-reinforced CBSs [21].

3.2.1.1 Ultrasonic dispersion

The ultrasonic duration and energy are critical factors to influence the dispersion efficiency of carbon materials. Chaipanich et al. [22] first mixed the CNT with a proportion of water under 10 min ultrasonic dispersion and then mixed the solution with cement materials, followed by the residual water adding into the mixer once the blends formed. They observed denser microstructures and improved mechanical properties of cementitious composites. Nochaiya et al. [23] dispersed the CNT in mixing water for 5 min sonication, and obtained similar CNT–cement composites with improved mechanical properties. Different from the previous ultrasonic duration, sonication time up to 2 h was investigated by Du and Dai Pang [8]. In Fig. 3.1, the sedimentation of black precipitates gradually increased from the well-dispersed GNP suspension over time. The increase of sonication time reduced the sedimentation rate indicating the better dispersion of GNP suspension. However, they did not test the structural integrity of GNP and the mechanical properties of the cementitious composites. Similarly, the effect of sonication time from 2 h to 5 h on the electrical conductivity of CNT–polyaniline composites was investigated [24]. The longer duration increased the structure toughness but decreased the electrical conductivity because of the conversion of conductive state. It could be seen that the sonication time mainly depended on the matrix/base of the carbon materials. The nanoparticle tends to disperse more easily in water rather than in a polymer-based matrix. Moreover, for the cement matrix, the differences in the ultrasonic duration might relate

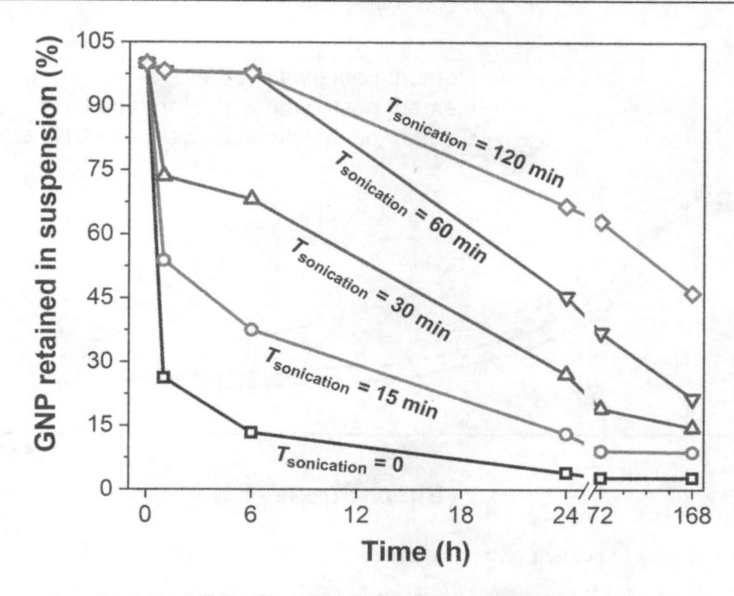

Figure 3.1 Effect of ultrasonic duration on stability of GNP solution [8].

to the water-to-binder (w/b) ratios, since it was concluded by Kim et al. [25] that cement composites at low w/b ratios benefited CNT dispersion and improved piezoresistivity. Furthermore, the shapes and dosages of fillers also relate to the sonication time, given different ultrasonic durations based on the concentrations and aspect ratios of carbon fillers carried out in studies [26]. In addition to the ultrasonic duration, the effect of ultrasonic energy on the dispersion of CNT was studied by Zou et al. [27]. They found that the dispersion efficiency increased with the increase of ultrasonic energy until reaching a plateau. Similar results were obtained by Metaxa et al. [28], with the first increased and then decreased flexural strength of the cementitious composites prepared with increasing ultrasonic energy. These results indicate optimal ultrasonic energy to achieve the optimal mechanical properties of cementitious composites because the high energy might destroy the CNT and deteriorate the reinforcement effect.

3.2.1.2 Surface modification

The surface modification to improve the dispersion of carbon nanomaterials includes the application of surfactants in suspension and the surface coating of nanomaterials. Collins et al. [29] compared five different surfactants on the efficiency of CNT dispersion in cement paste, and found that the polycarboxylate could strengthen the CNT–cement composites with improved compressive strength. Chuah et al. [6] investigated the effect of polycarboxylate, air-entrainment, and gum Arabic admixtures on the dispersion of GO in cement composites. They observed an identical conclusion to Collins' results that polycarboxylate was the most promising surfactant to disperse the carbon nanomaterials. Similarly, Luo et al. [30] compared another five surfactants on the CNT dispersion in aqueous solutions. Not only was the CNT solution with sodium dodecylbenzene sulfonate found with the largest delaminating time, but also the composites were observed with the best mechanical properties and electrical conductivity. By the UV spectroscopy

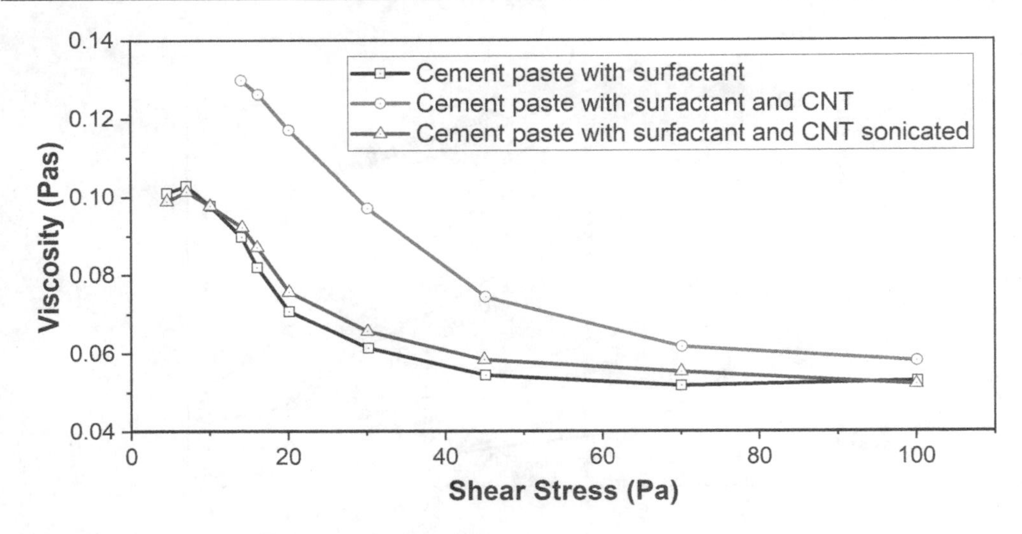

Figure 3.2 Shear viscosity of cement paste reinforced with MWCNTs [21].

observation, Shuang and Baomin [31] compared four surfactants and drew the conclusion that the polyoxyethylene nonylphenylether at the concentration of 0.7 g/l exhibited the best dispersion efficiency. Among these surfactants, it was recommended that the ionic surfactants are more suitable for the CNT/water-soluble solutions [32]. For the identical surfactant with various concentrations, Konsta-Gdoutos et al. [21] investigated their effects on the dispersion efficiency of CNT. Given the CNT agglomerates increased the viscosity under small shear strength, as shown in Fig. 3.2, it has been found that the surfactant-to-CNT weight ratio of 4.0 led to the suspension with the smallest viscosity and illustrated the best CNT dispersion. Metaxa et al. [28] found that the surfactant to CNF weight ratio could be 2.0 or 4.0. Generally, the surfactant can decrease the van der Waals forces, modify the hydrophobicity of carbon materials, and enhance their dissolubility. Although an effective surfactant can successfully disperse carbon nanomaterials, uneven distribution of fillers or poor bond among fillers to matrix might still occur in the cementitious composites.

Similar to the effect of surfactants, special treatment of previous surface decoration on the conductive fillers have been proposed for better uniform dispersion. By investigating the impact of polyacrylic acid polymers on CNT surface functionalization and the followed dispersion in water, Cwirzen et al. [33] observed acceptable workability and mechanical strengths of the CNT–cement composites. As plotted in Fig. 3.3, it revealed the available ionic or covalent bonding between CNT and polymers that prevented the reagglomeration of CNT suspension. However, for the CNT coated by gum Arabic, it should be noted that lower cement hydration was observed for the CNT–cement composites that hindered the final mechanical strengths. Li et al. [34] pretreated the CNT by a mixed solution of H_2SO_4 and HNO_3, and they found that the cementitious composites with treated CNT were provided with both high mechanical properties and piezoresistivity than the counterparts with untreated CNT. They proposed that the smoother surface of treated CNT was more beneficial to electron movements and clung to the cement matrix. More like a mechanical modification, the modified CNF after debulking process was proved more efficient to improve the mechanical performance of cementitious composites, implying the good dispersion of modified CNF in the nanocomposites.

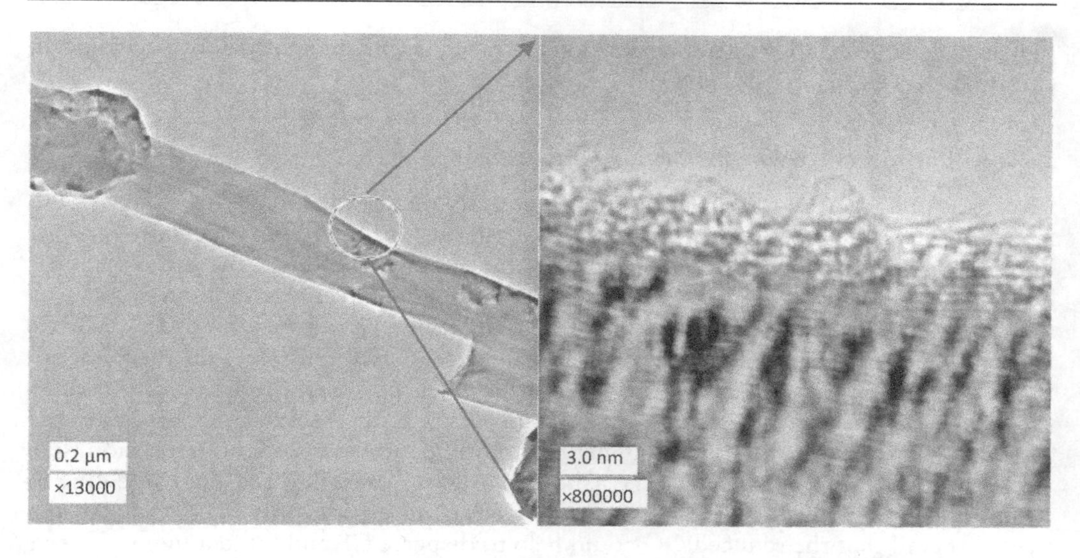

Figure 3.3 Transmission electron microscope (TEM) images of CNT covered by polymers.

3.2.1.3 *Mineral additives*

Chung, Sanchez, and Dong and et al. [35–37] deemed that SF could improve the dispersion of CF, CB, and CNF and improve the mechanical, electrical, and piezoresistive properties of cementitious composites, while they did not conduct further experiments on the specific influences by SF. Kim et al. [38] tested the CNT cementitious composites filled with various contents of SF and achieved the composites with improved compressive strength and electrical conductivity. The number and size of CNT agglomerations was reduced because the SF intermixed with CNT and broke the formation of agglomerations. The results showed the SF dosage of 10% by weight of cement exhibiting best performances. Similar results emerged for the SF-incorporated CNT cementitious composites for the application of EMI shielding [39]. In addition, it had been proposed that the SF could disperse the plate-like structures of carbon materials, such as GNP and GO. Li et al. [40] observed that the GO could be covered with SF and thus have less tendency to get agglomerated. Bai et al. [7] concluded similar results on the GNP cementitious composites filled with SF, and they proposed that the appropriate dosage of SF should be 5–10% by the weight of cement. In general, the main reason for improved dispersion is that the SF particles insert into the conductive fillers to separate the agglomerates. On the other hand, not all mineral additives have the capacity to improve the dispersion of carbon nanoparticles. The effect of nanosilica on the dispersion and stability of CNT aqueous solution relied on the CNT concentration. For the small dosage of CNT, the nanosilica nearly has no influence on the CNT dispersion. However, in the case of high CNT concentration, the dispersion and stability would be decreased with the addition of nanosilica [41]. Moreover, there exist mineral additives that can block the dispersion of carbon materials. According to the studies by Mendoza et al. [42], the presence of calcium hydroxide ($Ca(OH)_2$) in the cementitious composites could affect the dispersion of CNT, since it hindered the electrostatic repulsion between CNT and superplasticizer molecules and resulted in the reagglomeration. Similarly, mineral additives such as hydrated lime that can generate $Ca(OH)_2$ in cementitious composites might retard the dispersion of carbon materials.

3.2.1.4 Carbon materials in various dimensions

Investigations have revealed that carbon materials themselves possess the functionality to improve the dispersion efficiency of other carbon materials, especially for carbon materials with different dimensions. Öztürk et al. [43] applied the hybrid CF and NCB in the self-sensing reinforced beams and obtained a lower electrical resistivity and higher deflection sensing capacity compared to the counterparts with single CF or NCB. Lu et al. [44] employed GO to assist the dispersion of CF and CNT in the cementitious composites; the results of UV spectroscopy and improved mechanical properties indicated that the CF and CNT were well dispersed with the existence of GO. They attributed the phenomenon to the higher electrostatic repulsion of GO suspension than that of water. In the polymer matrix, Chen et al. [45] found that the EMI shielding of composites with reduced GO and CF exceeded the counterparts only filled with CF, and they observed that the reduced GO could help to disperse CF and build a more effective conductive network. It could be observed in Fig. 3.4a and b that the reduced GO was successfully immobilized on the surface of CF. Since the GNP has similar structures to reduced GO, studies concluded similar results on the composites filled with CF and GNP [46]. Apart from the 2D carbon materials, the 0D carbon black is also found beneficial for the dispersion of 1D carbon materials. Zhang et al. [47] observed the grape bunch structures of CNT and CB in the cementitious composites, and concluded that the addition of CB could improve the functionality of composites. As shown in Fig. 3.4c, the addition of CB reduced the distance between fillers and increased the possibility of conductive passages generation. Moreover, studies have mentioned the usage of CB and CF in the cementitious composites for conductivity development [48], which implied the application potential of the mixture of carbon materials at different dimensions.

3.2.1.5 Hybrid dispersion methods

Hybrid dispersion methods could more efficiently homogenize the carbonic nanoparticles in aqueous solutions and cement matrix. Table 3.2 lists different dispersion ways incorporating at least two of the aforementioned methods. Parveen et al. [4] proposed a novel dispersant of Pluronic F-127 incorporating 1 h ultrasonication to disperse CNT in cementitious composites, which expressed a significant enhancement on the mechanical strengths and modules. To reduce the sizes of agglomerations, Musso et al. [49] dispersed CNT suspension in acetone by means of an ultrasonic probe. Konsta-Gdoutos et al. [21] controlled the ultrasonic energy and surfactant concentration to disperse CNT and found that the application of ultrasonication was critical to obtain the well-dispersed CNT suspensions. Wang et al. [50] utilized the gum Arabic to premix with CNT solution, followed by 30 min ultrasonication and defoamer treatment to reduce the generated air bubbles. Subsequently, they combined the surfactant methylcellulose and ultrasonication to disperse the GNP in cementitious composites [51]. With different mixing orders, Wang et al. [52] first premixed the CF with water and then added the hydroxyethyl cellulose for another stirring; the deformer was the last added to eliminate air bubbles. In both studies, they achieved denser microstructures and improved the mechanical properties of

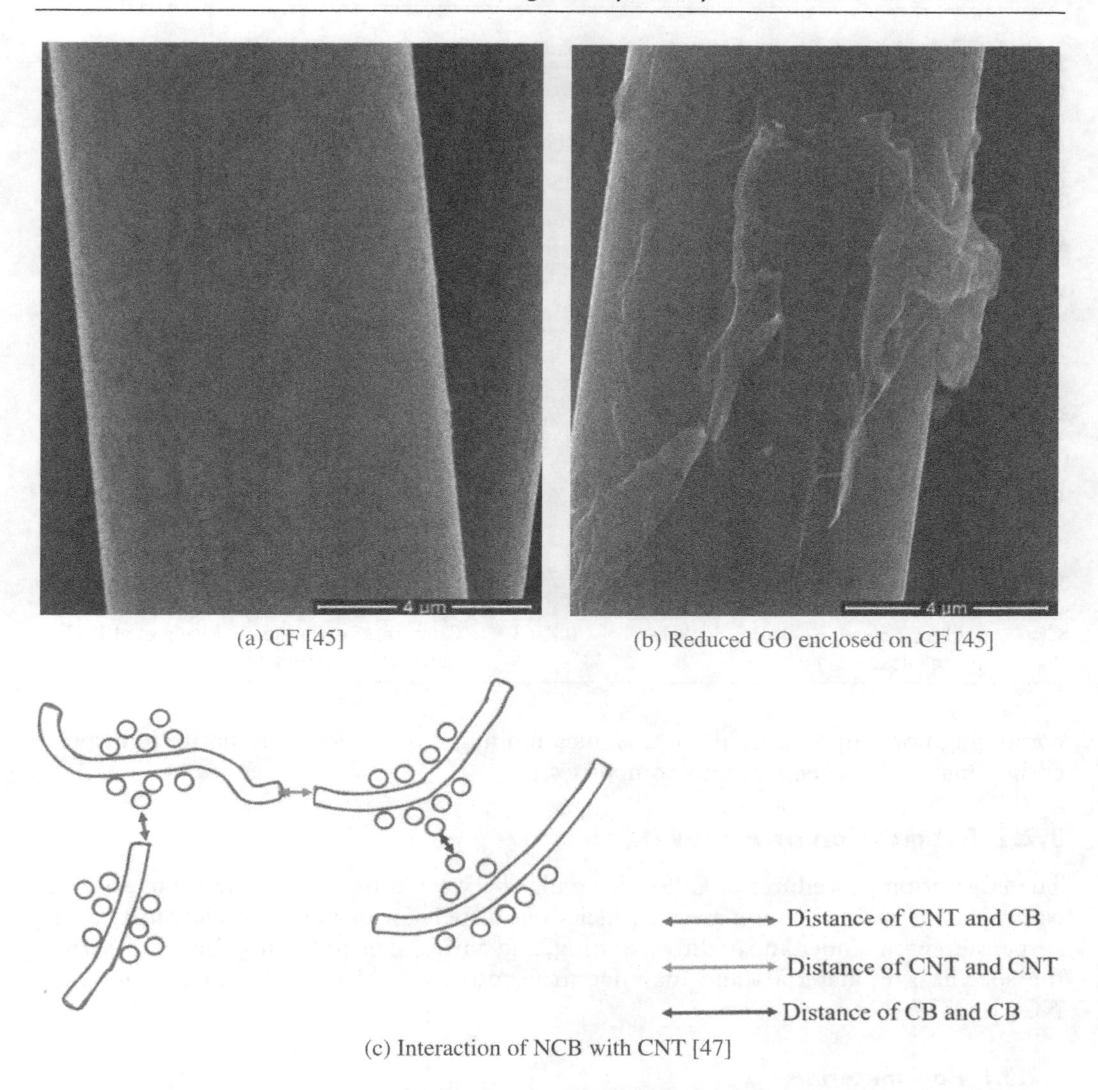

Figure 3.4 Relative positions among mixed carbon materials with different dimensions.

cementitious composites. However, long-term sonication could raise the temperature of suspension and affect the dispersion of CNT. To solve the problem, Xu et al. [53] applied a surfactant (TNWDIS) with an aromatic ring and hydrophilic group to mix with CNT suspension, and then the ultrasonication was conducted six times with breaks. Similar to the studies which carried out the ultrasonication in ice/water bath [27], this method could avoid the temperature increasing of suspension during sonication. As mentioned above, long-term ultrasonication might destroy the structural integrity and increase the temperature of suspensions, which strongly affects the mechanical properties of cementitious composites. The application of surfactant could overcome these disadvantages, but the dispersion efficiency was always limited, let alone the potential damage to the mechanical properties of cementitious composites. As a consequence, the hybrid dispersion methods

Table 3.2 Hybrid dispersion methods of carbon materials in cementitious composites

Carbon materials types	*Dispersion methods*	*Surfactant-to-carbon materials ratio*	*Characterization*	*Reference*
CNT	Ultrasonication + Pluronic F-127	25–30	Microstructures Flexural strength	Parveen et al. [4]
CNT	Ultrasonication + superplasticizer + acetone	2.2	Microstructures	Musso et al. [49]
CNT	Ultrasonication + polymer	15–91	Microstructures Compressive strength	Cwirzen et al. [33]
CNT	Ultrasonication + surfactant	4.0	Rheology	Konsta-Gdoutos et al. [21]
CNT	Ultrasonication + surfactant (TNWDIS)	0.2	Microstructures Flexural strength Porosity	Xu et al. [53]
CNT	Ultrasonication + Gum Arabic	6.0	Porosity Microstructure Fracture energy	Wang et al. [50]
GNP	Ultrasonication + methylcellulose	7.0	UV spectroscopy Microstructure	Wang et al. [51]
CF	Ultrasonication + hydroxyethyl cellulose	0.75–3	Compressive, tensile, and flexural strengths	Wang et al. [52]

combining both surfactants and ultrasonication might be a better alternative to disperse carbon materials in cementitious composites.

3.2.2 Fabrication procedure

The fabrication procedures of CBSs are similar between different conductive fillers especially among various nanomaterials, including dispersion of nanoparticles, mixing the suspension with binder and additive, casting, and curing. The following content describes the specific raw materials and manufacturing process regarding the CBSs containing NCB or MWCNT.

3.2.2.1 *Raw materials*

Fig. 3.5 illustrates the general-purpose cement, SF, and NCB particles applied in this test [54]. The general-purpose cement sourced from Independent Cement & Lime

Figure 3.5 Morphology of raw materials [54]. (a) General-purpose cement. (b) SF. (c) NCB.

Table 3.3 Physical and chemical properties of general-purpose cement [54]

Fineness Index (m^2/kg)	*Initial setting time (h)*	*Final setting time (h)*	*Chloride (%)*	*Portland clinker (%)*	*Gypsum (%)*	*Mineral addition (%)*
370–430	1.5	3	0.01	85–94	5–7	Up to 7.5

Table 3.4 Detailed components of commercial SF [54]

Silicon as SiO_2	*Sodium as Na_2O*	*Potassium as K_2O*	*Available alkali*	*Chloride as Cl*	*Sulfuric anhydride*	*Moisture content*	*Bulk density*
89.6%	0.11%	0.23%	0.25%	0.16%	0.83%	1.5%	625 kg/m^3

Pty Ltd., Australia, was used throughout the experiments. To improve the density and pore structures of cement mixture, SF purchased from Concrete Waterproofing Manufacturing Pty. Ltd., Australia, was applied as binder to partially substitute cement. NCB from Xinxiang Deron Chemical Co. Ltd., China, was used as conductive fillers for the fabrication of CB/CBSs (CBCS). The MWCNT is considered to possess much more electrical contact points than single-walled CNTs, thus performing better piezoresistive sensitivity to forces and deformations. Their specific chemical and physical parameters can be found from Tables 3.3–3.6. In addition, the polycarboxylic acid-based superplasticizer was used to improve the dispersion of CB particles and enhance the workability of composites. Lab tap water was directly used throughout the experiments. Tap water was used throughout the concrete mixing in UTS Tech Lab.

3.2.2.2 Process without dispersion

The specific fabrication process of NCB-filled CBS without ultrasonic dispersion is illustrated in Fig. 3.6, which includes the well-dispersed NCB solution preparation and the mixing of NCB–cement composites. There are seven major steps for the manufacturing of piezoresistive NCB–cement composite. First, half of the water was added into the mixer, then gently poured in the water reducer, and stirred for 5 min at low speed to avoid the formation of air bubble. Second, weighted NCB was added into the solution,

Table 3.5 Physical properties of carbon black [54]

Particle size (nm)	*Resistivity (Ω·cm)*	*Pour density (g/l)*	*DBP (ml/100g)*	*Surface area (m^2/g)*	*pH value*	*Ash content (%)*
20	<0.43	0.375	280	254	7.5	<0.3

Table 3.6 Physical and electrical properties of commercial MWCNT [54]

CNTs purity (wt.%)	*Inner diameter (nm)*	*Outer diameter (nm)*	*Length (μm)*	*Loose density (g/cm^3)*	*Electrical resistivity (Ω·cm)*	*Specific area (m^2/g)*
>95%	3–5	8–15	3–12	2.1	10.2	233

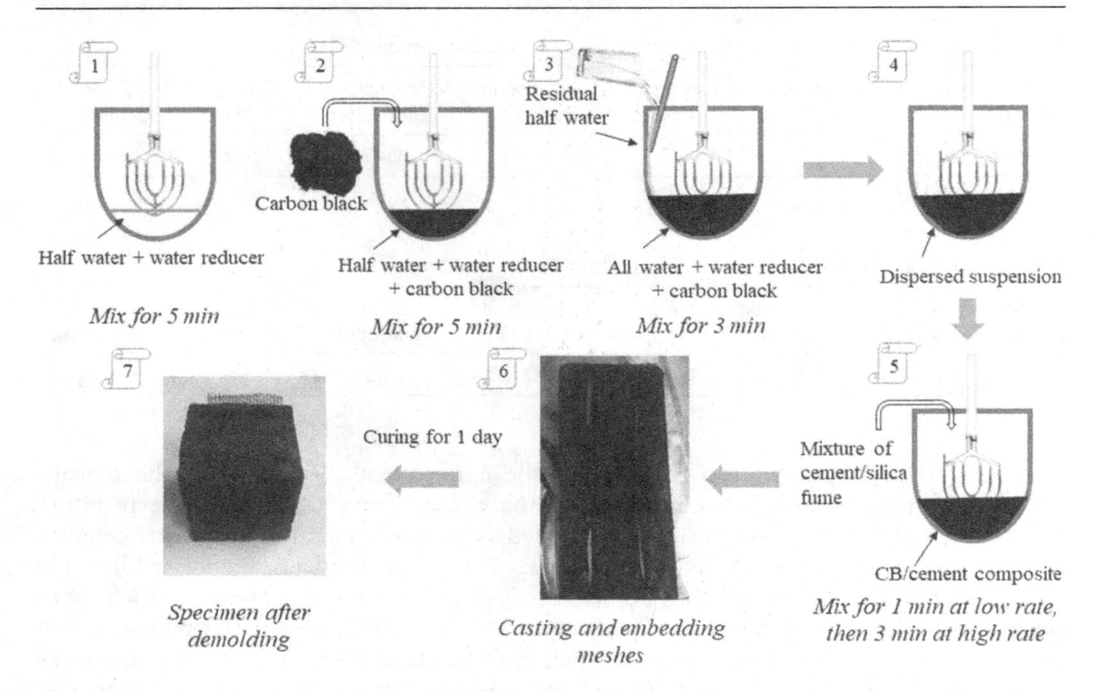

Figure 3.6 Fabrication procedures of NCB-filled cementitious composites and the treatments after curing.

followed by another 5 min mixing. The superplasticizer plays the key role in enhancing the dispersion efficiency of NCB. Third, slowly decant the rest of the water down a stirring rod to wash away the attached carbon black, then another 3 min stirring was needed. The mixing rate of second and third procedures was kept at a low speed, owing to the solution might splash out at high mixing rate. Fourth, the well-dispersed NCB solution is prepared for NCB–cement composites. Fifth, pour in the mixture of cement and silica fume, and it is first stirred at the low rate for 1 min and then switched to the high rate for another 3 min mixing. Sixth, the mixture is poured into the mold and copper meshes are inserted. Mechanical vibration is followed before placing the specimens into curing chamber. Finally, specimens were demolded after one day curing in the chamber, and then moved to the curing chamber for another 28 days, with the controlled temperature at 20±5°C and relative humidity of 95%.

3.2.2.3 Process with dispersion

As shown in Fig. 3.7, the water reducer was first added into the water and the solution was stirred to reach uniformity. Then, the weighted MWCNT was added into the solution with another gentle stirring, followed by the sonication treatment for 1 h with the ultrasound frequency of 40 kHz. During the ultrasonic dispersion process, the solution temperature increases, necessitating periodic replacement of the ultrasonic water to prevent the elevated temperature from affecting the dispersion of MWCNT. Afterward, the well-dispersed MWCNT solution was poured into the Hobart mixer, and then cement and silica fume were added into the mixer. The following procedures are identical to the aforementioned.

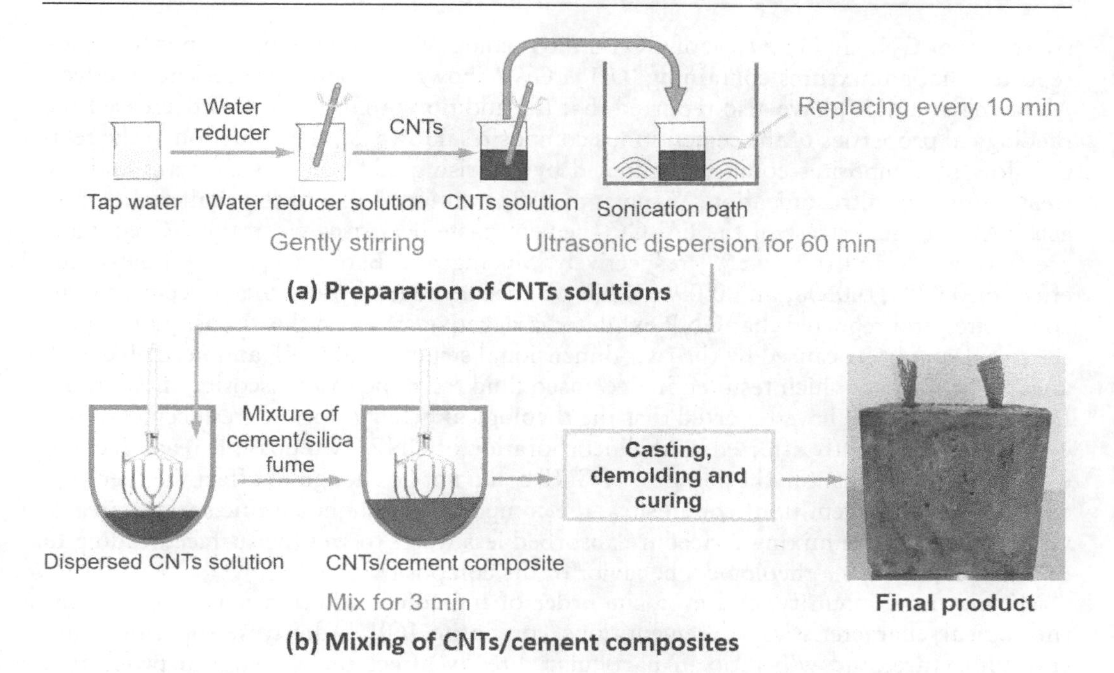

Figure 3.7 Preparation of MWCNTs solutions and MWCNT–cementitious composites.

3.2.3 Rheological properties of CBS

3.2.3.1 Rheological properties with conductive filler

The rheological properties were usually adopted to evaluate the effects of nanomaterials on the fresh properties of cementitious composites. Fig. 3.8 displays yield stress and viscosity of cementitious composites modified with different contents of GNP and GO [55]. The yield stress shows that the paste can overcome the resistance of plastic flow and has a direct relation with the working property of the paste. The smaller the yield stress, the better the performance of the paste. Chougan et al. [56] compared the effects of

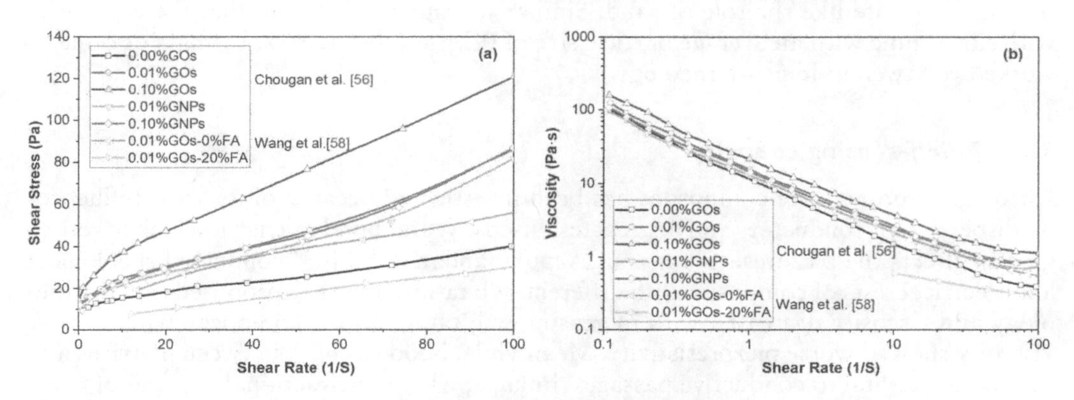

Figure 3.8 Rheology of the cementitious composites modified with different contents of GNPs and GOs. (a) Shear rate versus shear stress. (b) Shear rate versus viscosity.

two types of GNP and one type of GO on the rheology of cementitious composites. They reported that admixtures containing 0.01% GNP showed a dominant lubrication effect. Alatawna et al. [57] have also reported that the addition of GNP and GO decreased the rheological properties of the cementitious composites. However, the reduction in the relative flow of composites could be lightened by the rising addition of surfactant and the treating time of ultrasonication. Wang et al. [58] mentioned that with the increase of fly ash (FA), the yield stress of the FA–GO–cement paste decreases when the GO contents are 0.01 wt.% and 0.03 wt.%, respectively. Shuang and Baomin [31] investigated the effect of 0.03%, 0.06%, and 0.09% GNP on the rheological properties of cementitious composites and reported that GNP exhibited a negative effect on the viscosity of cement slurries. It might be caused by the two-dimensional structure of GNP and needlelike AFt at early hydration, which resulted in decreased fluid resistance and viscosity. In addition, Lamastra et al. [59] have reported that the rheological behavior of the fresh cementitious composite was slightly affected by the incorporation of GNP. Two possible reasons could explain the phenomenon: the addition of GNP could not be enough to affect the rheological behavior of cementitious composites; the comparatively slight specific surface area of GNP utilized in the mixing procedure absorbed less water to wet the surface, leading to minimal effect on the rheological behavior of the composite.

Mixing time, intensity, and even the order of ingredient addition may influence the rheological characteristics of cementitious composite [60]. Additives, superfine mineral admixture, and w/b ratio in particular directly affect the rheological properties. Normally a higher w/b ratio leads to lower apparent viscosity and worse rheology, while insufficient water could prohibit the hydration of cement and influence the final strength of composite. Therefore, plasticizer is always used to ensure the composite having both satisfactory mechanical and rheological characteristics. Papo and Piani [61] investigated the effect of superplasticizers on the rheological property of cement paste and concluded that the polyacrylate superplasticizer was the best to meliorate the rheology of composite. This is mainly due to the hydrophobic groups attached on the surfaces of cement particles and slowing down the hydration and agglomeration rate of mixture. As for using mineral admixture for rheological modification, Ferraris et al. [62] used the ultrafine fly ash to improve rheology of cement paste and proposed that the modified concrete rheology could maintain the mechanical properties and reduce cost, thereby eliminating the use of more water and superplasticizer. It is believed that mineral admixture affects the rheology of composite like the role of sand. Similarly, Chung [35] found that the composite with silica fume with an average particle size of 0.1 μm (used at 15% by mass of cement) worked very well to improve rheology.

3.2.3.2 Self-sensing capacity

Rheological property of composites has been investigated because of its great influences on dispersion of conductors and aggregates, viscosity, and bond strength, which in return greatly affects the piezoresistive output. Vipulanandan et al. [63] conducted rheological and electrical tests of composite with different w/b ratios, and they proposed that the rheology and viscosity decreased with increasing w/b ratio and the composite with a lower viscosity showed worse piezoresistivity. Meanwhile, bond strength between matrices and conductors is vital to conductive passages' linkage, where the fractional changes of resistivity and piezoresistivity directly affect the strain-sensing ability. The more addition of SP and higher w/b ratio led to a higher slump flow and lower viscosity of cement paste.

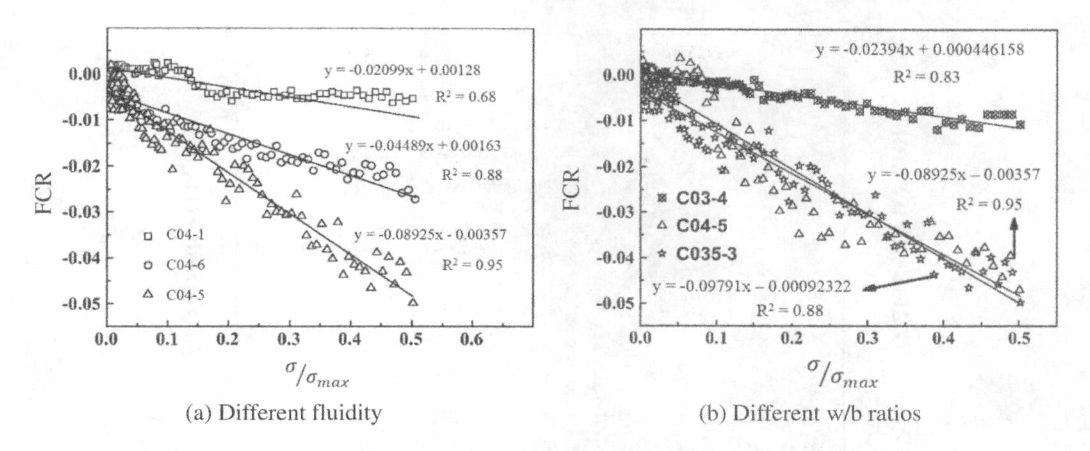

Figure 3.9 Self-sensing performance for the cementitious composites with CNF.

Fig. 3.9 shows the CNF-filled cementitious composites in different slump flow and viscosity. In the context of cementitious composites, the electrical resistivity exhibited a nonlinear relationship with changes in slump flow and viscosity. Specifically, the resistivity initially decreased as the slump flow increased and the viscosity decreased. This indicated that an optimal increase in slump flow and reduction in viscosity facilitated the dispersion of CNF within the composite. However, excessively high fluidity had an adverse effect on the electrical resistivity of the cementitious composites [64].

3.2.4 Electrode configuration

Electrode materials are required to possess two fundamental characteristics: low electrical resistance and a stable electrically conductive property. The predominant choices for electrode materials consist of copper, silver, or stainless steel bars, meshes, or plates. Normally, the electrodes should be attached to the surface, or embedded inside of the CBSs. The influence of electrode configuration on the electrical resistance of CF-reinforced cement paste was investigated by Reza et al. [65]. The first configuration used parallel silver plates as electrodes on both ends of the specimens, which linked with an ammeter and a voltmeter. In the second configuration, a separate voltmeter was connected to two silver paste circles mounted on the surface of the specimens, coupling with the same silver plates to access the applied current. The third one adopted the commonly used four-probe method with four silver paste circles to measure electrical resistivity of cement sensor. It is worth mentioning that there were no embedded copper grids to substitute the conductive plates and paste as electrodes to measure the electrical resistance, so the measured resistance refers to the surface resistance rather than the volume resistance. It was found that the indicator changed sensitively with contact pressures on the interface of plate and specimens for the methods of parallel plate, while negligible alteration was observed for the four-probe method. This is mainly owing to the loose contact between the silver plates and the surface of cement where the interface was not smooth enough. Han et al. [66] draw the conclusion that the embedded electrodes are better than that on the interface of cement sensor, because the electrodes can fully contact with matrix to reduce contact resistivity. However, the use of embedded electrodes will influence the

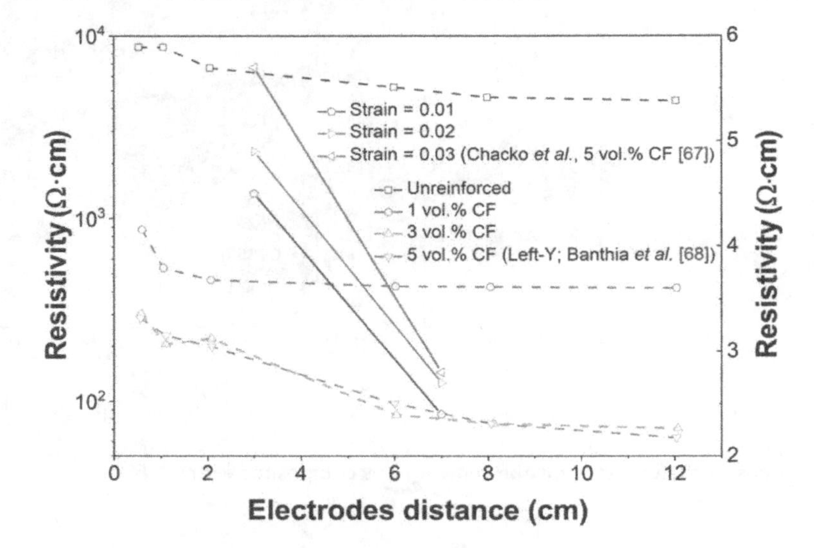

Figure 3.10 Effect of electrodes spacing on the resistivity of cement composite [69].

mechanical strength and durability characteristics of cement-based materials, leading to stress concentration in the electrode region and resulting in premature cracking.

As indicated in Fig. 3.10, for both studies, that is, References [67, 68] the measured electrical resistivity decreased with increasing electrode space. Although there were only two chosen distances analyzed by the former, the results by the latter did demonstrate that the electrical resistivity would become constant when the space of electrodes reached the threshold. According to the hypothesis by Banthia et al. [68], this phenomenon perhaps depends on the weakened capacitance effect with increasing distance. In general, increasing the distance of electrodes is better than increasing the size of copper grids to minimize the effect of capacitance. Besides, for volume resistance measurement, a larger space of electrodes could also decrease the interference from surface resistance [69].

3.2.5 Curing regime

3.2.5.1 Curing methods

Steam curing, moist curing, and air curing are mostly used for cement composite curing. Concrete stream-cured at early age can achieve high mechanical strength and durability. At high curing temperatures, the hydration of cement and induced drying shrinkage are accelerated. The rapidly formed hydration products at an early age, however, may wrap the unhydrated cement particles and prevent water from entering them, resulting in a nonuniform microstructure of the paste that affects the conductive framework formation. Moist curing improves cement hydration and allows hydration products gradually to fill the spaces between unhydrated cement grains, thus providing a denser microstructure and higher strength, especially long-term strength for the paste. However, nonconductive hydration products in moisture environment enhance the resistivity of cementitious composite. Besides, moisture may more easily induce the effects of capacitance and inductance than dry condition, which may cause larger electrical resistivity of CBS. As depicted

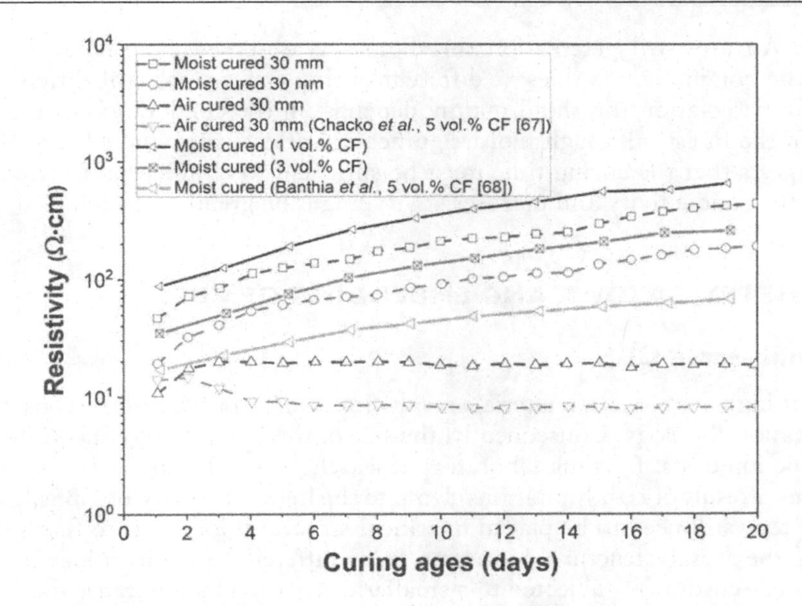

Figure 3.11 Effect of air curing and moist curing on the resistivity of cement mortar with curing ages.

in Fig. 3.11, results from both References [67, 68] showed that the moist-cured conductive mortar has higher resistivity than the air-cured mortar. There is also a tendency for the resistivity of moist-cured sensors to increase with the increasing curing ages while the value remains stable for the counterparts cured in air; such differences possibly are due to the huge water content alteration during the moist curing process.

Air curing with lower moisture means a slower hydration rate at an early age. According to the research by Whiting and Nagi [70], the resistivity of air-cured plain concrete is larger than that of moist cured. The discrepancy is probably due to their different conductive mechanisms. For the CBS, the conductors in composite may be enwrapped tightly by the moist-stimulated hydration products, leading to the decreased conductivity. But for the plain composite, the conductivity mainly comes from the residual moisture in composite, which explains that the air-cured specimens with insufficient moisture exhibit higher electrical resistivity than the moist-cured specimens.

3.2.5.2 Curing time

The piezoresistivity of CBS is affected by the curing age as the bonding forces between the conductors and matrixes are a function of the curing age. According to the research on CNT-reinforced cement sensor by Galao et al. [71], the fractional changes of resistivity to cyclic stress at an age of 28 days showed better periodicity when compared with the performance at 7 days or 14 days. Because of the poorer bonding at the early curing period, the composite possesses lower resistance to external stress when subjected to loading. Furthermore, due to the swift hydration of cement mixture at the early age, the void ratio in the conductor–matrix interface increases owing to the intensive drying shrinkage, which also contributes to the resistivity growth. On the other hand, Chen et al. [72] investigated the effect of curing time on the conductivity of cementitious composite. An unvaried percolation threshold was observed at different curing ages, although the conductivity values at different curing times were still different. This is because the percolation threshold mainly depends on the conductors in the composite rather than the moist, although moist significantly affects the conductivity of composites. It suggests that the curing time must be sufficient to eliminate the effect of moist, otherwise the conductivity and piezoresistivity might be greatly affected.

3.3 GEOMETRY, LAYOUT, AND EMBEDDING OF CBSs

3.3.1 Small-scale CBS

The current high cost of nanomaterials results in elevated production expenses for CBSs based on nanotechnology. Consequently, the size of these sensors is constrained to minimize production costs. Existing laboratory research primarily emphasizes small embedded CBSs as a result of cost limitations. Despite the high sensitivity of CBSs, due to their small size, they still need to be placed in critical structural locations to fulfill the role of monitoring the overall structure. Moreover, being different from direct load on the CBSs, the embedded sensors are subjected to a smaller distributed load because the surrounded cement matrix bears the majority of the load, which affects the force transmission and sensing efficiency of CBSs. As a consequence, the CBSs connected in series are proposed to magnify the sensing ranges and electrical resistivity changes through collecting the resistivity changes from all CBSs in multiple locations under compression. As shown in Fig. 3.12, multiple small CBSs are connected through electrical soldering on the joints, which were then coated by insulated glue to eliminate the influence from pore solutions after being embedded into concrete structures [73].

3.3.2 Sensors in mortar slab

3.3.2.1 Sensors configuration

The connection among CBSs and mortar slab is critical to both the piezoresistive performance of CBSs and the mechanical properties of the mortar slab because the interfacial transition zones (ITZ) can influence the force transmission from the slab to the sensors. To certify the excellent connection, macro- and micromorphology of ITZ are displayed

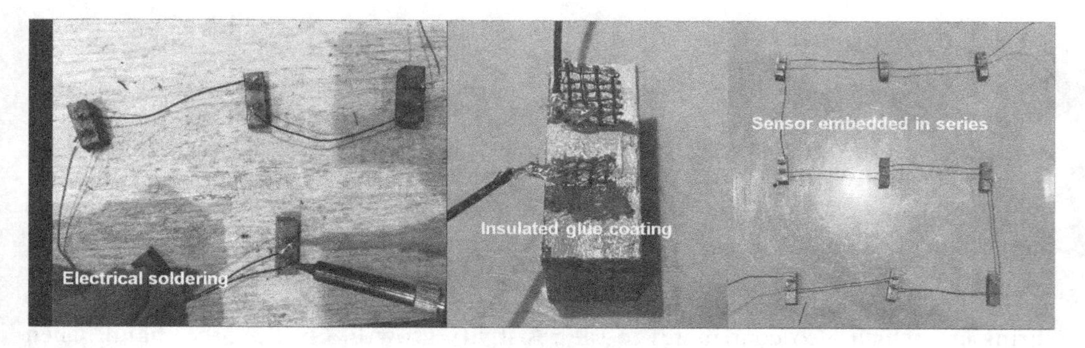

Figure 3.12 The joint connection, insulated glue coating, and CBSs in series [73].

Figure 3.13 Morphology of the interface between embedded CBSs and mortar slab [73].

and analyzed. Fig. 3.13 shows the macro-morphology of CBSs during embedding into cement mortar slab and after crushing [73]. Given the mortar casting and sensors' embedding procedures were carried out on a vibration table, it could be observed that no significant air pores or gaps were generated around the ITZ. Moreover, even after crushing, the bonding between the CBSs and the mortar slab is firm and dense, indicating an excellent and stable ITZ under external forces. Therefore, the force applied on the surface of mortar slab can be successfully transferred to the sensors and induce piezoresistivity.

3.3.2.2 Boundary between sensors with slab

Fig. 3.14 shows the micromorphology of ITZ between mortar slab and CBSs. Three different regions of multiple connections between CBSs and mortar slab were exhibited in Fig. 3.14a– h. For the CBSs with intact surface, there were less porous structures and gaps than that of the deteriorated surface with exposed CNF. But generally, it could be found that all surficial conditions of CNF-filled CBSs connected well to the cement matrix. On the one hand, one of the critical advantages of CBSs compared to other traditional sensing

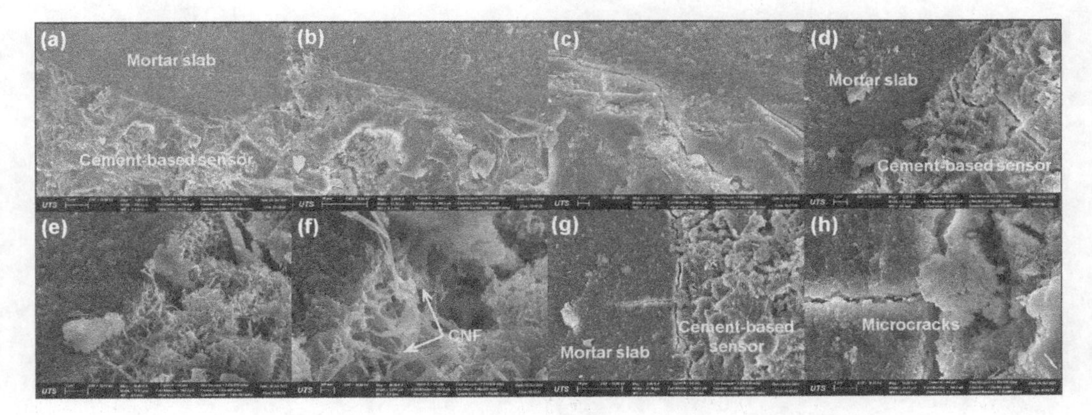

Figure 3.14 Microstructures of the interfaces between embedded CBSs and the slab mortar [73].

techniques is the "intrinsic properties" to concrete structures that will never deteriorate their durable and mechanical properties. The good connection implied that the mechanical properties of mortar slab with embedded CBSs could be well maintained, which is fundamental for practical application. On the other hand, the applied force on the slab can be successfully transmitted to the CBSs, so that the piezoresistive performance can be expressed subsequently. Still, in some specific regions, microcracks along ITZ extending to the mortar slab can be observed. Even though the cement mortar slabs were manufactured with similar raw materials and mix proportion to that of CBSs, the microcracks were inevitable due to their different shrinkages [74]. This will definitely affect the stress transmission from the slab to the CBSs but given the microcracks are limited and the widths are less than 1.0 μm, it is believed that the negativity will not greatly affect the self-sensing performance of mortar slab with embedded CBSs.

3.3.2.3 Self-sensing under compression

Fig. 3.15 shows the fractional changes of electrical resistivity (FCR) of cement mortar slab with embedded CBSs under deformation-controlled compression. It was found that the mortar slab showed excellent self-sensing performance, with the first decreasing and then returning electrical resistivity, respectively, when the force was loaded or unloaded. It demonstrated that the embedded CBSs have satisfactory piezoresistivity and good connection with the mortar slab. In addition, compared to the FCR under 4 kN, the electrical resistivity changes almost doubled for the mortar slab subjected to 8 kN forces, indicating an excellent linearity between electrical resistivity changes and applied static forces.

It should be noted that there exists a gradual decrease of FCR in each cycle of compression. Previous studies also found similar phenomenon because of the permanent deformation of CBSs [75]. In this study, in addition to the compressed CBSs, the micro-defects generated in cement mortar slabs during compression can absorb partial energy that result in reduced FCR values. Moreover, the connections among CBSs and mortar

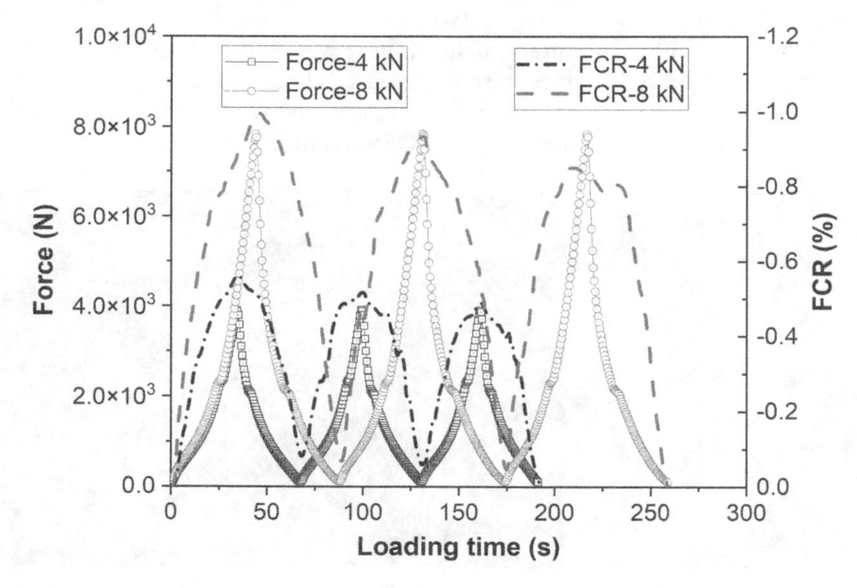

Figure 3.15 Self-sensing behavior of mortar slab under experimental compression [73].

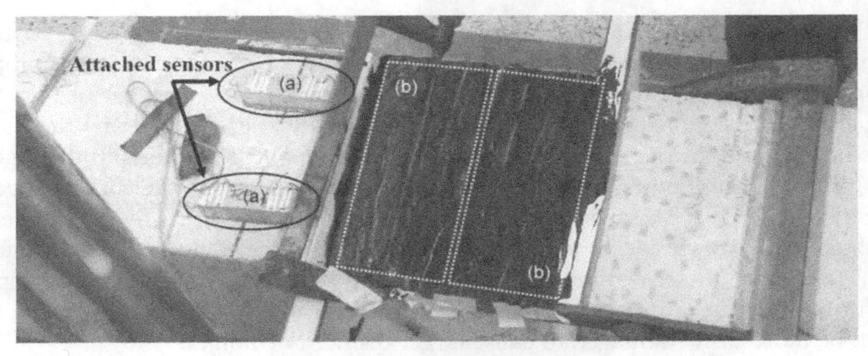

(a) Attached sensors in the concrete beam

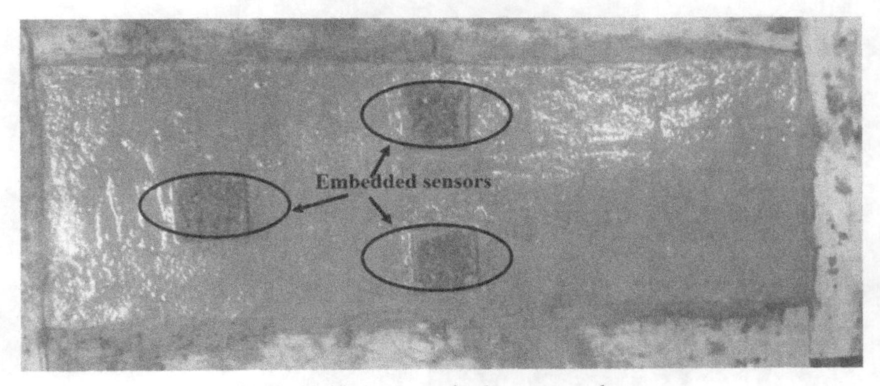

(b) Embedded sensors in the concrete beam

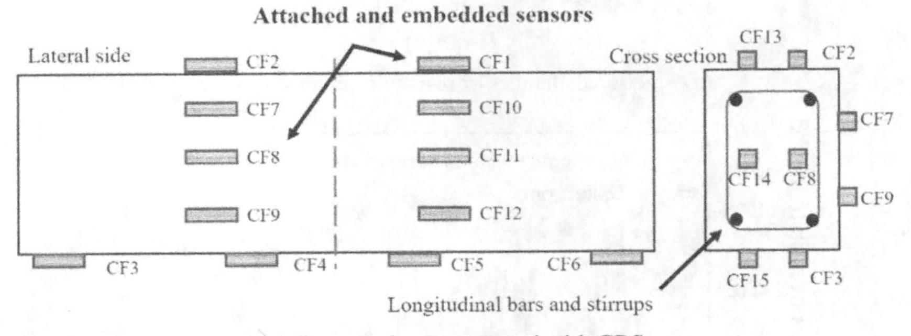

(c) Concrete beam mounted with CBSs

Figure 3.16 Schematic plot of the CBSs applied to the concrete beams [69].

slab might be influenced to decrease the FCR values. To overcome these shortcomings, the preloading treatment on the mortar slab with embedded CBSs might be an option to stabilize the electrical resistivity changes.

3.3.3 Sensors in reinforced concrete beam

As shown in Fig. 3.16, for self-sensing concrete beams with coexistence of compression and tension zones, the compressive and tensile loadings on CBS lead to different fractional changes of resistivity with usually a negative value for the former and a positive value for the latter. This mainly depends on the different fracture mechanisms as electrical resistance of the compression zone is decreased by closer space of matrixes, conductors, and compressed pores, while contrarily the increasing resistance is from the development of microcracks and larger space of conductors in the tension zone. Meanwhile, it seems that the fractional changes of resistivity in the tension zone is more than one order of magnitude larger than its counterpart in the compressive zone under the same deflection of the concrete beams, as shown in Fig. 3.17 [76]. This figure also shows that the irreversible resistivity is higher in the tension zone with worse repeatability, while much lower in the compression zone with both excellent sensitivity and repeatability. This might be owing to the much higher compressive strength of concrete than its tensile strength, which means that the sensor in the tension zone can be more easily damaged to cause larger resistivity changes.

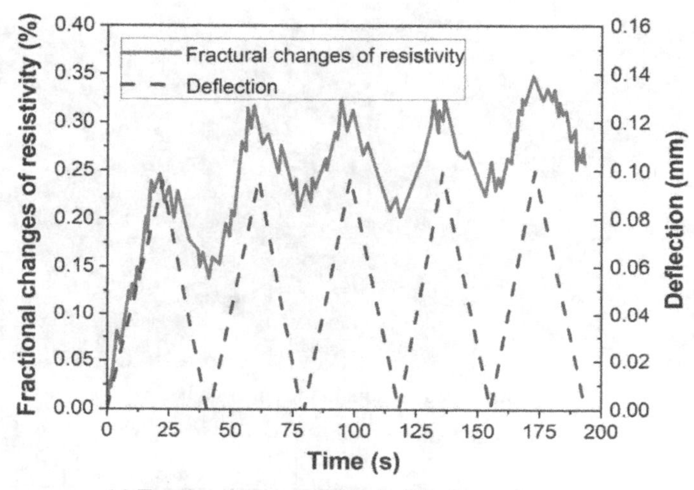

(a) Fractional changes of resistivity in tension zone

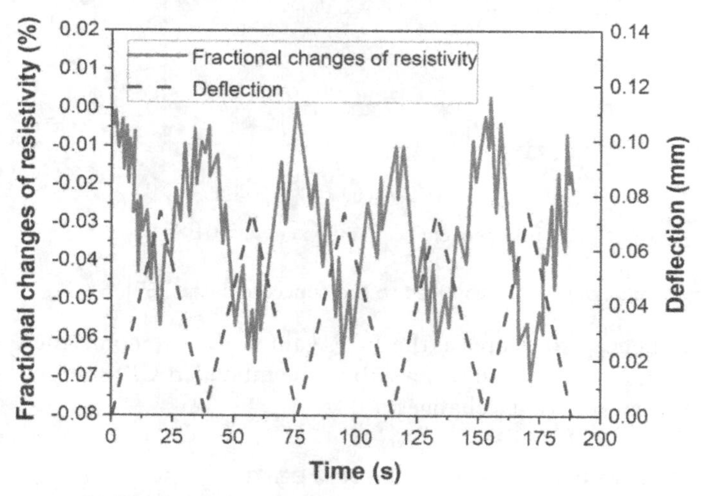

(b) Fractional changes of resistivity in compression zone

Figure 3.17 Fractional changes of resistivity of concrete beam under flexural loading with strain-sensing coating [76].

On the other hand, the experiments by Baeza et al. [77] demonstrated a higher gauge factor and better sensing capacity in the tension zone. The discrepancy is possibly owing to their different testing conditions where the CFs in the tests were oxidation treated before mixing. This may strengthen the bonding between the fibers and matrixes and increase the tensile strength of the CBSs, making them well preserved even in the tensile zone and thus increasing their electrical sensing ability.

3.3.4 Sensors in unreinforced beam

3.3.4.1 Sensors configuration

Fig. 3.18 depicts three embedding configurations of NCB-filled CBSs (CBCS) in the small-scale unreinforced beams (40 mm × 40 mm × 160 mm) for three-point bending test [54]. The piezoresistive-based sensing performances of CBCS were investigated in both the compression and tension zones. Since the distribution of stress was varied based on the distance to load point, the effects of embedding locations of sensors in unreinforced beams were also considered. Fig. 3.18a-1 shows the first sensors configuration in compression zone, where the distance of CBCS-1 and CBCS-2 to center of beams is 10 mm, and

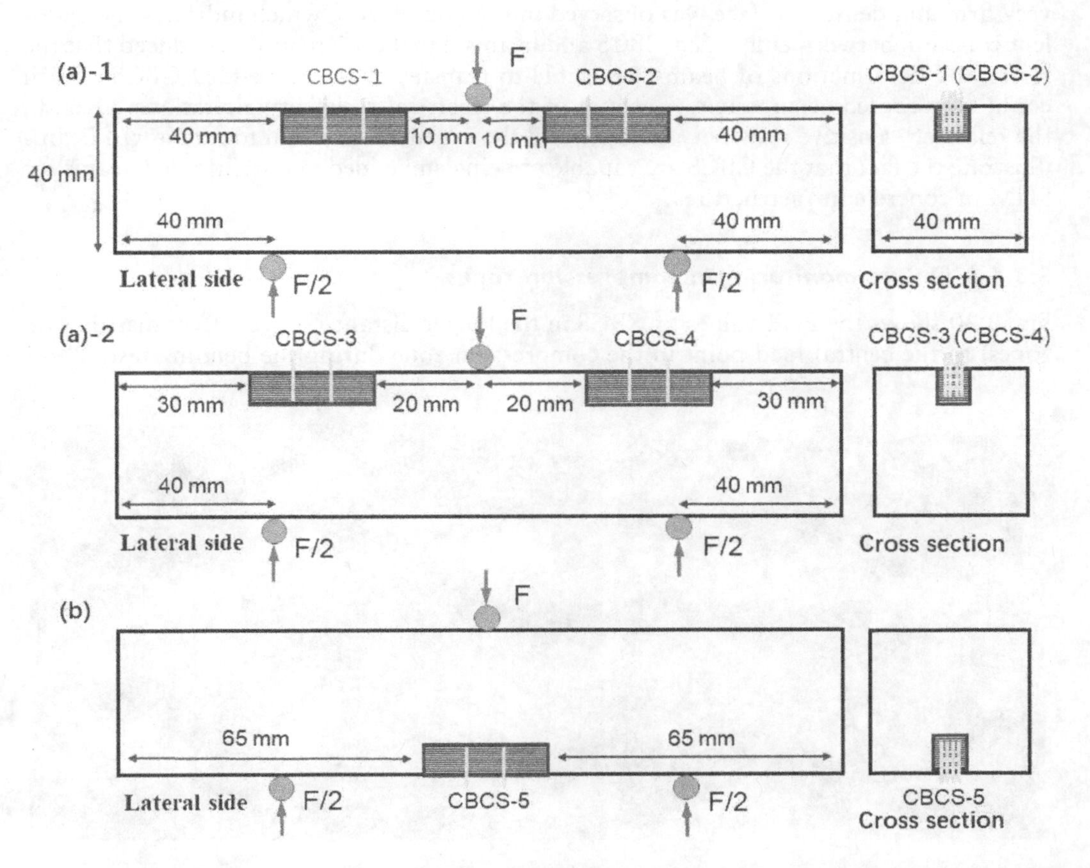

Figure 3.18 Configuration of CBSs in different zones of unreinforced beam under three-point-bending [54].

very close to the load applied point. Another sensor configuration (CBCS-3 and CBCS-4) in the compression zone is conducted by longer distance of 20 mm to the load point, as shown in Fig. 3.18a-2. Generally, it could be considered that the former one was in the high stress concentration zone, while the latter one was in the low stress distribution zone. The sensors in these regions were subjected to both compressive and tensile stresses. As for the CBCS in the tension zone, the sensors only embedded in the central section (CBCS-5) were investigated in Fig. 3.18b, because of the largest tensile stress in the beams and the relatively weak sensing ability of CBSs for the tensile stress monitoring.

3.3.4.2 Boundary between sensors with beam

Previous studies usually assumed a good cohesion and dense boundaries between embedded CBSs when embedding CBSs to monitor concrete structures [78]. However, boundaries between CBCS and unreinforced beams are significant to the piezoresistive performance of CBSs, owing to the stress transfer from the beams to CBCS. To obtain the characteristics of their boundaries, the concrete beams cooperating embedded CBCS were cut in half to expose the boundaries. Fig. 3.19 illustrates the morphology and the microstructures of CBCS, unreinforced beams, and their boundaries. It was found that a very firm and dense interface was observed in the boundaries, which indicates the excellent cohesion between embedded CBCS and beams. Furthermore, it was deduced that the stress and deformations of beams were able to transfer to the embedded CBCS, which could be recorded accurately on account of the electrical resistivity alterations. Overall, the relatively sensitive resistivity changes and the excellent cohesion to monitored beams illustrate the fact that the CBCS are capable of being embedded in unreinforced beams for SHM in concrete infrastructures.

3.3.4.3 Failure monitoring in compression zones

Fig. 3.20 shows the FCR values of CBCS in the longer distance (lower stress distribution zones) to the central load point in the compression zone during the bending test. Three

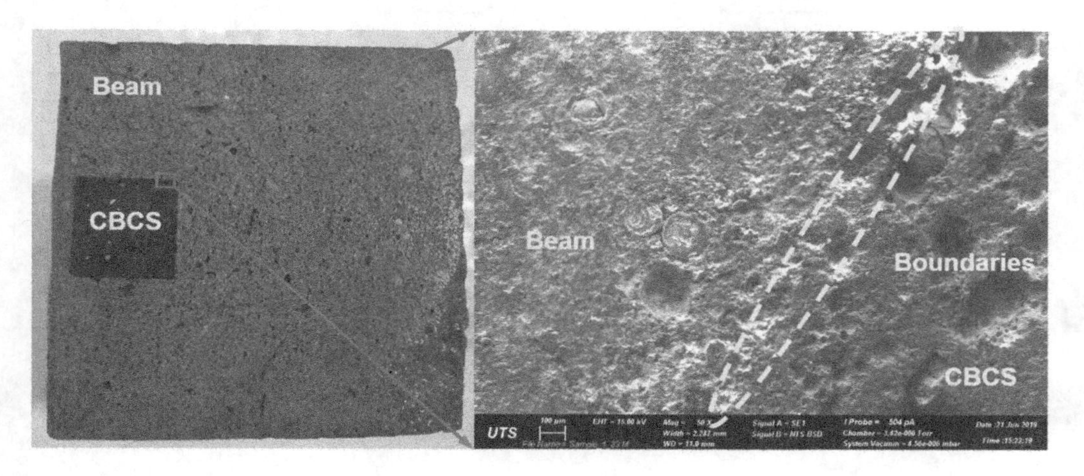

Figure 3.19 Morphology and microstructures of interfaces between embedded CBCS and unreinforced beam [54].

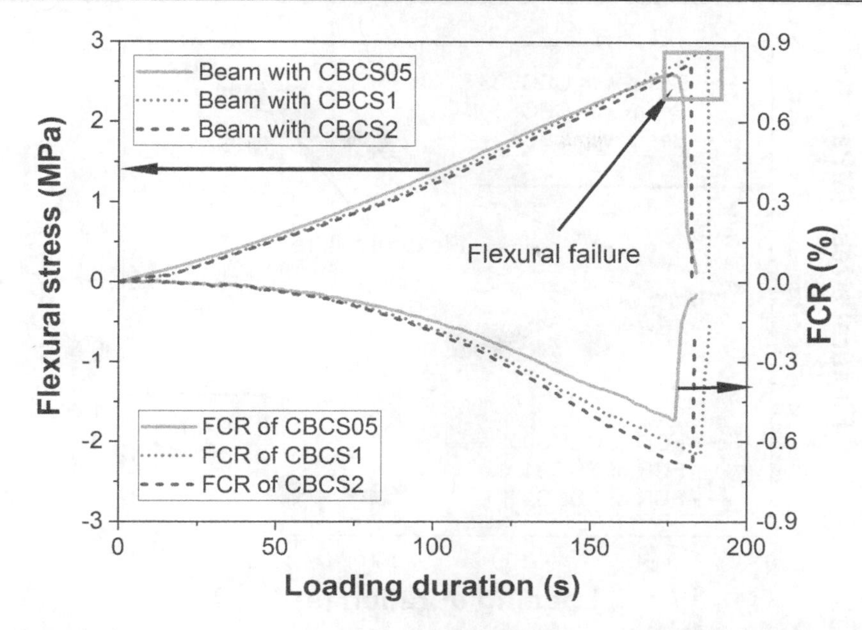

Figure 3.20 Resistivity of CBCS containing various NCB contents with long distance to the central loading point in compression zone [54].

different kinds of CBCS with various NCB contents of 0.5% (CBCS05), 1% (CBCS1), and 2% (CBCS2) illustrated similar tendencies to the FCR decreases as to the increase of loading. Since the intrinsic embedded NCB/cementitious sensors had much less negative impacts on the mechanical properties of monitored concrete beams, it was observed that the beams with CBCS05, CBCS1, and CBCS2 displayed very close ultimate flexural strength, reaching approximately 2.59 MPa, 2.87 MPa, and 2.73 MPa, respectively. After the peak flexural strength, the stress was dramatically decreased to almost zero which implied the flexural failure of unreinforced concrete beams.

In terms of electrical resistivity variations for different CBCS, it seems that the FCR values decreased consistently to the flexural stress and increased almost at the same time when the flexural failure occurred. This is due to the CBCS under compression in the compression zone of unreinforced concrete beams to reduce their electrical resistivity and the excellent cohesion and dense interface between CBCS and beams. Also, it was seen that both the increasing rate and ultimate values of FCR increased with the increase of NCB content, with the ultimate FCR at the flexural failure point rising from 0.51%, to 0.65%, to 0.71% for the CBCS05, CBCS1, and CBCS2, respectively. However, even with the highest piezoresistive sensitivity to bending failure monitoring, it was observed that the CBCS2 exhibited the worst resistivity repeatability compared to those of CBCS05 and CBCS1. The slightly decreased electrical resistivity mainly originated from the permanent compressed CBCS and the constant stress situation in the compression zones even at the bending moment of flexural failure. As previous studies found that the NCB particles could lower the mechanical properties of cementitious composites [79], the reason for the worst resistivity repeatability of CBCS2 is due to its relatively lower bearing capacity and the more compression deformations in the bending process. Overall, with the CBCS embedded at a longer distance (lower stress distribution

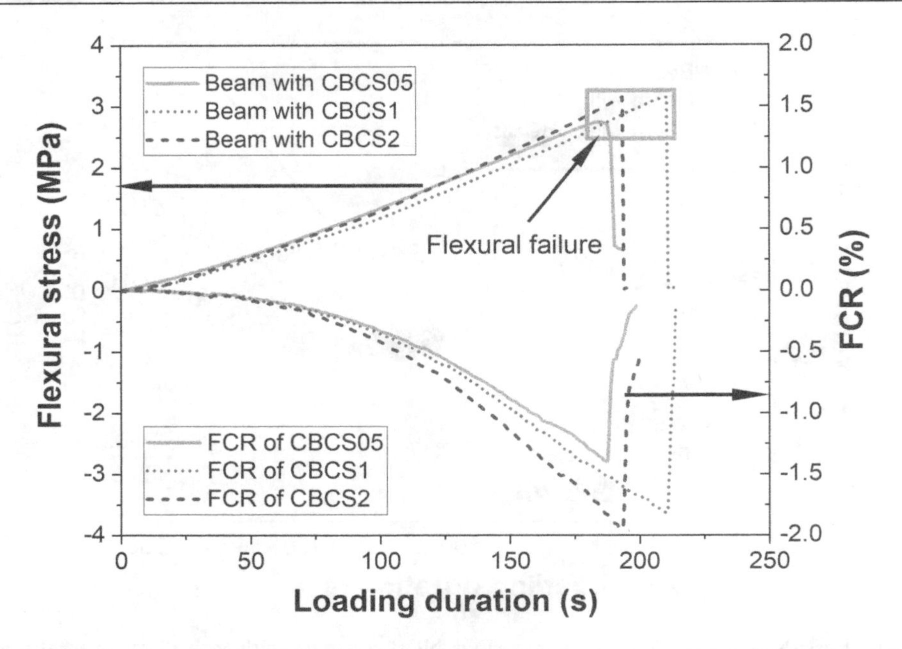

Figure 3.21 Resistivity of CBCS containing various CB contents with close distance to the central loading point in compression zone [54].

zone) to the loading point in the three-point-bending concrete beams, the minor changes of electrical resistivity can still be alternatively monitored for the health of the beams in their service life.

For the CBCS embedded in the closer distance (stress concentration zone) to the load point, much higher values of FCR were observed in Fig. 3.21 compared to the above-mentioned ones with longer distance from the loading point. In general, the CBCS2 once again performed the best piezoresistive sensitivity by the larger changes of FCR, which was followed by the moderate sensitivity of CBCS1 and the worst piezoresistive performances by CBCS05. Compared to the counterparts embedded in the long-distance zone, the FCR values of CBCS05 nearly trebled and surged to 1.44%, and the values increased sharply to 1.82% and 1.96% for the CBCS1 and CBCS2, respectively, in the flexural failure. It indicates that the close distance areas were subjected to higher stress and deformation in the three-point-bending test, and thus the CBCS embedded in these zones can be more efficient to monitor the damage and health conditions of concrete structures to achieve excellent performances of CBSs. However, it should be emphasized that the irreversible electrical resistivity also increased significantly, especially for the CBCS2, whose irreversibility increased from less than 0.3% to a value of higher than 0.5%. This may be attributed to the lower bearing capacity and the larger compression deformations induced by higher stress and closer distances among conductive NCB particles, which could permanently decrease the electrical resistivity of CBCS. In comparison to the CBCS under uniaxial compression, whose final resistivity were increased after reaching 4.0 MPa cyclic compression, the results demonstrated that the embedded CBCS in unreinforced concrete beams never generated irremediable damages or cracks due to the permanent increase in electrical resistivity.

3.3.4.4 Failure monitoring in tension zones

The CBCS embedded in the tension zone of three-point-bending beams were right below the loading point where presented the largest tensile stress concentration. Fig. 3.22 displays the electrical resistivity changes of embedded CBCS with the flexural stress development of unreinforced concrete beams. Compared to the CBCS in the compression zones, the embedded NCB/CBSs in the tension zone caused severe negativities on the ultimate flexural strength of the unreinforced concrete beams. The flexural strength of the beams embedded with CBCS05 reached 2.91 MPa, which considerably decreased to only 2.65 MPa and 2.48 MPa for those beams with sensors of CBCS1 and CBCS2 embedded, respectively. The low tensile capacity of cementitious composites and the stress concentration in the tension zones during three-point-bending test [80], demonstrated the need of high bearing capacity of beams in the tension zones, such as adding longitudinal reinforcements in concrete beams. Therefore, since the NCB/CBSs had poor mechanical properties due to the increase of NCB content, the CBCS with higher NCB content were probably responsible for the reduced bending strength of the beams because of the poor tensile strength of CBCS themselves.

As for the electrical resistivity changes of CBCS in the tension zones, it showed completely opposite trend of FCR to the sensors in the compression zones. It was observed that the electrical resistivity increased with the flexural stress, especially with a sharp increase on FCR at the moment of flexural failure of beams. This can be explicated by the conductive fillers of CBCS in tension, whose distances among NCB particles are widened and thus increase the electrical resistivity. Similar to that of the CBCS in compression zones, the sensors with 2% NCB (CBCS2) were provided with the largest piezoresistivity of nearly 4.2% increase of FCR, and values of 3.4% and 3.1% were observed for the CBCS1

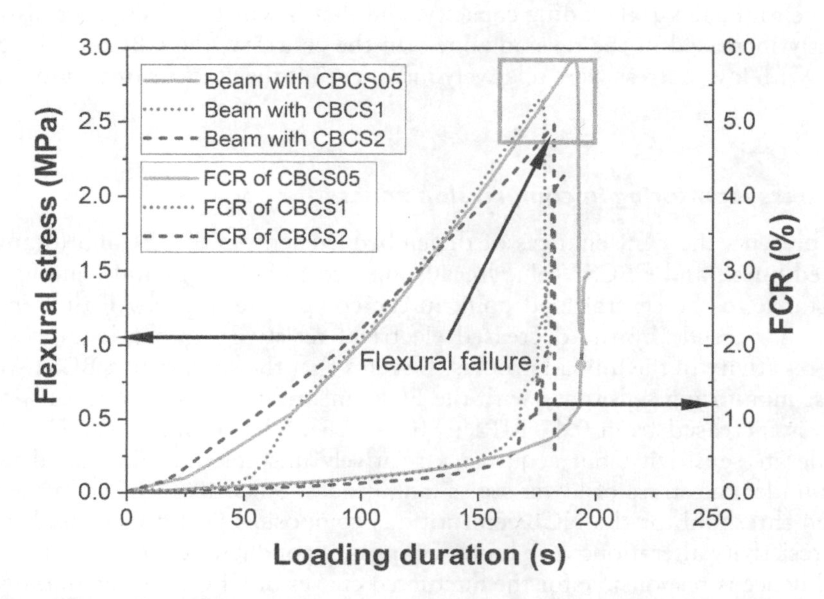

Figure 3.22 Resistivity of CBCS containing different NCB contents in tension zone during three-point bending [54].

Table 3.7 Factor *G* of the CBCS embedding in various regions of concrete beams for monitoring [54]

G (MPa-1)	*Compression zone*		*Tension zone*
	Longer distance	*Closer distance*	
CBCS05	1.97×10^{-3}	5.26×10^{-3}	1.07×10^{-2}
CBCS1	2.26×10^{-3}	5.78×10^{-3}	1.28×10^{-2}
CBCS2	2.60×10^{-3}	6.14×10^{-3}	1.69×10^{-2}

and CBCS05, respectively. The factor *G* to assess the failure monitoring sensitivity of CBCS was proposed according to Eq. (3.1):

$$G = \frac{\text{FCR}}{\sigma} \tag{3.1}$$

where σ is the flexural strength of the concrete beams. The factor *G* of CBCS embedded in various regions of concrete beams for failure monitoring is listed in Table 3.7. It demonstrated that the monitoring sensitivity of CBCS in tension zone was doubled than those in compression zone with close distance to the load point, and even five times higher than those embedded in the compression zone with long distance to the load point. This phenomenon is simply due to the much higher stress concentration in the tension zone than that in the compression zones.

Another finding is that the electrical resistivity of CBCS never returned to the original values, once the beams were damaged by cracks. As mentioned above, the bending strength of unreinforced concrete beams partially depended on the bearing capacity of CBCS in the tension zone. In this circumstance, the CBCS were first destroyed when the beams came to the flexural loading capacity, and that is why the electrical resistivity was permanently increased in the tension zone. On the contrary, the CBCS in the compression zones with lower stress were relatively intact even if the cracks throughout the beams occurred.

3.3.4.5 Stress monitoring in compression zones

Fig. 3.23 presents the FCR changes of the embedded sensors of CBCS05 (orange lines), CBCS1 (red lines), and CBCS2 (blue lines) subjected to cyclic flexural bending with the longer distance to the central load point in the compression zone. All the sensors performed similar trends by the decreased electrical resistivity in the loading stage and increased resistivity in the unload process. It seems that the sensors of CBCS2 were given the highest monitoring sensitivity with the FCR linearly increased from 0.1% to 0.21% as the stress increased from 0.47 MPa to 0.94 MPa. The second is the CBCS1 which got a moderate sensitivity, but acquired a relatively unstable and fluctuated resistivity, especially under the stress of lower stress magnitudes. Since 1% NCB is in the range of percolation threshold for the NCB/cementitious composites [79], this caused significant electrical resistivity alterations with minor changes of conductor contents and microstructures, and hence is responsible for the fluctuated curves of CBCS1 in the initial stage. As for the sensors of CBCS05, the lowest sensitivity with FCR values of 0.08% and 1.5% for the stress magnitudes of 0.47 MPa and 0.94 MPa were observed, respectively. Generally,

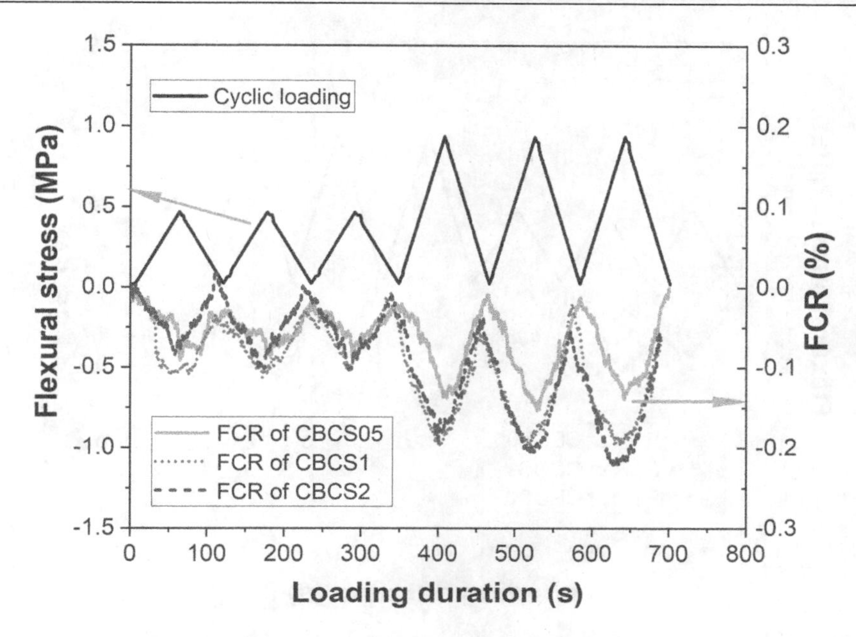

Figure 3.23 The FCR variations of embedded CBCS with the long distance to central load point in compression zones under cyclic flexural loading [54].

the stress-monitoring ability of these sensors embedded in unreinforced concrete beams were competent for their self-sensing behaviors and flexural failure monitoring capacity.

On the other hand, all the sensors expressed cyclic resistivity alterations on account of applied flexural stress on concrete beams. Despite possessing the worst sensitivity, the sensors CBCS05 were observed with best resistivity repeatability as the electrical resistivity almost returned to the initial values, while the electrical resistivity for both the CBCS1 and CBCS2 sensors permanently decreased by approximately 0.5% after the cyclic loading. This can be explained by the decreased mechanical properties of NCB cementitious composites with increase of NCB content, which became "softer" and easier to cause more deformations of sensors and closer distances among conductive NCB particles to decrease the resistivity. Meanwhile, the phenomenon of reduced rather than increased electrical resistivity demonstrated that these sensors were not destroyed to generate cracks in the compression. Moreover, slight electrical resistivity increases occurred for the CBCS2 in the initial loading stage. Since the high content of NCB in cementitious composites is more likely to agglomerate together to produce defects or weak zones in the composites, the initial resistivity increases were probably due to the densification on the NCB agglomeration defects and weak zones by compression.

In terms of the sensors embedded with close distance to the load point in the compression zones of unreinforced concrete beams, their electrical resistivity changes under cyclic flexural stress are displayed in Fig. 3.24. Compared to the sensors long distance to the load point, higher FCR values were observed for all CBCS and the FCR values of CBCS2 gradually increased from 0.2% to 0.4%, which were also doubled for the sensors CBCS05 and CBCS1. Similar to the CBCS1 with the long distance to load point, much more severe resistivity fluctuations were observed for the same sensors embedded at a close distance to the loading point. The reason is identical to the above-mentioned instability since

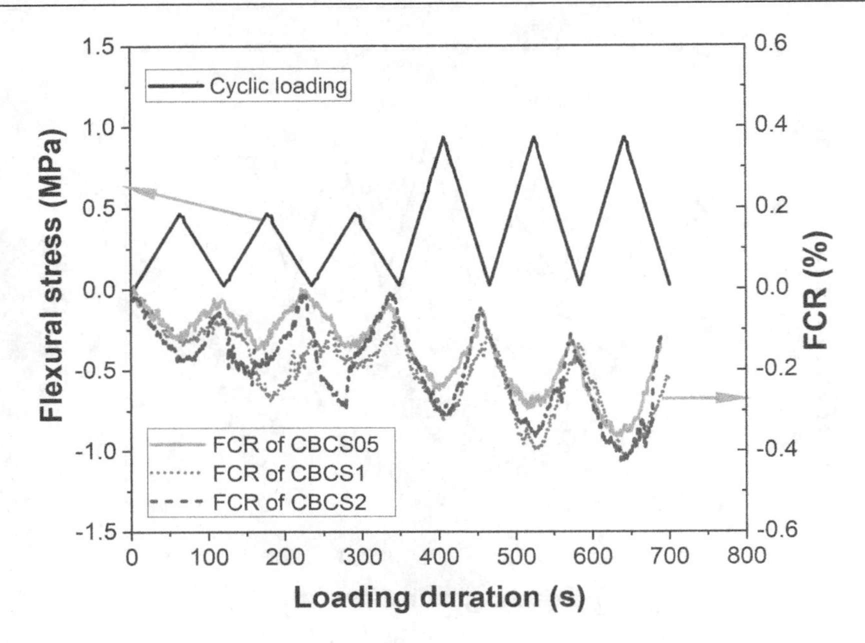

Figure 3.24 FCR alterations of embedded CBCS with close distance to the central loading point in compression zones under cyclic flexural loading [54].

the 1% NCB is in the range of percolation threshold for NCB/cementitious composites. In addition, Fig. 3.24 represents that the sensors that went through small stress magnitude exhibited excellent resistivity repeatability, while clear FCR irreversibility occurred when the stress magnitudes doubled. The irreversible FCR for CBCS1 reached 0.1% and for CBCS05 and CBCS02, it was higher than 0.2%. It demonstrates that the sensors maintained good performances under the stress magnitude of 0.47 MPa. However, high compressive stress, microcracks initiation, and connected conductors occurred close to the loading point by the doubled stress magnitude. Furthermore, severe irreversibility by CBCS1 is also probably due to the effect of slight percolation in the cementitious composites. Overall, for the majority of concrete structures with known load points, such as concrete beams and columns, the sensors close to the load point exhibit higher changes in electrical resistivity and seem easier to capture compared to the sensors far from the loading point. However, the drawbacks of slightly decreased repeatability and more fluctuated resistivity of sensors with high content of carbon black, especially for CBCS1 and CBCS2, should be particularly resolved in the future.

3.3.4.6 Stress monitoring in tension zones

Fig. 3.25 shows the electrical resistivity changes of various embedded CBCS in the tension zones under cyclic flexural stress. Contrary to the sensors in compression zones, those in tension zones showed increased electrical resistivity with the increase of flexural stress and decreased resistivity as the stress returned to the original level, indicating the tensile stress induced in the embedded sensors and the longer distance between carbon black particles in the tension zones. Generally, the sensors with 2% NCB (CBCS2) possessed the highest FCR values of slightly over 0.5% in the first three

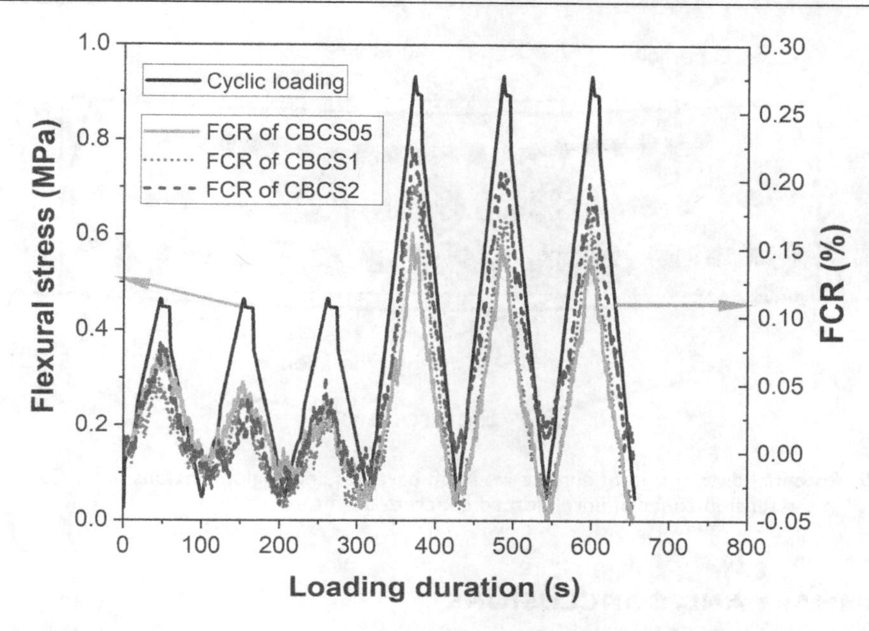

Figure 3.25 FCR alterations of embedded CBCS in tension zones of unreinforced concrete beams under cyclic flexural loading [54].

cycles and boomed to approximately 2% when the stress magnitudes doubled to 0.94 MPa. For the stress magnitude at 0.47 MPa, the sensors CBCS05 ranked the second sensitivity by the average FCR values reaching approximately 0.5%, which were superior to that of CBCS1. This phenomenon partially contributed to the gradually decreased resistivity in the loading process. However, when the applied stress increased to 0.94 MPa, the electrical resistivity changes of CBCS1 were consistent to the results in the compression zone and gently higher than the FCR of CBCS05, with both values approaching 0.18%.

Even with the highest average FCR values for CBCS2, the tendency observed was that the gradually decreased electrical resistivity occurred for the sensors under the stress in low magnitudes. Similar phenomenon was observed for the CBCS1, which were responsible for the aforementioned smaller FCR values than that of CBCS05. To explicate the decreased electrical resistivity in tension zones, Fig. 3.26 illustrates the stress conditions of sensors and the potential distributions of conductive NCB particles and their agglomerations in the tension zones under flexural stress. Unlike the sensors under axial tension where the electrical resistivity increased gradually [81], the CBSs in tension zone of beams subjected to bending moment still had partial regions under compression. It indicates that the electrical resistance of sensors was dependent on the coupled actions of both compression and tension zones between two electrodes, and the actual resistivity was the combination of the resistivity decreases in the region under compression and the resistivity increases in the region under tension. For the sensors with high NCB contents (CBCS1 and CBCS2), the compression zone was relatively easy to compress to cause microcracks. Thus, the decreased electrical resistivity might compensate for the resistivity increase in the tension zones and decrease the bulk electrical resistivity.

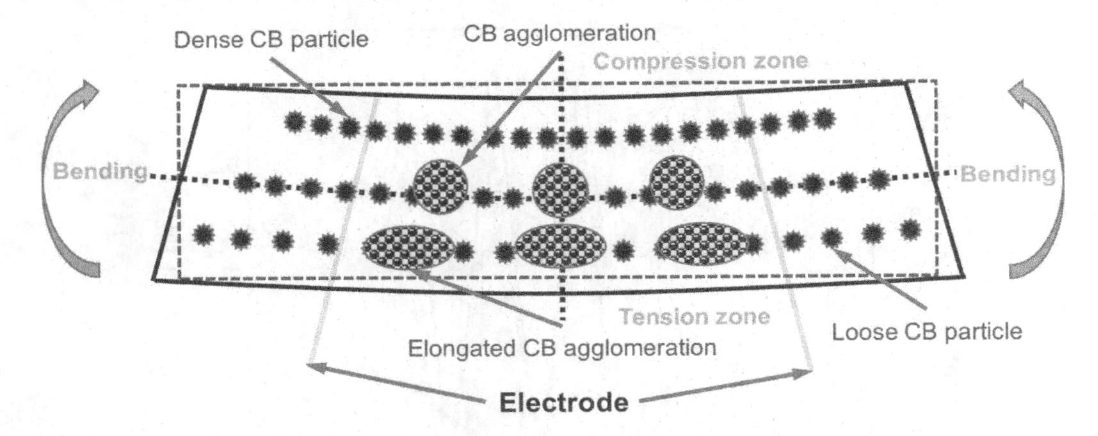

Figure 3.26 Potential distribution of conductive NCB particles and agglomerations in CBCS embedded in the tension zones of unreinforced concrete beams.

3.4 SUMMARY AND CONCLUSIONS

This chapter summarizes the fabrication, geometry, layout, and challenges of the CBSs. The fabrication of CBSs involves dispersion of conductive fillers, mixing, casting, electrode configuration, and curing regime to ensure optimal performance. The geometry and layout involve the individual CBSs or the embedded sensors in multiple concrete structures. The challenges exist in achieving uniformity, dispersion of nanomaterials, environmental interference, and scalability for large-scale production. There are some key conclusions:

1. Achieving uniform dispersion of nanomaterials in CBSs is a critical aspect of sensor fabrication that directly impacts their sensing capabilities and overall performance. The hybrid utilization of mechanical ultrasonication, chemical additives, and coupling of nanomaterials with different shapes can be an efficient way of dispersion. The choice of electrode configuration depends on the specific application and the desired sensitivity and accuracy of the sensor; compared to the electrodes attached on the surface, the embedded ones have more direct and intimate interaction with the matrix, resulting in better sensitivity and accuracy. The CBSs should be installed in the key location of concrete structures, ensuring the monitoring of these areas provides valuable data for assessing structural health and identifying potential issues. The curing process relates to the curing duration, method, moisture, and temperature control to ensure the CBSs develop the desired mechanical strength. It is a critical step in ensuring their optimal performance and long-term durability.
2. The self-sensing performance of CBSs varies depending on whether they are embedded within the structure or left unembedded, the different forms of concrete structure seem to affect the performance as well. The strategic layout of these sensors is essential for effectively monitoring various parameters in concrete structures. It is suggested to connect the CBSs in series to amplify the sensing ranges and electrical resistivity changes. This approach involves gathering the resistivity changes from all CBSs placed at multiple locations under compression, thus enhancing the overall

sensitivity and measurement capabilities. Additionally, since there are conductive materials such as longitudinal bars and stirrups in concrete structures, when using CBSs, it is necessary to avoid direct contact between them, especially when the CBSs exhibit good conductivity. Specific isolation measures should be implemented to address this concern.

3. CBSs offer promising capabilities for structural health monitoring in civil infrastructures, until several challenges need to be addressed to fully realize their potential. In order to expand commercialization, it is essential to reduce the manufacturing cost of CBSs, which also means lowering the production and dispersion costs of nanomaterials. Standardizing the sensor manufacturing process is necessary for unified calibration and performance comparison. The sensitivity of sensors to environmental factors, including temperature, humidity, freeze-thaw effects, etc., needs to be addressed. Additionally, the development of data acquisition systems is crucial as it forms the foundation for accurate utilization of CBSs.

REFERENCES

1. W.-J. Long, H.-D. Li, C.-L. Fang, F. Xing, Uniformly dispersed and re-agglomerated graphene oxide-based cement pastes: A comparison of rheological properties, mechanical properties and microstructure, Nanomaterials 8(1) (2018) 31.
2. W. Li, W. Dong, Y. Guo, K. Wang, S.P. Shah, Advances in multifunctional cementitious composites with conductive carbon nanomaterials for smart infrastructure, Cement and Concrete Composites 128 (2022) 104454.
3. W. Li, X. Li, J. Chen Shu, G. Long, M. Liu Yan, H. Duan Wen, Effects of nanoalumina and graphene oxide on early-age hydration and mechanical properties of cement paste, Journal of Materials in Civil Engineering 29(9) (2017) 04017087.
4. S. Parveen, S. Rana, R. Fangueiro, M.C. Paiva, Microstructure and mechanical properties of carbon nanotube reinforced cementitious composites developed using a novel dispersion technique, Cement and Concrete Research 73 (2015) 215–227.
5. X. Li, L. Wang, Y. Liu, W. Li, B. Dong, W.H. Duan, Dispersion of graphene oxide agglomerates in cement paste and its effects on electrical resistivity and flexural strength, Cement and Concrete Composites 92 (2018) 145–154.
6. S. Chuah, W. Li, S.J. Chen, J.G. Sanjayan, W.H. Duan, Investigation on dispersion of graphene oxide in cement composite using different surfactant treatments, Construction and Building Materials 161 (2018) 519–527.
7. S. Bai, L. Jiang, N. Xu, M. Jin, S. Jiang, Enhancement of mechanical and electrical properties of graphene/cement composite due to improved dispersion of graphene by addition of silica fume, Construction and Building Materials 164 (2018) 433–441.
8. H. Du, S. Dai Pang, Dispersion and stability of graphene nanoplatelet in water and its influence on cement composites, Construction and Building Materials 167 (2018) 403–413.
9. L. Guo, J. Wu, H. Wang, Mechanical and perceptual characterization of ultra-high-performance cement-based composites with silane-treated graphene nano-platelets, Construction and Building Materials 240 (2020) 117926.
10. S. Imanian Ghazanlou, M. Jalaly, S. Sadeghzadeh, A. Habibnejad Korayem, High-performance cement containing nanosized Fe_3O_4-decorated graphene oxide, Construction and Building Materials 260 (2020) 120454.
11. G. Jing, K. Xu, H. Feng, J. Wu, S. Wang, Q. Li, X. Cheng, Z. Ye, The non-uniform spatial dispersion of graphene oxide: A step forward to understand the inconsistent properties of cement composites, Construction and Building Materials 264 (2020) 120729.
12. J. Lin, E. Shamsaei, F. Basquiroto de Souza, K. Sagoe-Crentsil, W.H. Duan, Dispersion of graphene oxide–silica nanohybrids in alkaline environment for improving ordinary Portland cement composites, Cement and Concrete Composites 106 (2020) 103488.

13. J. Liu, J. Fu, Y. Yang, C. Gu, Study on dispersion, mechanical and microstructure properties of cement paste incorporating graphene sheets, Construction and Building Materials 199 (2019) 1–11.
14. W. Qin, Q. Guodong, Z. Dafu, W. Yue, Z. Haiyu, Influence of the molecular structure of a polycarboxylate superplasticiser on the dispersion of graphene oxide in cement pore solutions and cement-based composites, Construction and Building Materials 272 (2021) 121969.
15. B. Şimşek, S. Doruk, Ö.B. Ceran, T. Uygunoğlu, Principal component analysis approach to dispersed graphene oxide decorated with sodium dodecyl sulfate cement pastes, Journal of Building Engineering 38 (2021) 102234.
16. W. Dong, W. Li, K. Wang, S.P. Shah, Physicochemical and piezoresistive properties of smart cementitious composites with graphene nanoplates and graphite plates, Construction and Building Materials 286 (2021) 122943.
17. W. Dong, W. Li, K. Vessalas, X. He, Z. Sun, D. Sheng, Piezoresistivity deterioration of smart graphene nanoplate/cement-based sensors subjected to sulphuric acid attack, Composites Communications 23 (2021) 100563.
18. X. Yan, D. Zheng, H. Yang, H. Cui, M. Monasterio, Y. Lo, Study of optimizing graphene oxide dispersion and properties of the resulting cement mortars, Construction and Building Materials 257 (2020) 119477.
19. M. Abedi, R. Fangueiro, A. Camões, A. Gomes Correia, Evaluation of CNT/GNP's synergic effects on the mechanical, microstructural, and durability properties of a cementitious composite by the novel dispersion method, Construction and Building Materials 260 (2020) 120486.
20. S.K. Adhikary, Z. Rudzionis, R. Ghosh, Influence of CNT, graphene nanoplate and CNT-graphene nanoplate hybrid on the properties of lightweight concrete, Materials Today: Proceedings 44 (2021) 1979–1982.
21. M.S. Konsta-Gdoutos, Z.S. Metaxa, S.P. Shah, Highly dispersed carbon nanotube reinforced cement based materials, Cement and Concrete Research 40(7) (2010) 1052–1059.
22. A. Chaipanich, T. Nochaiya, W. Wongkeo, P. Torkittikul, Compressive strength and microstructure of carbon nanotubes–fly ash cement composites, Materials Science and Engineering: A 527(4-5) (2010) 1063–1067.
23. T. Nochaiya, P. Tolkidtikul, P. Singjai, A. Chaipanich, Microstructure and characterizations of Portland–carbon nanotubes pastes, Advanced Materials Research 55 (2008) 549–552.
24. S. Yun, J. Kim, Sonication time effect on MWNT/PANI-EB composite for hybrid electro-active paper actuator, Synthetic Metals 157(13-15) (2007) 523–528.
25. H. Kim, I. Park, H.-K. Lee, Improved piezoresistive sensitivity and stability of CNT/cement mortar composites with low water–binder ratio, Composite Structures 116 (2014) 713–719.
26. R.K. Abu Al-Rub, A.I. Ashour, B.M. Tyson, On the aspect ratio effect of multi-walled carbon nanotube reinforcements on the mechanical properties of cementitious nanocomposites, Construction and Building Materials 35 (2012) 647–655.
27. B. Zou, S.J. Chen, A.H. Korayem, F. Collins, C.M. Wang, W.H. Duan, Effect of ultrasonication energy on engineering properties of carbon nanotube reinforced cement pastes, Carbon 85 (2015) 212–220.
28. Z.S. Metaxa, M.S. Konsta-Gdoutos, S.P. Shah, Carbon nanofiber cementitious composites: Effect of debulking procedure on dispersion and reinforcing efficiency, Cement and Concrete Composites 36 (2013) 25–32.
29. F. Collins, J. Lambert, W.H. Duan, The influences of admixtures on the dispersion, workability, and strength of carbon nanotube–OPC paste mixtures, Cement and Concrete Composites 34(2) (2012) 201–207.
30. J. Luo, Z. Duan, H. Li, The influence of surfactants on the processing of multi-walled carbon nanotubes in reinforced cement matrix composites, Physica Status Solidi (a) 206(12) (2009) 2783–2790.
31. D. Shuang, W. Baomin, Study on dispersion of graphene nanoplates and rheological properties, early hydration of cement composites, Materials Research Express 6(9) (2019) 095086.
32. L. Vaisman, H.D. Wagner, G. Marom, The role of surfactants in dispersion of carbon nanotubes, Advances in Colloid and Interface Science 128–130 (2006) 37–46.
33. A. Cwirzen, K. Habermehl-Cwirzen, V. Penttala, Surface decoration of carbon nanotubes and mechanical properties of cement/carbon nanotube composites, Advances in Cement Research 20(2) (2008) 65–73.

34. G.Y. Li, P.M. Wang, X. Zhao, Pressure-sensitive properties and microstructure of carbon nanotube reinforced cement composites, Cement and Concrete Composites 29(5) (2007) 377–382.
35. D.D. Chung, Dispersion of Short Fibers in Cement, Journal of Materials in Civil Engineering 17(4) (2005) 379–383.
36. F. Sanchez, C. Ince, Microstructure and macroscopic properties of hybrid carbon nanofiber/silica fume cement composites, Composites Science and Technology 69(7) (2009) 1310–1318.
37. W. Dong, W. Li, Y. Guo, X. He, D. Sheng, Effects of silica fume on physicochemical properties and piezoresistivity of intelligent carbon black-cementitious composites, Construction and Building Materials 259 (2020) 120399.
38. H. Kim, I.W. Nam, H.-K. Lee, Enhanced effect of carbon nanotube on mechanical and electrical properties of cement composites by incorporation of silica fume, Composite Structures 107 (2014) 60–69.
39. I.-W. Nam, H.-K. Kim, H.-K. Lee, Influence of silica fume additions on electromagnetic interference shielding effectiveness of multi-walled carbon nanotube/cement composites, Construction and Building Materials 30 (2012) 480–487.
40. X. Li, A.H. Korayem, C. Li, Y. Liu, H. He, J.G. Sanjayan, W.H. Duan, Incorporation of graphene oxide and silica fume into cement paste: A study of dispersion and compressive strength, Construction and Building Materials 123 (2016) 327–335.
41. W. Li, W. Ji, F. Torabian Isfahani, Y. Wang, G. Li, Y. Liu, F. Xing, Nano-silica sol-gel and carbon nanotube coupling effect on the performance of cement-based materials, Nanomaterials 7(7) (2017) 185.
42. O. Mendoza, G. Sierra, J.I. Tobón, Influence of super plasticizer and $Ca(OH)_2$ on the stability of functionalized multi-walled carbon nanotubes dispersions for cement composites applications, Construction and Building Materials 47 (2013) 771–778.
43. O. Öztürk, M. Koçer, A. Ünal, Multifunctional behavior of composite beams incorporating hybridized carbon-based materials under cyclic loadings, Engineering Structures 250 (2022) 113429.
44. Z. Lu, A. Hanif, G. Sun, R. Liang, P. Parthasarathy, Z. Li, Highly dispersed graphene oxide electrodeposited carbon fiber reinforced cement-based materials with enhanced mechanical properties, Cement and Concrete Composites 87 (2018) 220–228.
45. J. Chen, J. Wu, H. Ge, D. Zhao, C. Liu, X. Hong, Reduced graphene oxide deposited carbon fiber reinforced polymer composites for electromagnetic interference shielding, Composites Part A: Applied Science and Manufacturing 82 (2016) 141–150.
46. W. Qin, F. Vautard, L.T. Drzal, J. Yu, Mechanical and electrical properties of carbon fiber composites with incorporation of graphene nanoplatelets at the fiber–matrix interphase, Composites Part B: Engineering 69 (2015) 335–341.
47. L. Zhang, B. Han, J. Ouyang, X. Yu, S. Sun, J. Ou, Multifunctionality of cement based composite with electrostatic self-assembled CNT/NCB composite filler, Archives of Civil and Mechanical Engineering 17(2) (2017) 354–364.
48. Y. Ding, Z. Chen, Z. Han, Y. Zhang, F. Pacheco-Torgal, Nano-carbon black and carbon fiber as conductive materials for the diagnosing of the damage of concrete beam, Construction and Building Materials 43 (2013) 233–241.
49. S. Musso, J.-M. Tulliani, G. Ferro, A. Tagliaferro, Influence of carbon nanotubes structure on the mechanical behavior of cement composites, Composites Science and Technology 69(11–12) (2009) 1985–1990.
50. B. Wang, Y. Han, S. Liu, Effect of highly dispersed carbon nanotubes on the flexural toughness of cement-based composites, Construction and Building Materials 46 (2013) 8–12.
51. B. Wang, R. Jiang, Z. Wu, Investigation of the mechanical properties and microstructure of graphene nanoplatelet–cement composite, Nanomaterials 6(11) (2016) 200.
52. C. Wang, K.-Z. Li, H.-J. Li, G.-S. Jiao, J. Lu, D.-S. Hou, Effect of carbon fiber dispersion on the mechanical properties of carbon fiber-reinforced cement-based composites, Materials Science and Engineering: A 487(1–2) (2008) 52–57.
53. S. Xu, J. Liu, Q. Li, Mechanical properties and microstructure of multi-walled carbon nanotube-reinforced cement paste, Construction and Building Materials 76 (2015) 16–23.
54. W. Dong, W. Li, Z. Luo, G. Long, K. Vessalas, D. Sheng, Structural response monitoring of concrete beam under flexural loading using smart carbon black/cement-based sensors, Smart Materials and Structures 29(6) (2020) 065001.

55. W. Li, F. Qu, W. Dong, G. Mishra, S.P. Shah, A comprehensive review on self-sensing graphene/cementitious composites: A pathway toward next-generation smart concrete, Construction and Building Materials 331 (2022) 127284.
56. M. Chougan, E. Marotta, F.R. Lamastra, F. Vivio, G. Montesperelli, U. Ianniruberto, A. Bianco, A systematic study on EN-998-2 premixed mortars modified with graphene-based materials, Construction and Building Materials 227 (2019) 116701.
57. A. Alatawna, M. Birenboim, R. Nadiv, M. Buzaglo, S. Peretz-Damari, A. Peled, O. Regev, R. Sripada, The effect of compatibility and dimensionality of carbon nanofillers on cement composites, Construction and Building Materials 232 (2020) 117141.
58. Q. Wang, X. Cui, J. Wang, S. Li, C. Lv, Y. Dong, Effect of fly ash on rheological properties of graphene oxide cement paste, Construction and Building Materials 138 (2017) 35–44.
59. F.R. Lamastra, M. Chougan, E. Marotta, S. Ciattini, S.H. Ghaffar, S. Caporali, F. Vivio, G. Montesperelli, U. Ianniruberto, M.J. Al-Kheetan, Toward a better understanding of multifunctional cement-based materials: The impact of graphite nanoplatelets (GNPs), Ceramics International 47(14) (2021) 20019–20031.
60. M. Yang, H.M. Jennings, Influences of mixing methods on the microstructure and rheological behavior of cement paste, Advanced Cement Based Materials 2(2) (1995) 70–78.
61. A. Papo, L. Piani, Effect of various superplasticizers on the rheological properties of Portland cement pastes, Cement and Concrete Research 34(11) (2004) 2097–2101.
62. C.F. Ferraris, K.H. Obla, R. Hill, The influence of mineral admixtures on the rheology of cement paste and concrete, Cement and Concrete Research 31(2) (2001) 245–255.
63. C. Vipulanandan, A. Mohammed, Smart cement rheological and piezoresistive behavior for oil well applications, Journal of Petroleum Science and Engineering 135 (2015) 50–58.
64. H. Wang, X. Gao, R. Wang, The influence of rheological parameters of cement paste on the dispersion of carbon nanofibers and self-sensing performance, Construction and Building Materials 134 (2017) 673–683.
65. F. Reza, G.B. Batson, J.A. Yamamuro, J.S. Lee, Volume electrical resistivity of carbon fiber cement composites, Materials Journal 98(1) (2001) 25–35.
66. B. Han, X. Guan, J. Ou, Electrode design, measuring method and data acquisition system of carbon fiber cement paste piezoresistive sensors, Sensors and Actuators A: Physical 135(2) (2007) 360–369.
67. R.M. Chacko, N. Banthia, A.A. Mufti, Carbon-fiber-reinforced cement-based sensors, Canadian Journal of Civil Engineering 34(3) (2007) 284–290.
68. N. Banthia, S. Djeridane, M. Pigeon, Electrical resistivity of carbon and steel micro-fiber reinforced cements, Cement and Concrete Research 22(5) (1992) 804–814.
69. W. Dong, W. Li, Z. Tao, K. Wang, Piezoresistive properties of cement-based sensors: Review and perspective, Construction and Building Materials 203 (2019) 146–163.
70. D.A. Whiting, M.A. Nagi, Electrical resistivity of concrete: A literature review, R&D Serial 2457 (2003) 1078.
71. O. Galao, F.J. Baeza, E. Zornoza, P. Garcés, Strain and damage sensing properties on multifunctional cement composites with CNF admixture, Cement and Concrete Composites 46 (2014) 90–98.
72. B. Chen, K. Wu, W. Yao, Conductivity of carbon fiber reinforced cement-based composites, Cement and Concrete Composites 26(4) (2004) 291–297.
73. W. Dong, W. Li, Y. Guo, Z. Sun, F. Qu, R. Liang, S.P. Shah, Application of intrinsic cement-based sensor for traffic detections of human motion and vehicle speed, Construction and Building Materials 355 (2022) 129130.
74. I. Maruyama, H. Sasano, M. Lin, Impact of aggregate properties on the development of shrinkage-induced cracking in concrete under restraint conditions, Cement and Concrete Research 85 (2016) 82–101.
75. W. Dong, W. Li, L. Shen, Z. Sun, D. Sheng, Piezoresistivity of smart carbon nanotubes (CNTs) reinforced cementitious composite under integrated cyclic compression and impact, Composite Structures 241 (2020) 112106.
76. S. Wen, D.D.L. Chung, Carbon fiber-reinforced cement as a strain-sensing coating, Cement and Concrete Research 31(4) (2001) 665–667.
77. F.J. Baeza, O. Galao, E. Zornoza, P. Garcés, Multifunctional cement composites strain and damage sensors applied on reinforced concrete (RC) structural elements, Materials 6(3) (2013) 841–855.

78. M.-q. Sun, R.J.Y. Liew, M.-H. Zhang, W. Li, Development of cement-based strain sensor for health monitoring of ultra high strength concrete, Construction and Building Materials 65 (2014) 630–637.
79. Y. Dai, M. Sun, C. Liu, Z. Li, Electromagnetic wave absorbing characteristics of carbon black cement-based composites, Cement and Concrete Composites 32(7) (2010) 508–513.
80. J. Li, B. Samali, L. Ye, S. Bakoss, Behaviour of concrete beam–column connections reinforced with hybrid FRP sheet, Composite Structures 57(1–4) (2002) 357–365.
81. X. Fu, W. Lu, D.D.L. Chung, Improving the strain-sensing ability of carbon fiber-reinforced cement by ozone treatment of the fibers, Cement and Concrete Research 28(2) (1998) 183–187.

Chapter 4

Signal collection and processing of cement-based sensors

4.1 INTRODUCTION

The most common fundamental measurement principle of CBSs is to measure electrical signals that can be mapped to changes in the external environmental parameters or intrinsic geometric shape of the material. Signal processing for data acquisition and interpretation for CBSs refers to the set of techniques and methods used to collect, analyze, and make sense of the raw signals generated by various electronic receiving devices. For CBSs, they primarily respond to external factors such as temperature, humidity, and physical–mechanical changes by monitoring electrical signals. Therefore, accurate reception of electrical signals is crucial for the practical application of CBSs. Currently, there are three main approaches used to receive and analyze these electrical signals for CBSs. The first approach involves electrical resistance and resistivity changes [1–4], the second involves electrochemical impedance spectroscopy analysis [5–8], and the third involves monitoring changes in capacitance [9–12]. This chapter comprehensively introduces the electrical resistivity measurements under the alternative current (AC) or direct current (DC), the influence of current intensity, impedance spectrum responses, the equivalent circuit model, its parameters analysis, and the capacitance measurement. With specific cases, it is expected to be an essential guide for students, researchers, engineers, and enthusiasts seeking to comprehend the signal collection and processing of CBSs.

4.2 ELECTRICAL RESISTIVITY/CONDUCTIVITY MEASUREMENT

4.2.1 Basic connection and formulation

The methods for electrical resistivity/conductivity measurement of CBSs consist of the two-probe method and the four-probe method. The four-probe method, also referred to as the Wenner method, is a commonly employed approach for determining the electrical resistivity of concrete. This method involves the insertion of four evenly spaced electrodes into the concrete specimen, arranged in a linear array. By applying a constant current between the outer electrodes and measuring the voltage drop across the inner electrodes, the electrical resistivity of the concrete can be measured. The two-probe method is another technique used for measuring the electrical resistivity of CBSs. In this method, only two electrodes are used to serve as the current-carrying electrode and voltage-measuring electrode synchronously. As shown in Fig. 4.1, the function generator was connected to the series circuit of resistor (R) and cement specimen as power supply,

DOI: 10.1201/9781032663685-4

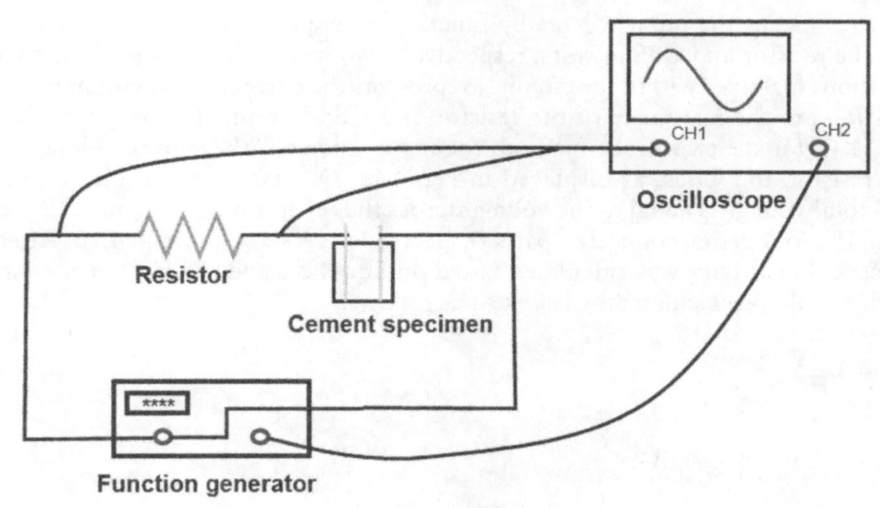

(a) Two-probe method

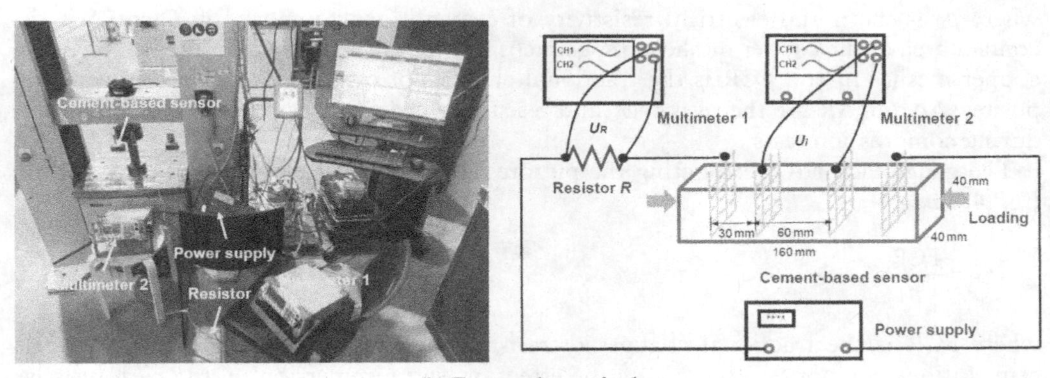

(b) Four-probe method

Figure 4.1 **Experimental configuration of electrical resistivity measurement under two-probe and four-point methods.**

and an oscilloscope was attached to measure the voltages of the resistor and generator. According to the circuit, the electrical resistance R_c of the CBS could be calculated by Eqs. (4.1)–(4.4) as follows:

$$V_{\text{input}} = V_R + V_{C1} \tag{4.1}$$

$$V_{\text{input}} = V_R + V_{C1} + V_{C2} \tag{4.2}$$

$$I = \frac{V_R}{R_R} \tag{4.3}$$

$$R_{C1} = \frac{V_{C1}}{I} = \frac{V_{C1}}{\frac{V_R}{R_R}} = \frac{R_R V_{C1}}{V_R} \tag{4.4}$$

where V_{input} means the input voltage by function generator, and V_R and V_{C1} are the voltages of the resistor and CBS in volts, respectively; V_{C2} means the voltage of the two outer cementitious matrixes within the circuit. I represents the current in the circuit in amperes; R_R and R_{C1} are the resistance of the resistor and CBS in ohms (Ω), respectively. Based on Eq. (4.1), for the two-probe method, the input voltage is the sum of voltages applied on the resistor and CBSs, as displayed in Fig. 4.1a. Otherwise, based on Eq. (4.2), the applied total voltage is equal to the voltage across the resistor plus the sum of the voltages between the four series-connected pairs of electrodes, as shown in Fig. 4.1b. Afterward, the electrical resistivity was calculated based on Eq. (4.5), and the fractional changes of resistivity could be calculated by Eq. (4.6) as follows:

$$\rho_C = \frac{R_{C1}S}{L} \tag{4.5}$$

$$\text{FCR} = \frac{\Delta\rho}{\rho_C} \times 100\% \approx \frac{\Delta R}{R_C} \times 100\% \tag{4.6}$$

where ρ_C is the initial electrical resistivity of cementitious composite in Ω·cm; S is the contact area of the copper meshes to specimens in cm^2; and L is the distance between two copper meshes in cm. FCR is the fractional changes of resistivity of cementitious composites; $\Delta\rho$ and ΔR are the resistivity and resistance changes, respectively, of composites during compression tests.

The gauge factor (G) evaluating the piezoresistive sensitivity can be drawn based on Eq. (4.7):

$$G = \frac{\text{FCR}}{\varepsilon} \tag{4.7}$$

where FCR is the fractional changes of resistivity and ε is the compressive strain of cementitious composite. To evaluate the effect of water immersion on the piezoresistive performances, water immersion treatment on the specimen has been carried out. The specimens are immersed in water with the upper surface slightly lower than the water level.

4.2.2 Direct current (DC)

The direct current for electrical resistivity measurement is considered the most straightforward approach, but its limitation lies in the induced ion movement, causing considerable electrical polarization within the CBSs and greatly affecting the measurement of electrical resistance. Fig. 4.2 shows a typical changing pattern of electrical resistivity of the CBSs with DC electrical measurement under cyclic loading. In this case, polarization can lead to an apparent increase in resistance due to the redistribution of charges and the movement of free ions in residual solutions under the electric field. As a result, the electrical resistivity of the CBSs could gradually increase with the duration of the measurement and affect the self-sensing performance.

In spite of the application of DC for electrical resistance measurement, some special treatment can be employed to mitigate the effects of polarization in the self-sensing performance of CBSs. The first approach entails prolonging the duration of measurement

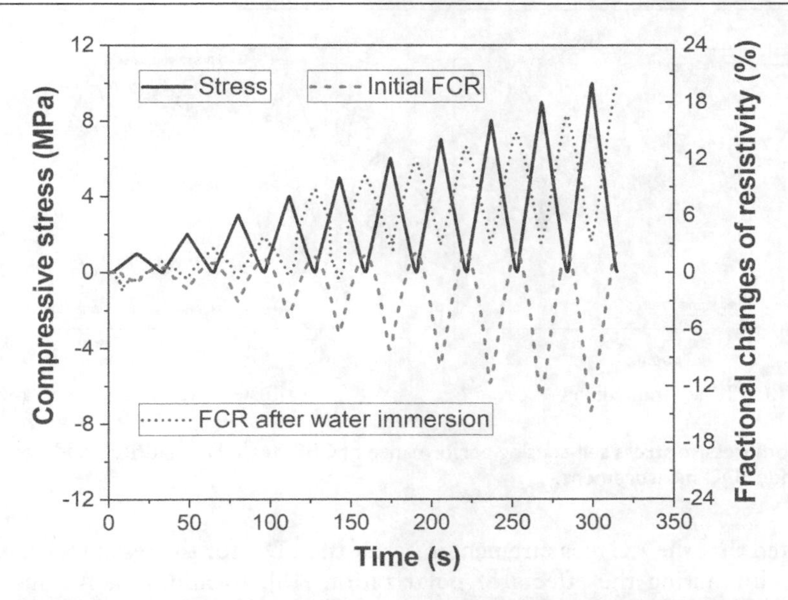

Figure 4.2 Electrical resistivity measured by DC of CBSs under cyclic loading with severe polarization.

while maintaining a constant voltage. This extension of measuring time allows the polarization effect to progressively diminish, eventually leading to a consistent and stable flow of electric current. By maintaining a steady input voltage, the resultant output resistance becomes stable and can be observed as pressure variations occur. In particular, as shown in Fig. 4.3, even though the measured resistance value may not correspond directly to the material's intrinsic electrical resistance, the consistent aspect lies in the unchanging rate at which this resistance value responds to alterations in pressure.

Drying treatment can have a significant impact on the polarization of CBSs. Polarization in cementitious materials can occur due to the movement of ions within the pore solution, which is a liquid phase that exists within the CBSs' pore spaces. Therefore, with the reduction of the pore solution during the drying process, the concentration of ions in the pore solution can increase. This higher concentration of ions can lead to more pronounced polarization when an external electric field is applied. However, with continued drying, the entirety of the pore solution is eliminated, leading to a complete confinement of ion mobility. Consequently, the polarization effect experiences a notable attenuation, resulting in a substantial weakening of its influence. The drying treatment for CBSs is conveniently carried out by placing them in an oven set at temperatures between 40°C and 60°C for one to three days. The specific drying conditions may vary depending on the individual CBS with different w/b ratios, curing conditions, and additives. In current research on CBSs, many of them employ DC measurements to assess resistance, which is also feasible to precisely evaluate the self-sensing performance of CBSs under drying conditions [13–17].

4.2.3 Alternative current (AC)

Unlike the aforementioned DC, where polarization can accumulate over time and affect measurement accuracy, AC reverses the direction of current flow periodically, preventing significant accumulation of charges and reducing polarization effects. Many studies have

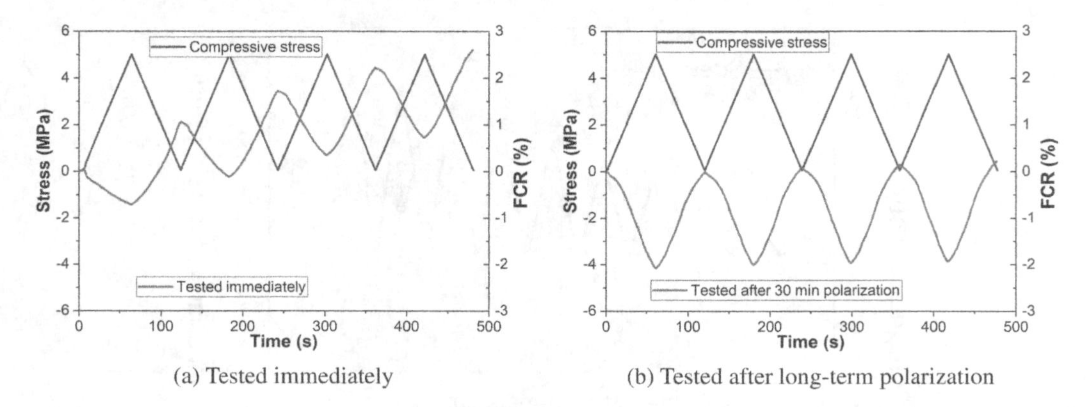

(a) Tested immediately (b) Tested after long-term polarization

Figure 4.3 Compressive stress self-sensing performance of CBSs tested immediately and after polarization under DC measurement.

demonstrated that the AC measurement is better than DC for the resistance measurement by greatly eliminating the effects of polarization [18]. In addition, AC measurements allow for frequency-dependent analysis. Different frequencies can reveal variations in the material's electrical properties at different scales or conditions. This can provide insights into the material's microstructure, moisture content, and other characteristics. More specific descriptions will be covered in the following section of electrochemical impedance spectroscopy. Researchers have concluded that the electrical resistance of cementitious materials behaved as resistor–capacitor circuits, resulting in a different impedance at different frequencies, as shown in Fig. 4.4 [19].

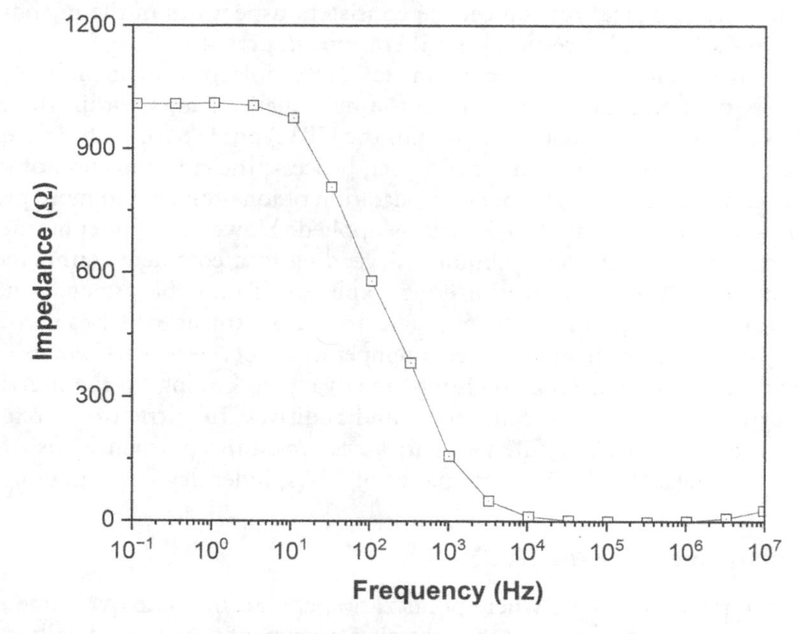

Figure 4.4 Influence of frequency on total impedance illustrated by equivalent circuit on mortar cylinder at age 45 days [19].

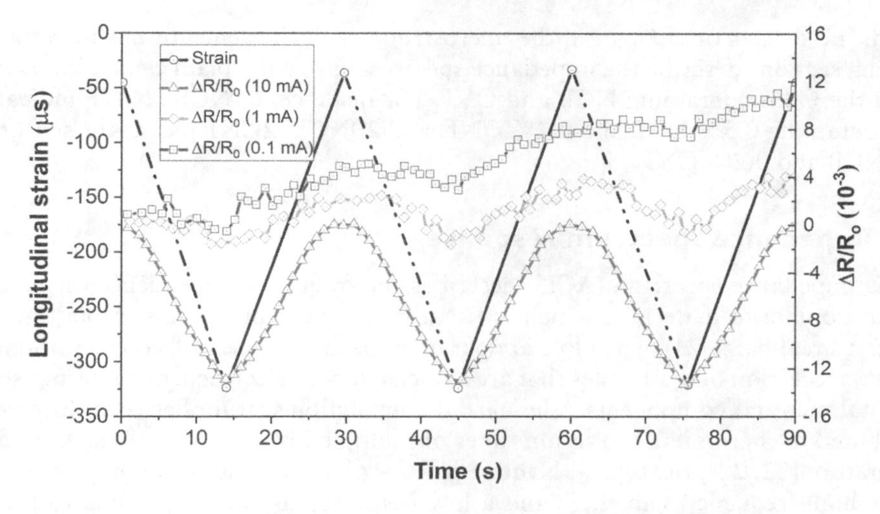

Figure 4.5 Influence of current intensity on piezoresistive responses of CBSs [20].

4.2.4 Current intensity

As for the intensity of applied current/voltage, Galao et al. [20] applied a current of 0.1 mA, 1.0 mA, and 10 mA, respectively, to CBSs to observe the altered resistivity with longitudinal strain, and the results are shown in Fig. 4.5. With increasing current intensity from 0.1 mA to 10 mA, the repeatability of resistivity is gradually improved, and the current of 10 mA shows the best consistency, demonstrating a correlation with the longitudinal strain. However, the research did not check the piezoresistive performance of CBS with current intensity higher than 10 mA. Nevertheless, it can be predicted that the resistivity of sensor might decrease, since large current intensity may be detrimental to the piezoresistivity of cement sensor. Almost at the same time, Konsta-Gdoutos and Aza [17] applied voltages of 10 V, 20 V, and 30 V, respectively, to the composite to pursue the optimum applied voltage. The voltages of 10 V and 30 V led to poor electrical property with up to 12% deviation in the measured resistivity, while the voltage of 20 V was the best to achieve stability in resistivity. From the above investigations, it can be inferred that a proper intensity of applied current/voltage is significant to achieve the piezoresistivity of sensor with good repeatability and sensitivity. Perhaps the input current/voltage intensity depends on the types and amount of cement matrix and the coupled conductors. Due to limited research, no conclusive recommendations can be made in this aspect.

4.3 ELECTROCHEMICAL IMPEDANCE SPECTROSCOPY

Impedance spectroscopy is a technique used to analyze the electrical response of CBSs over a range of frequencies. Impedance spectroscopy involves applying AC signals with varying frequencies to the sensor and measuring the electrical impedance, which is a complex quantity, including both resistance (real part) and reactance (imaginary part). Different frequencies can penetrate different depths of the material to analyze the distribution of water in the pores. In addition, changes in the impedance spectrum at specific frequencies can indicate defects or changes in the microstructure of the cementitious materials. The following sections describe the impedance spectroscopy of the plain cementitious materials and the CBSs containing NCB and CNF. For instance, 05NCB02CNF indicates the CBSs containing 0.5% NCB and 0.2% CNF, and 20NCB02CNF means the sensors with 2.0% NCB and 0.2% CNF.

4.3.1 Impedance spectrum response

The AC impedance spectrum (ACIS) describes the frequency dependence of the impedance for composite materials, which is usually presented by Nyquist plots (imaginary part versus real part) [21]. The plots are parameterized in terms of frequency and usually involve a succession of semicircles that are associated with electrochemical response of the individual material components. The partial convolution feature between different arcs is attributed to the similar relaxation times of multiple involved responses. According to the literature [22, 23], the typical Nyquist plot for plain cementitious composite includes a single high-frequency semicircle and a low-frequency arc corresponding to the effect of external electrode, whereas the cementitious composite with the inclusion of conductive fillers presents two separated semicircles in addition to electrode arc. The origin of dual semicircles characteristic was explained by a "frequency-switchable coating model" in some studies on carbon fiber-reinforced composites [23]. The schematic diagram for typical Nyquist plots for plain cementitious composites and CBSs containing conductive fillers are shown in Fig. 4.6. For the case of CBS, the left-side R_{cusp} is directly a result of short circuit of conductive materials and accounts for the combined transfer of electron current through the conductive materials and ionic current through the electrolyte [24],

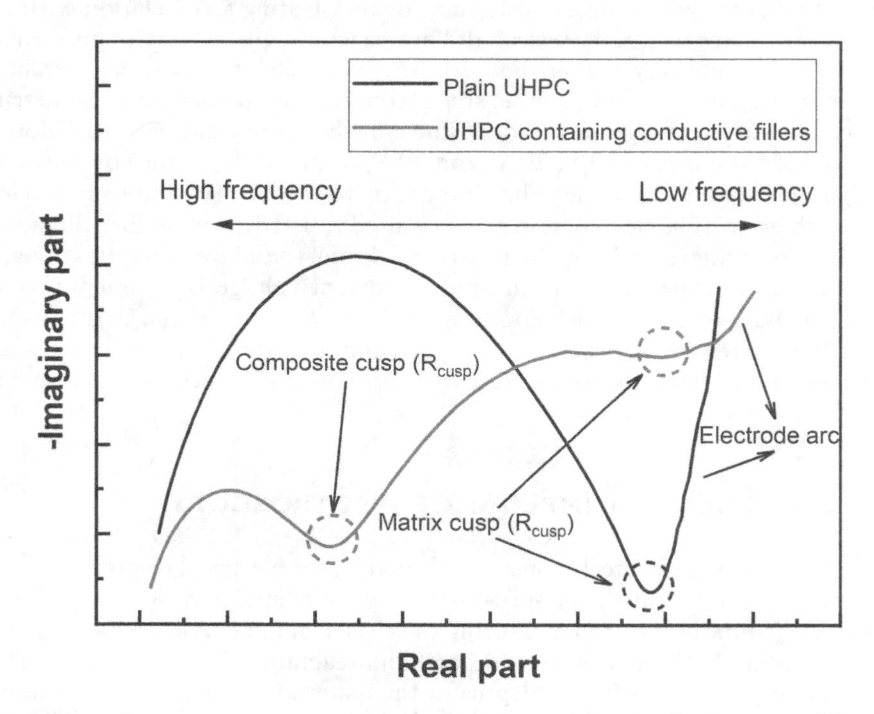

Figure 4.6 Typical Nyquist plots for plain cementitious materials and CBS-containing conductive fillers.

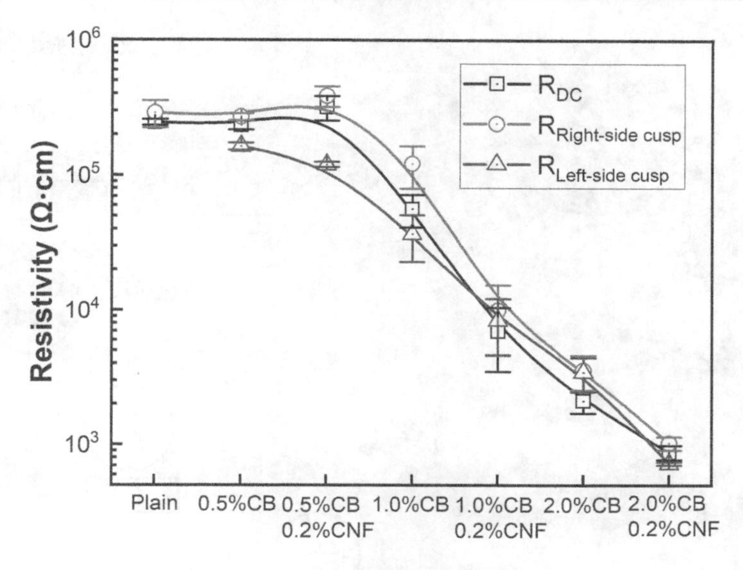

Figure 4.7 Comparisons of RDC, left-side R_{cusp}, and right-side R_{cusp}.

whereas right-side R_{cusp} was attributed to AC resistance of the cement matrix and coincided with the four-point DC resistance of composite [25, 26]. Fig. 4.7 compares the R_{cusp} values extracted from ACIS with R_{DC} values measured from four-point method [24]. It is indicated that right-side R_{cusp} generally coincides with the R_{DC}, which supports the percolation zone analysis in the previous section. In addition, the left-side R_{cusp} of CBS is relatively smaller than right-side R_{cusp}/R_{DC} with the increase of NCB/CNF content, and it gradually almost coincides when the NCB/CNF content enters the percolation zone, in which the number and continuity of conductive passages greatly improved. This phenomenon should further substantiate the fact that the left-side R_{cusp} is a direct consequence of short circuit of conductive materials. Therefore, the relationship between the left-side R_{cusp} and the right-side R_{cusp}/R_{DC} can be potentially used to identify the conductive states inside the cementitious composites.

4.3.2 Equivalent circuit model

The equivalent circuit model comprised of a series of discrete electrical components can be established based on ACIS mentioned in the last section. In general, equivalent circuit model is divided into several parts, corresponding to the different conductive behaviors of multiple phases in CBSs. Considering previous relevant studies [27–32], a new equivalent model was proposed based on different conducive phases in the microstructures of cementitious materials. The schematic diagram of microstructure and their corresponding equivalent electrical components are illustrated in Fig. 4.8. For the carbon nanomaterials–filled cementitious composites, tunneling conduction contributed by very closed adjacent NCB/CNF, contact conduction by interconnected NCB/CNF, and ionic conduction contributed by pore electrolytes mainly govern the conductive behavior [33].

Individual NCB/CNF or its agglomerates are tightly embedded in cement matrix, which means there is inevitable coating of matrix phases enclosing the NCB/CNF except for those directly connected NCB/CNF. The NCB/CNF distributed within the matrix is

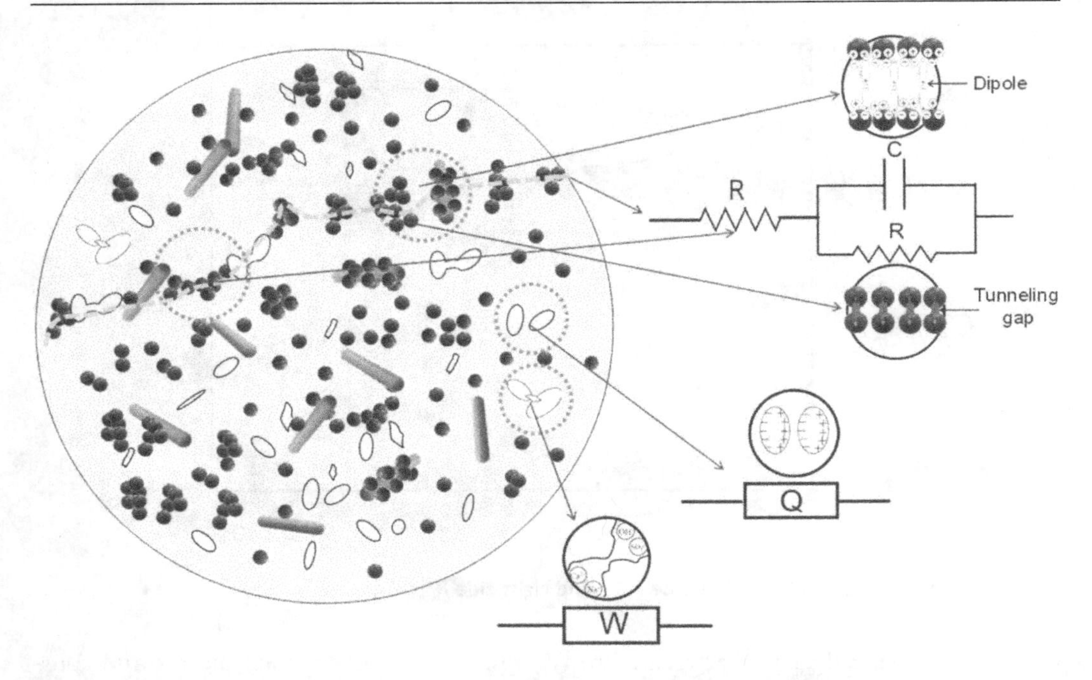

Figure 4.8 Schematic diagram of microstructure and equivalent circuit components.

in the form of a partially conductive path, which contains a certain portion of continuous conductive path and a certain portion of discontinuous conductive path. To explicate electrical behavior of partially conductive path, Jiang et al. [31] adopted the $R_0(C_1R_1)$ circuit component to represent partially contacted graphene in cement matrix, where R_0 represented resistance of composite and directly contacted graphene; C_1 and R_1 represented capacitance and resistance of GNP/matrix/GNP structures, respectively. Similarly, Wang et al. [29] adopted a $R_{as}(C_sR_{cs})$ circuit component to explicate the conductive path of partially contacted NCB aggregates where polymer matrix is as the dielectric, where R_{as} represents the resistance of aggregate, C_s and R_{cs} represent the capacitance and resistance of NCB/matrix/NCB structures, respectively. In this study, the electrical behavior of directly connected NCB/CNF structures and electrolyte resistance by continuously connected micropores is represented by a resistor [34]. The electrical behavior of unconnected NCB/CNF structures and unconnected pores is explicated by a parallel R–C circuit, showing both the capacitive and resistive behaviors. For the plain cementitious materials, the resistor represents the electrolyte resistance by unconnected micropores, whereas the capacitor represents the capacitance of insulated cement matrix.

For the case of single NCB and hybrid NCB/CNF-filled CBS, the resistor in parallel R–C circuit additionally represents the nonohmic resistance of NCB(CNF)/matrix/NCB(CNF) structures due to the tunneling effect, in which the adjacent NCB/CNF particles are very closely neighboring the electrons to conquer a potential barrier and allow the tunneling to take place. Previous studies proposed that the maximum tunneling distance for electrical current transfer in cement matrix is typically less than several nanometers [35]. The capacitor in the parallel R–C circuit represents the sum of the capacitance of insulated matrix and the double-layer capacitance generated by NCB(CNF)/matrix/NCB(CNF) structures. For plain cementitious material, the capacitor is only related to the capacitance of insulated matrix. Under the external electric field, the electrical charges will be orientationally deposited along the NCB/CNF "plates" due to the Maxwell–Wagner interfacial polarization, where the matrix between the NCB/CNF structures is considered as dielectric, resulting in the formation of a series of tiny electrical double-layer capacitor [36]. Similar capacitive feature has also been discussed in the graphite-filled cementitious material [37]. Nevertheless, the capacitive behaviors only occur in the region where the contact between NCB/CNF grains is poor [26], with relatively large matrix gap between them that can enable electric charges to be stored. Consequently, the parallel *R–C* circuit component is essentially controlled by the "contact state" of NCB/CNF structures that are related to the partially conductive paths inside the single NCB and hybrid NCB/CNF-filled CBSs. As the NCB content increases, the distance between adjacent NCB/CNF structures reduces, accordingly resulting in intensified tunneling effect and weakened NCB(CNF)/matrix/NCB(CNF) capacitance.

Based on the discussion, the conductive path contributed by NCB/CNF structures and connected pores can be represented by a *R*(*CR*) circuit component, that is resistance and capacitive elements conduct electricity in series and parallel. This physical explanation of the parallel *R*(*CR*) circuit components is also consistent with the equivalent response models of carbon nanotubes (CNT) cement-based composites [38]. In addition, the ions chemically or physically absorbed in the interfacial area between pores and matrix, the opposite charges at the solid layer and the solid–liquid bilayer also act as double-layer capacitance unit, as so-called "solid–liquid interface capacitance" under AC [39, 40]. However, owing to the complex nature of unconnected pores such as different shape, size, pore structure, and their moisture content, their time constant response to different frequency ranges are different, leading to "dispersion effect" in ACIS [41]. Consequently, the complex unconnected pores cannot be considered as a perfect capacitor but instead are represented by a constant phase element (CPE, expressed by Q), and its formula is illustrated in Eq. (4.8):

$$Z_Q = \frac{1}{Y_0\omega^n}\left(\cos\frac{n\pi}{2} - j\sin\frac{n\pi}{2}\right) \tag{4.8}$$

where Z_Q denotes the impedance (Ω), $\omega = 2\pi f$ denotes the angular frequency, Y_0 represents the capacitance or resistance, when $n = 1$, CPE behaves as ideal capacitor and with the capacitance of $C = Y_0$, when $n = 0$, CPE behaves as ideal resistor with the resistance of $R = 1/Y_0$, when $0 < n < 1$, CPE has both the capacitive and resistive behaviors.

It is worth noting that although the CBS was dried at 60°C in oven for 24 hours to eliminate the ionic conductivity, there is still certain moisture inside sensor, and the ionic conductivity still plays an important role. This is probably related to the complex types of water in pores and the drying treatment only evaporate part the of bulk capillary water [42]. The water evaporation alters the ionic centration and internal diffusion inside the cement composites [43]. For the partially saturated specimens, the charge transfer and ion diffusion not only go through the interconnected solids or continuous fluids but also through the interface region of solid–liquid phases [44]. Thus, the diffusion impedance of ion diffusion in the solution stemming from gradient of ion concentration should also be

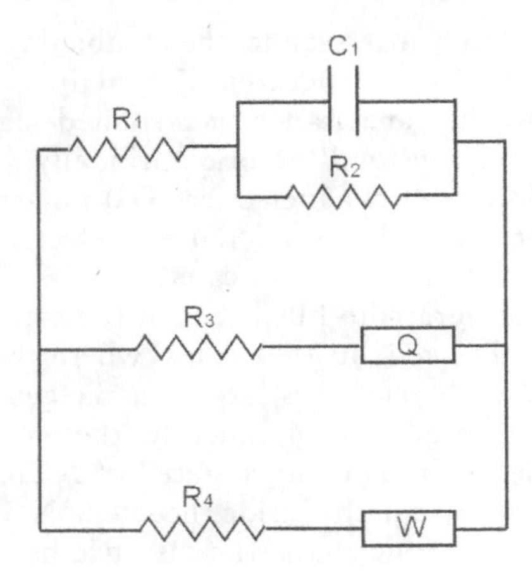

Figure 4.9 Equivalent circuit model for the CBSs.

taken into consideration. The diffusion effect is often represented by Warburg impedance [45], and its formula is shown in Eq. (4.9):

$$Z_w = \frac{\sigma}{\sqrt{\omega}} - j\frac{\sigma}{\sqrt{\omega}} \tag{4.9}$$

where Z_w denotes Warburg impedance and σ denotes Warburg coefficient.

Based on the analysis of the microstructure and the involved conductive behaviors, the overall equivalent circuit model of CBS is illustrated in Fig. 4.9, which is expressed as $((R_1(C_1R_2))(R_3Q)(R_4W))$ and its impedance formula is shown in Eq. (4.10):

$$\frac{1}{Z} = \frac{1}{R_1 + \frac{R_2}{1+\omega^2 R_2^2 C_1^2} - j \cdot \frac{\omega R_2^2 C_1}{1+\omega^2 R_2^2 C_1^2}} + \frac{1}{R_3 + Z_Q} + \frac{1}{R_4 + Z_w} \tag{4.10}$$

where Z denotes bulk impedance of CBS; C_1 represents the capacitive behavior of the insulated matrix and NCB(CNF)/matrix/NCB(CNF) structures; R_1 represents the resistive behavior of NCB(CNF)/matrix/NCB(CNF) structures; R_2 represents the electrical behavior of directly connected NCB/CNF structures and connected pore solution; Z_Q represents the electrical behavior of complex unconnected pore structure; R_3 represents the bulk resistance in series with the unconnected pore in the direction of electrical current and "R_3 + Z_Q" is presented as "oblique line" in the Nyquist plot; Z_w represents the electrical behavior of ions diffusion; R_4 represents the bulk resistance in series with charge-diffusion layers in the direction of electrical current, and "R_4 + Z_w" is presented as a "depressed semicircle" in the Nyquist plot.

4.3.3 Analysis of equivalent circuit parameters

Table 4.1 shows the fitting parameters of the proposed model $((R_1(C_1R_2))(R_3Q)(R_4W)$. As shown in Fig. 4.10, the resistance value R_1 of plain samples to 05NCB is nearly unchanged,

Table 4.1 Fitting parameters of equivalent circuit model for CBSs

Groups	*Parameters* / *Samples*	R_1 (Ω)	R_2 (Ω)	C_1 (F)	R_3 (Ω)	Z_W (Ω)	R_4 (Ω)	Z_Q		*Regression SS*
								Y_0	*n*	
Plain	S_1	1660	8.56E+5	2.24E–11	11465	48797	22542	7.03E–8	0.416	0.00140
	S_2	1737	5.72E+5	1.87E–11	8754	1.02E+5	1.24E+5	2.38E–7	0.192	8.76E–4
05CB	S_1	1563	2.82E+5	3.00E–11	**0.0493**	90647	47179	2.33E–7	0.448	1.87E–3
	S_2	1853	2.40E+5	2.69E–11	**2.06E–4**	86283	58795	2.27E–7	0.484	8.64E–4
05CB02CNF	S_1	3654	63241	3.93E–11	95941	7.56E+6	8374	7.52E–7	0.292	0.0292
	S_2	2970	57294	4.24E–11	1.04E+5	3.71E+6	5896	5.13E–7	0.332	0.0209
10CB	S_1	500.6	50422	4.88E–11	**2.26E–5**	3.43E+5	8431	1.74E–5	0.354	7.59E–3
	S_2	680.2	75568	4.70E–11	1411	1.64E+6	10607	1.34E–5	0.324	6.33E–3
10CB02CNF	S_1	1082	5180	5.30E–11	**2.85E–4**	107674	5795	6.38E–5	0.258	3.69 E–4
	S_2	613.1	5267	6.67E–11	419.5	1468	9185	6.85E–6	0.483	7.41E–6
20CB	S_1	2099	267.9	6.34E–09	2482	**3.80E–6**	754.1	1.95E–4	0.0405	2.03E–4
	S_2	2276	241.9	6.02E-09	239.8	693	603.3	5.29E–5	0.0804	4.40E–5
20CB02CNF	S_1	273.1	232.2	3.47E-09	402.3	288	230.9	3.19E–4	0.219	3.32E–3
	S_2	302.3	220.3	1.95E-09	533.4	797.7	154.5	9.24E–4	0.166	2.75E–3

Note: Regression SS means the regression sum of squares, which reflects how well a fitting curve represents the fitting data. A higher regression SS indicates the fitting curve does not fit the data well. The bolds values denote the abnormal values.

the resistance value R_2 drops by about two-thirds, and the dielectric capacitance value C_1 slightly increases. These variations are well consistent with the fact that the dosage of 0.5% NCB is at the starting percolation threshold. There is no continuous conductive path formed by NCB particles/agglomerates as a consequence of inadequate concentration of conductive fillers [46–48]. The addition of 0.5% NCB does not contribute to additional continuous conductive path compared to connected pores filled with electrolyte in plain samples. In such a case, the resistance value R_1 is merely varied and only resistance value R_2 drops since 0.5% NCB particles/agglomerates dispersed in cement matrix in discontinuous form and only contribute the tunneling conduction. From 05NCB to 05NCB02CNF, the resistance value R_1 almost doubles, the resistance value R_2 drops by about three-fourths. The increase in R_1 should be attributed to refined pore structure, which blocks the continuous conductive paths in certain range [49, 50]. The increase in R_1 is also observed in the case from 10NCB to 10NCB02CNF, which probably reveals the beneficial role of CNF in modifying the microstructures. The unexpected larger R_1 value for 20NCB compared to 10NCB is probably owing to higher occurrence of localized defects due to increasing NCB agglomerates in 20NCB, which blocks the formation of continuous pathways to certain extent [51]. As R_1 increases, there are obvious decrease in R_2 for both cases from 05NCB to 05NCB02CNF and from 10NCB to 10NCB02CNF. Especially for the case from 10NCB to 10NCB02CNF, there is a significant decrease in R_2 by nearly one order of magnitude.

These variations indicate that the incorporation of 0.2% CNF mainly contributes to the conductive network in the form of increasing tunneling conduction (discontinuous conductive paths) for the cases of 05NCB02CNF and 10NCB02CNF. However, different trends are observed from 20NCB to 20NCB02CNF, in which case R_2 is nearly unaltered while R_1 critically drops to near one order of magnitude. This implies that the incorporation of 0.2% CNF mainly contributes to the conductive network in the form of increasing

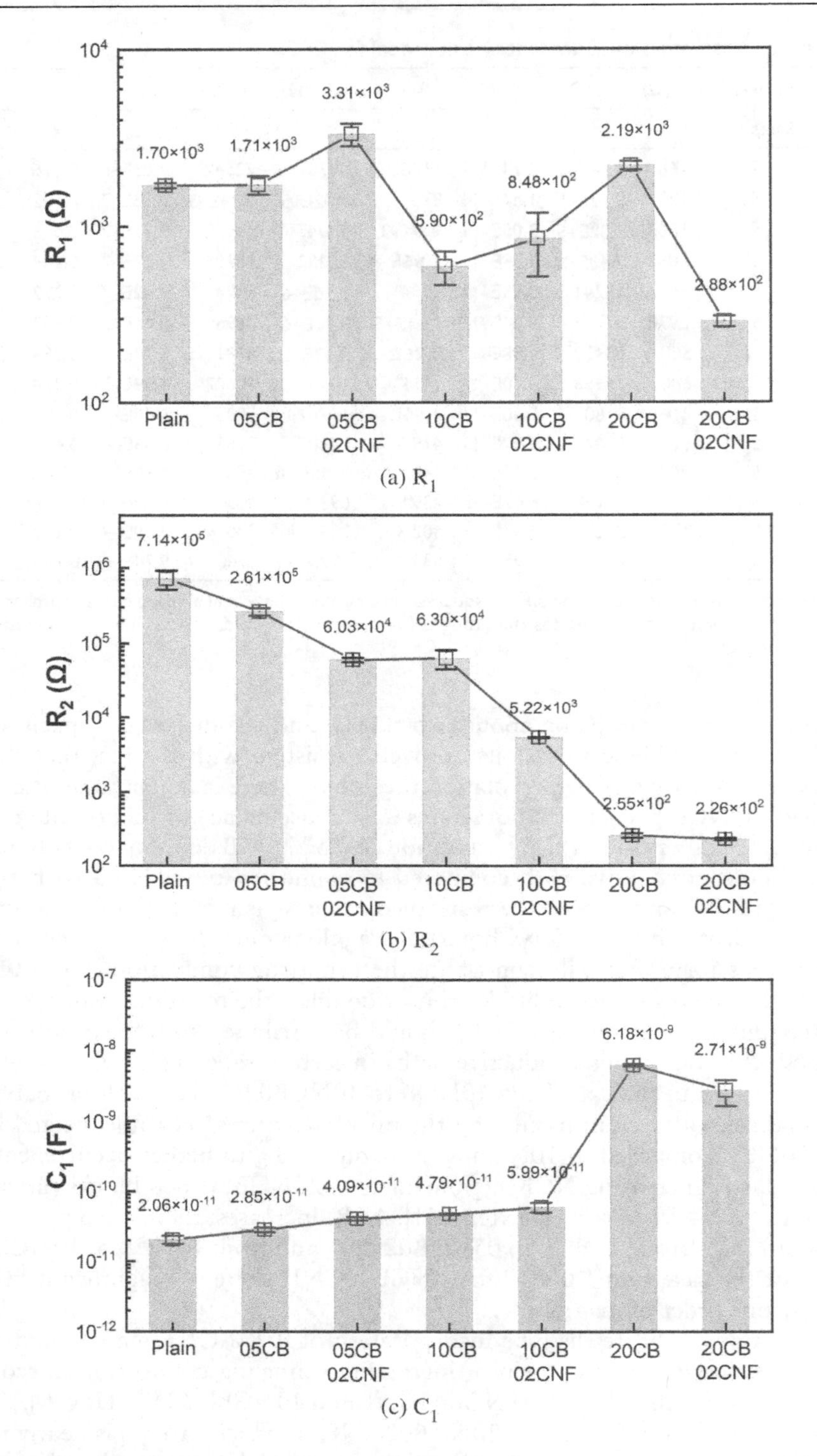

Figure 4.10 Variations of equivalent circuit model fitting parameters for single NCB and hybrid NCB/CNF filled CBSs.

contact conduction (continuous conductive paths) for the case of 20NCB02CNF. These variation trends agree well with conductive characteristics shown in microstructures. For the cases of 05NCB02CNF and 10NCB02CNF, NCB particles/agglomerates are sparsely distributed in the matrix without forming a continuous percolating network. Although the fibrous CNF with high aspect ratio can play a crucial role as long-range conductor and fill the dielectric gap between NCB, they mainly achieve the cooperative improvement effect in electrical conductivity by narrowing the dielectric gap between adjacent NCB particles/agglomerates and enhancing the tunneling conduction.

When compared with the case from 05NCB to 05NCB02CNF, the significant drop in R_2 for the case from 10NCB to 10NCB02CNF evidences more efficient improvement in tunneling conduction, which is consistent with the fact that more conductive contact points provided by higher dosage of 1.0% NCB [52]. The distinctive synergetic improvement efficiency in electrical properties brought by CNF on the CBSs filled with different dosage of NCB can be related to the percolation characteristics. When NCB content approaches the start of percolation threshold (i.e., 0.5% NCB), the NCB marginally distributed in matrix can provide limited conductive contact point. When the NCB content is within the percolation zone (i.e., 1.0% NCB) where the tunneling distance becomes the dominant factor [53], the assistance of CNF shows the most efficient synergetic improvement in electrical properties. The contribution of tunneling conduction provided by CNF in 05NCB02CNF cannot even overcome the reduced conduction by refined porosity structure, which leads to increase in bulk resistivity from 05NCB to 05NCB02CNF. When the NCB content approaches to the end of percolation zone (i.e., 2.0% NCB), the NCB particles/agglomerates already broadly distribute and form percolating conductive network in the matrix, in this case CNF mainly helps form the continuous conductive path and contribute to contact conduction rather than tunneling conduction.

The above hypothetical analysis is further supported by the variation trend of capacitance value C_1. From plain samples to 10NCB02CNF, there is a continuous increase in C_1, but the increasing range is limited within the same order of magnitude. As previously discussed, C_1 is assigned to the capacitance of insulated matrix and the double-layer capacitance of NCB(CNF)/matrix/NCB(CNF) structures due to polarization at NCB(CNF)/matrix interfaces. The incorporation of conductive fillers into humidified matrix generates numerous "tiny double layers" with high specific capacitance and adds into cement matrix. It is worth mentioning that although the characteristic time constant corresponding to such two electrochemical responses is different, which is evidenced by the nonsymmetrical variations in electrical parameters and can lead to changes in the time constant, the variation trend of C_1 is directly linked to the effects of the incorporating CNF/NCB. An example is the variation of C_1 that is far remarkably less than the counterpart of R_1 from plain samples to 10NCB02CNF. When the dosage of NCB/CNF is low, before reaching the percolation threshold, the numerous "tiny double layer capacitor" by individual NCB(CNF)/matrix/NCB(CNF) structure can be anticipated to be connected in parallel, which is stemmed from the fact that the current from one local NCB(CNF)/matrix/NCB(CNF) capacitor to the adjacent populated one does not follow a single channel, but follows multiple channels as shown in Fig. 4.11a [25]. In this case, the capacitance contributed by NCB(CNF)/matrix/NCB(CNF) structures can be evaluated by Eq. (4.11):

$$C_p = \sum_{i=1}^{n} C_i \tag{4.11}$$

where C_i is the capacitance of localized capacitor and can be evaluated based on Eq. (4.12):

$$C_i = A \cdot k \frac{\varepsilon}{d} \tag{4.12}$$

where A is the area of individual NCB/CNF structure; k is the relative permittivity of dielectric mortar; ε is the permittivity of space; d is the distance between two NCB/CNF structures. According to the equations, C_i increases with the increase of NCB/CNF dosage since A increases and d decreases. In addition, since the number of NCB(CNF)/matrix/NCB(CNF) capacitors increases with the increase of NCB(CNF) dosage, C_p increases. Therefore, there is a gradual and steady increase in C_1 from plain samples to 10NCB02CNF. Interestingly, there is a dramatic increase in C_1 by approximately two orders of magnitude from 10NCB02CNF to 20NCB. This is probably related to the presence of increasing number of NCB agglomerates. Compared with dispersed NCB/CNF, the presence of NCB/CNF agglomerates form the bigger NCB(CNF) agglomerate/matrix/NCB(CNF) agglomerate capacitors that are associated with higher storage capacity and larger capacitance values. The increase in capacitance with the increasing concentration of conductive fillers before reaching the percolation threshold is also detected in previous studies [25]. In addition, there is a sudden decrease in C_1 from 20NCB to 20NCB02CNF, which should be related to the fact that the number of NCB(CNF)/matrix/NCB(CNF) capacitors should be assumed connected in series when dosage of NCB/CNF approach the percolation threshold as illustrated in Fig. 4.11b. In this case, the charge can only be stored in the local regime where the contact of NCB/CNF structures is poor and the capacitance C_s can be evaluated by Eq. (4.13):

$$C_s = \frac{1}{\sum_{i=1}^{n} \frac{1}{C_i}} \tag{4.13}$$

According to Eq. (4.13), the increase of individual local C_i is equivalent to decrease of C_s. Consequently, the capacitive character weakened from 20NCB to 20NCB02CNF reveals that the incorporation of CNF reduces the regimens where the contact of NCB/CNF is poor. It agrees well with the previous hypothesis that synergetic improvement of CNF on the electrical conductivity of 20NCB mainly helps to make up the continuous conductive paths and contributes to the contact conduction. The consistent findings by analyzing the variations of different electrical parameters should validate the proposed equivalent circuit model.

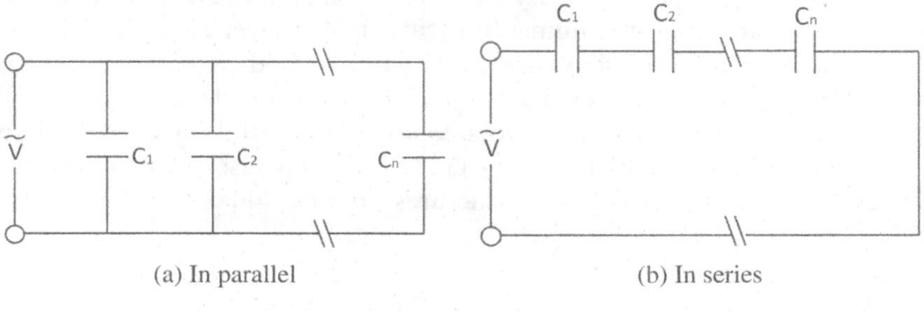

(a) In parallel

(b) In series

Figure 4.11 Circuit of NCB(CNF)/matrix/NCB(CNF) capacitors.

Overall, the proposed model based on conductive phase involved in microstructures can provide reliable information to characterize the conductive behaviors, showing close agreement with the results of microstructural characterization. The inherent behaviors of CNF, as long-range conductor to play the role to achieve the synergetic improvement effect in electrical properties for single NCB-filled CBSs with different NCB dosages, can be successfully analyzed based on the variations of equivalent circuit parameters. When the dosage of NCB approaches percolation threshold (i.e., for the cases containing 0.5% and 1.0% NCB), CNF mainly assists in forming discontinuous conductive paths, narrows the gap between neighboring NCB structures, and contributes to the tunneling conduction. These characteristics can be reflected by the increase in nearly unchanged/increase in R_1, the decrease in R_2, and the increase in C_1. Particularly, a comparably significant drop in R_2 and increase in C_1 for cases from 10NCB to 10NCB02CNF in comparison to the cases from 05NCB to 05NCB02CNF indicate much higher efficiency of synergetic improvement effect brought by CNF on 10NCB than 05NCB, which can be closely correlated to percolation characteristics and the microstructures.

When the NCB dosage approaches the percolation threshold (i.e., 2.0% NCB), CNF mainly assists in forming continuous path, directly connects the neighboring NCB structures, and contributes to contact conduction. These characteristics can be reflected by the decrease in R_1, nearly unchanged in R_2, and the reduction in C_1. Thus, the essential information concerning conductive behaviors of single NCB and hybrid NCB/CNF-filled CBSs can be properly extracted from a semi-qualitative point of view, based on the variations of equivalent circuit parameters that are assigned to different electrical phases involved in the microstructures. The obtained essential information such as the percolating state of NCB and the different forms of how CNF contribute to conductive network in hybrid NCB/CNF-filled CBSs can help understand the conductive behaviors and working mechanisms at microstructural level, which can provide new insight into the material design strategy for enhancing the electrical and sensing performances of CBSs.

4.4 CAPACITANCE MEASUREMENT

The capacitance-based self-sensing CBSs operate on the principle of capacitance, which means its capacity to store electrical charge following the application of voltage. Any alterations in mechanical attributes such as strain, stress, and deformation of CBSs can induce modifications in the intrinsic capacitance characteristics, that is why the measurement of capacitance could be another critical option for the CBSs. Traditional CBSs based on electrical resistivity measurements often require the assistance of conductive fillers to enhance the conductivity and piezoresistivity of the cement matrix. In contrast, capacitance-based CBSs do not necessitate additional conductive fillers, significantly reducing the manufacturing complexity and cost of CBSs. Despite these advantages, it can be imagined that the capacitance-based CBS is significantly more sensitive to external factors compared to the traditional CBSs with enhanced electrical conductivity. In addition to the effects of temperature, humidity, stress, and deformation, other factors such as material geometry, electrode properties, current frequency, and external electric fields can also cause certain interferences on the self-sensing performances.

Fig. 4.12 shows the capacitance and its fractional changes of plain cement paste subjected to cyclic loading in the low-stress and high-stress regimes, respectively [11]. The results showed that there is a reduction in capacitance under compressive stress. The

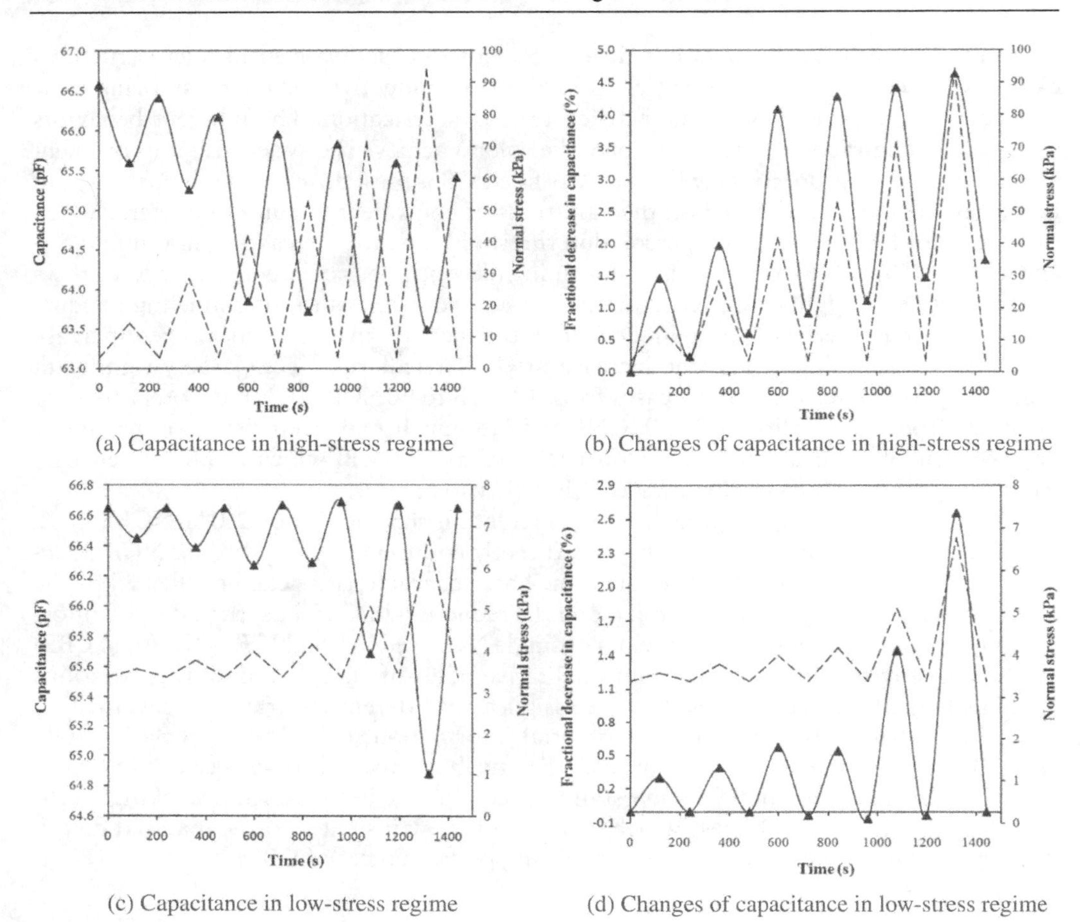

Figure 4.12 Fractional changes of capacitance for the plain cement paste: the dashed curve means the compressive stress and the solid curve is the capacitance [11].

magnitude of capacitance reduction is directly proportional to the magnitude of the applied stress. Under conditions of low stress, the variation in capacitance exhibits superior linearity and recoverability compared to high-stress conditions. This phenomenon is very similar to the traditional CBSs with conductive fillers because the high stress can disrupt the microstructure of cementitious matrix and influence capacitance. However, it must be noted that the applied compressive stress on the plain cement paste for capacitance-based self-monitoring is far below the stress applied on the traditional CBSs for resistivity-based self-monitoring. In this study, the applied stresses change in the range of a few to dozens of kPa [11], but the value can be thousand times higher reaching dozens of MPa from our previous studies and meanwhile maintain the excellent repeatability of CBSs [54, 55]. The primary reason lies in the low strength of pure cement paste, which prohibits the application of excessively high stress during capacitance measurements. Second, traditional CBSs incorporate nanofillers or fiber fillers to enhance porosity and microstructure, thereby improving the structure's recoverability upon compression. This prevents minor variations from inducing permanent changes in the resistive electrical characteristics. Considering the application of capacitance-based CBSs has more challenges compared to traditional CBSs, substantial research needs to be proposed in the future to address these challenges, including but not limited to the enhancement of its maximum applicable stress–strain range, repeatability, linearity, and capacitance variations under different frequencies.

Fig. 4.13 illustrates the flexural strength and Young's modulus of CNT-reinforced cement mortar as a function to the measured capacitance. It was found that the capacitance decreased with increased flexural strength and Young's modulus, and vice versa. Since the capacitance of cement mortar is firmly related to the CNT's state of dispersion,

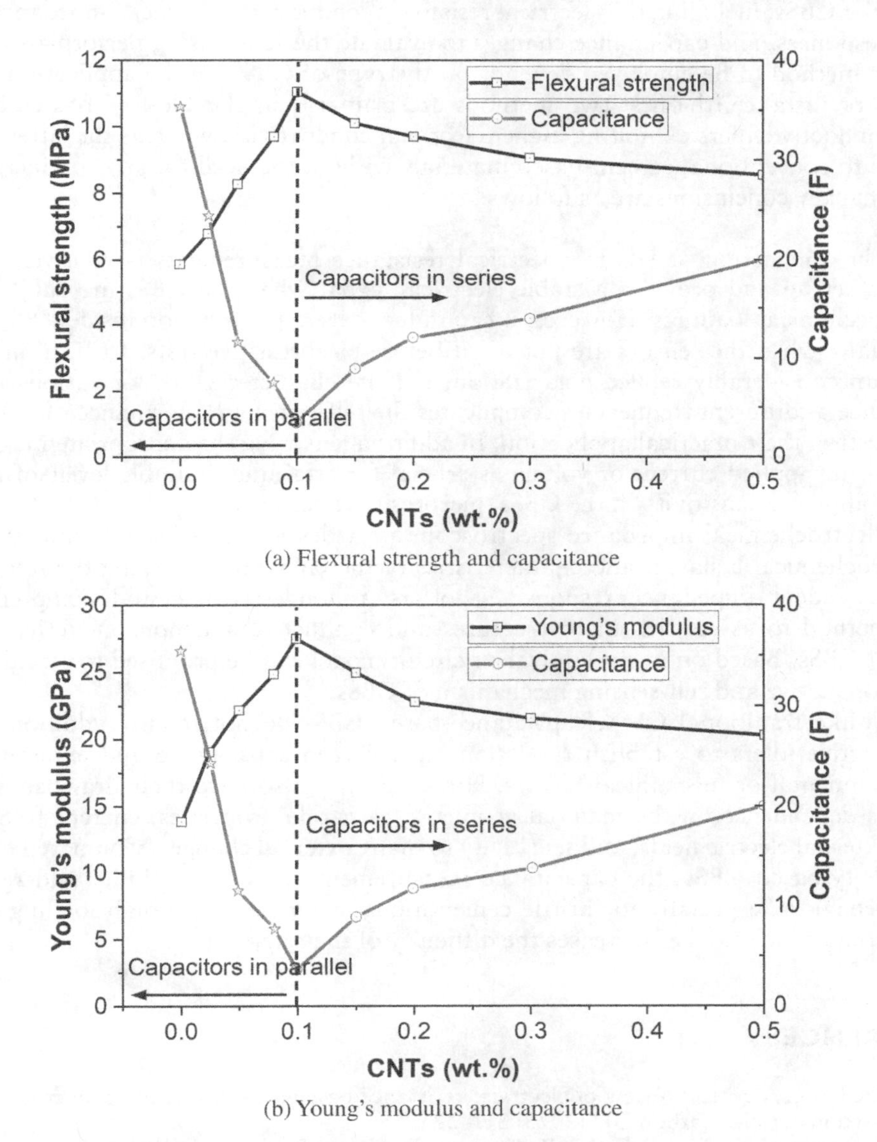

(a) Flexural strength and capacitance

(b) Young's modulus and capacitance

Figure 4.13 Relationship between (a) flexural strength and capacitance and (b) Young's modulus and capacitance of CNT-reinforced cement mortar [56].

the capacitance is able to be an indicator for the evaluation of nanomaterial's dispersion. Moreover, due to the linear relationship between capacitance to flexural strength and Young's modulus, the measurement of capacitance can be applied for the prediction of mechanical strengths of cement mortar reinforced with nanomaterials [56].

4.5 SUMMARY

This chapter systematically introduces the current methods for signal reception and processing of CBSs, including the electrical resistivity/conductivity changes, impedance spectrum responses, and capacitance changes to evaluate the self-sensing performances. The specific method to be employed depends on the type of CBS and the application conditions. For instance, the first two methods are primarily applicable to CBSs embedded with conductive fillers exhibiting higher electrical conductivity, whereas the latter can be applied to conventional cement-based materials without the need for any conductive fillers. Some key conclusions are as follows:

1. The application of DC for electrical resistance measurements is relatively simple to set up and provides a stable electrical signal when the CBSs are subjected to mechanical loadings. However, it can induce severe polarization inside CBSs, especially when the sensors are not dried before piezoresistive tests. Utilization of AC can considerably reduce polarization, but the characterization of various impedance at different frequencies complicates the self-sensing performance of CBSs and restrict their practical application. In addition, ensuring the appropriate magnitude of the applied current or voltage is crucial for attaining desirable levels of repeatability and sensitivity in the CBSs' piezoresistive responses.
2. Electrochemical impedance spectroscopy provides valuable insights into the electrochemical behavior and characteristics of the CBSs. By analyzing the frequency-dependent impedance response, it offers a nondestructive and comprehensive method to assess the microstructures and conductive and nonconductive phases of CBSs. Based on it, the equivalent circuit model can be proposed to elucidate the conductive and self-sensing mechanism of CBSs.
3. Unlike traditional CBSs, capacitance-based CBSs do not require additional conductive fillers to establish a relationship between capacitance and external environmental or mechanical factors. However, they also have their drawbacks, such as dependence on the material geometry, electrode properties, current frequency, external electric fields, and sensitivity to many external changes. Compared to resistivity-based CBSs, the capacitance measurement lacks repeatability and recovery behaviors, especially for brittle cementitious materials under high loading conditions, which further increases the difficulty of their application.

REFERENCES

1. D. Chung, A critical review of electrical-resistance-based self-sensing in conductive cement-based materials, Carbon 203 (2023) 311–325.
2. W. Dong, W. Li, N. Lu, F. Qu, K. Vessalas, D. Sheng, Piezoresistive behaviours of cement-based sensor with carbon black subjected to various temperature and water content, Composites Part B: Engineering 178 (2019) 107488.

3. S. Liang, H. Du, N. Zou, Y. Chen, Y. Liu, Measurement and simulation of electrical resistivity of cement-based materials by using embedded four-probe method, Construction and Building Materials 357 (2022) 129344.
4. W. Li, F. Qu, W. Dong, G. Mishra, S.P. Shah, A comprehensive review on self-sensing graphene/cementitious composites: A pathway toward next-generation smart concrete, Construction and Building Materials 331 (2022) 127284.
5. T.-C. Hou, J.P. Lynch, Electrical impedance tomographic methods for sensing strain fields and crack damage in cementitious structures, Journal of Intelligent Material Systems and Structures 20(11) (2009) 1363–1379.
6. H. Wang, F. Shi, J. Shen, A. Zhang, L. Zhang, H. Huang, J. Liu, K. Jin, L. Feng, Z. Tang, Research on the self-sensing and mechanical properties of aligned stainless steel fiber-reinforced reactive powder concrete, Cement and Concrete Composites 119 (2021) 104001.
7. L. Chi, Z. Xiaohong, S. Guoxing, Embedded resistivity sensor for compressive strength prediction of cement paste by electrochemical impedance spectroscopy, IEEE Sensors Letters 5(9) (2021) 1–4.
8. L. Zhang, S. Ding, L. Li, S. Dong, D. Wang, X. Yu, B. Han, Effect of characteristics of assembly unit of CNT/NCB composite fillers on properties of smart cement-based materials, Composites Part A: Applied Science and Manufacturing 109 (2018) 303–320.
9. D. Chung, Self-sensing concrete: From resistance-based sensing to capacitance-based sensing, International Journal of Smart and Nano Materials 12(1) (2021) 1–19.
10. D. Chung, First review of capacitance-based self-sensing in structural materials, Sensors and Actuators A: Physical 354 (2023) 114270.
11. D. Chung, Y. Wang, Capacitance-based stress self-sensing in cement paste without requiring any admixture, Cement and Concrete Composites 94 (2018) 255–263.
12. M. Ozturk, D.D.L. Chung, Capacitance-based stress self-sensing in asphalt without electrically conductive constituents, with relevance to smart pavements, Sensors and Actuators A: Physical 342 (2022) 113625.
13. L. Wang, F. Aslani, Mechanical properties, electrical resistivity and piezoresistivity of carbon fibre-based self-sensing cementitious composites, Ceramics International 47(6) (2021) 7864–7879.
14. J. Ou, B. Han, Piezoresistive cement-based strain sensors and self-sensing concrete components, Journal of Intelligent Material Systems and Structures 20(3) (2009) 329–336.
15. A. Belli, A. Mobili, T. Bellezze, F. Tittarelli, P. Cachim, Evaluating the self-sensing ability of cement mortars manufactured with graphene nanoplatelets, virgin or recycled carbon fibers through piezoresistivity tests, Sustainability 10(11) (2018) 4013.
16. W. Dong, W. Li, K. Wang, K. Vessalas, S. Zhang, Mechanical strength and self-sensing capacity of smart cementitious composite containing conductive rubber crumbs, Journal of Intelligent Material Systems and Structures 31(10) (2020) 1325–1340.
17. M.S. Konsta-Gdoutos, C.A. Aza, Self sensing carbon nanotube (CNT) and nanofiber (CNF) cementitious composites for real time damage assessment in smart structures, Cement and Concrete Composites 53 (2014) 162–169.
18. W. Dong, W. Li, Y. Guo, K. Wang, D. Sheng, Mechanical properties and piezoresistive performances of intrinsic graphene nanoplate/cement-based sensors subjected to impact load, Construction and Building Materials 327 (2022) 126978.
19. R. Spragg, C. Villani, K. Snyder, D. Bentz, J.W. Bullard, J. Weiss, Factors that influence electrical resistivity measurements in cementitious systems, Transportation Research Record 2342(1) (2013) 90–98.
20. O. Galao, F.J. Baeza, E. Zornoza, P. Garcés, Strain and damage sensing properties on multifunctional cement composites with CNF admixture, Cement and Concrete Composites 46 (2014) 90–98.
21. A.A. Elseady, I. Lee, Y. Zhuge, X. Ma, C.W. Chow, N. Gorjian, Piezoresistivity and AC impedance spectroscopy of cement-based sensors: Basic concepts, interpretation, and perspective, Materials 16(2) (2023) 768.
22. H.V. Le, M.K. Kim, D.J. Kim, J. Park, Electrical properties of smart ultra-high performance concrete under various temperatures, humidities, and age of concrete, Cement and Concrete Composites 118 (2021) 103979.
23. J. Torrents, T. Mason, E. Garboczi, Impedance spectra of fiber-reinforced cement-based composites: A modeling approach, Cement and Concrete Research 30(4) (2000) 585–592.

24. C.G. Berrocal, K. Hornbostel, M.R. Geiker, I. Löfgren, K. Lundgren, D.G. Bekas, Electrical resistivity measurements in steel fibre reinforced cementitious materials, Cement and Concrete Composites 89 (2018) 216–229.
25. H.V. Le, D.H. Lee, D.J. Kim, Effects of steel slag aggregate size and content on piezoresistive responses of smart ultra-high-performance fiber-reinforced concretes, Sensors and Actuators A: Physical 305 (2020) 111925.
26. S. Wansom, N. Kidner, L. Woo, T. Mason, AC-impedance response of multi-walled carbon nanotube/cement composites, Cement and Concrete Composites 28(6) (2006) 509–519.
27. X. Li, M. Li, Multifunctional self-sensing and ductile cementitious materials, Cement and Concrete Research 123 (2019) 105714.
28. G. Song, Equivalent circuit model for AC electrochemical impedance spectroscopy of concrete, Cement and Concrete Research 30(11) (2000) 1723–1730.
29. Y.J. Wang, Y. Pan, X.W. Zhang, K. Tan, Impedance spectra of carbon black filled high-density polyethylene composites, Journal of Applied Polymer Science 98(3) (2005) 1344–1350.
30. J. Han, J. Cai, J. Pan, Y. Sun, Study on the conductivity of carbon fiber self-sensing high ductility cementitious composite, Journal of Building Engineering 43 (2021) 103125.
31. L. Jiang, Z. Liu, Y. Yu, X. Ben, The effect of graphene on the conductivity of magnesium sulfate cement, Construction and Building Materials 312 (2021) 125342.
32. H. Allam, F. Duplan, S. Amziane, Y. Burtschell, Assessment of manufacturing process efficiency in the dispersion of carbon fibers in smart concrete by measuring AC impedance, Cement and Concrete Composites 127 (2022) 104394.
33. L. Liu, J. Xu, T. Yin, Y. Wang, H. Chu, Improving electrical and piezoresistive properties of cement-based composites by combined addition of nano carbon black and nickel nanofiber, Journal of Building Engineering 51 (2022) 104312.
34. D.A. Triana-Camacho, D.A. Miranda, E. García-Macías, O.A.M. Reales, J.H. Quintero-Orozco, Effective medium electrical response model of carbon nanotubes cement-based composites, Construction and Building Materials 344 (2022) 128293.
35. J. Xu, W. Zhong, W. Yao, Modeling of conductivity in carbon fiber-reinforced cement-based composite, Journal of Materials Science 45(13) (2010) 3538–3546.
36. G. Li, L. Wang, C. Leung, R. Hu, X. Zhao, B. Yan, J. Zhou, Effect of styrene–butadiene rubber on the electrical properties of carbon black/cement mortar, RSC Advances 5(86) (2015) 70229–70237.
37. W. Pichor, M. Frąc, M. Radecka, Determination of percolation threshold in cement composites with expanded graphite by impedance spectroscopy, Cement and Concrete Composites 125 (2022) 104328.
38. E. García-Macías, A. Downey, A. D'Alessandro, R. Castro-Triguero, S. Laflamme, F. Ubertini, Enhanced lumped circuit model for smart nanocomposite cement-based sensors under dynamic compressive loading conditions, Sensors and Actuators A: Physical 260 (2017) 45–57.
39. B. Dong, Y. Wu, X. Teng, Z. Zhuang, Z. Gu, J. Zhang, F. Xing, S. Hong, Investigation of the Cl– migration behavior of cement materials blended with fly ash or/and slag via the electrochemical impedance spectroscopy method, Construction and Building Materials 211 (2019) 261–270.
40. M. Cabeza, M. Keddam, X. Nóvoa, I. Sánchez, H. Takenouti, Impedance spectroscopy to characterize the pore structure during the hardening process of Portland cement paste, Electrochimica Acta 51(8–9) (2006) 1831–1841.
41. J. Cruz, I. Fita, L. Soriano, J. Payá, M. Borrachero, The use of electrical impedance spectroscopy for monitoring the hydration products of Portland cement mortars with high percentage of pozzolans, Cement and Concrete Research 50 (2013) 51–61.
42. C. Song, S. Choi, Moisture-dependent piezoresistive responses of CNT-embedded cementitious composites, Composite Structures 170 (2017) 103–110.
43. J. Zhang, A. Heath, H.M.T. Abdalgadir, R.J. Ball, K. Paine, Electrical impedance behaviour of carbon fibre reinforced cement-based sensors at different moisture contents, Construction and Building Materials 353 (2022) 129049.
44. B. Dong, Q. Qiu, J. Xiang, C. Huang, H. Sun, F. Xing, W. Liu, Electrochemical impedance interpretation of the carbonation behavior for fly ash–slag–cement materials, Construction and Building Materials 93 (2015) 933–942.

45. B. Dong, Q. Qiu, Z. Gu, J. Xiang, C. Huang, Y. Fang, F. Xing, W. Liu, Characterization of carbonation behavior of fly ash blended cement materials by the electrochemical impedance spectroscopy method, Cement and Concrete Composites 65 (2016) 118–127.
46. L. Wang, F. Aslani, A review on material design, performance, and practical application of electrically conductive cementitious composites, Construction and Building Materials 229 (2019) 116892.
47. W. Dong, W. Li, Z. Tao, K. Wang, Piezoresistive properties of cement-based sensors: Review and perspective, Construction and Building Materials 203 (2019) 146–163.
48. W. Li, W. Dong, Y. Guo, K. Wang, S.P. Shah, Advances in multifunctional cementitious composites with conductive carbon nanomaterials for smart infrastructure, Cement and Concrete Composites 128 (2022) 104454.
49. B. Díaz, B. Guitián, X.R. Nóvoa, C. Pérez, Conductivity assessment of multifunctional cement pastes by impedance spectroscopy, Corrosion Science 185 (2021) 109441.
50. B. Díaz, B. Guitián, X. Nóvoa, C. Pérez, Analysis of the microstructure of carbon fibre reinforced cement pastes by impedance spectroscopy, Construction and Building Materials 243 (2020) 118207.
51. Y. Guo, W. Li, W. Dong, Z. Luo, F. Qu, F. Yang, K. Wang, Self-sensing performance of cement-based sensor with carbon black and polypropylene fibre subjected to different loading conditions, Journal of Building Engineering 59 (2022) 105003.
52. W. Dong, W. Li, L. Shen, D. Sheng, Piezoresistive behaviours of carbon black cement-based sensors with layer-distributed conductive rubber fibres, Materials & Design 182 (2019) 108012.
53. L. Zhang, L. Li, Y. Wang, X. Yu, B. Han, Multifunctional cement-based materials modified with electrostatic self-assembled CNT/TiO_2 composite filler, Construction and Building Materials 238 (2020) 117787.
54. W. Dong, W. Li, K. Wang, Z. Luo, D. Sheng, Self-sensing capabilities of cement-based sensor with layer-distributed conductive rubber fibres, Sensors and Actuators A: Physical 301 (2020) 111763.
55. W. Dong, W. Li, K. Wang, B. Han, D. Sheng, S.P. Shah, Investigation on physicochemical and piezoresistive properties of smart MWCNT/cementitious composite exposed to elevated temperatures, Cement and Concrete Composites 112 (2020) 103675.
56. P.A. Danoglidis, M.S. Konsta-Gdoutos, S.P. Shah, Relationship between the carbon nanotube dispersion state, electrochemical impedance and capacitance and mechanical properties of percolative nanoreinforced OPC mortars, Carbon 145 (2019) 218–228.

Chapter 5

Mechanical properties and sensing performances

5.1 INTRODUCTION

The outstanding mechanical strength of cement-based sensors (CBSs) is a key factor that sets them apart from traditional sensors for SHM. Therefore, they can work well in conjunction with concrete structures without prematurely failing, as long as the concrete structure remains intact. The mechanical strength of conductive CBSs is greatly affected by the added conductive fillers, as well as other factors that also influence the strength of regular concrete, such as the water-to-cement ratio, additives, aggregate grading, and so on. Compared to 1D and 2D conductive fillers, 0D conductive fillers are less favorable for enhancing the mechanical strength of CBSs. For instance, nano-carbon black (NCB), which is commonly added to significantly improve the conductivity of cement-based materials and introduce piezoresistive performance, often diminishes the mechanical properties of CBSs due to its high specific surface area and its tendency to adsorb water particles [1]. Due to its high aspect ratio, carbon nanotube (CNT) and carbon nanofiber (CNF) can be considered as 1D conductive fillers. Their incorporation can significantly enhance the electrical and mechanical properties of CBSs through the bridging effect. However, a prerequisite for this enhancement is the proper dispersion of carbon nanofibers. Otherwise, the agglomerates formed by them could actually decrease the mechanical properties of the cement matrix [2]. Two-dimensional graphene nanoplate (GNP) can significantly enhance the mechanical performance of CBSs. The principle behind this enhancement lies in the superior mechanical properties of GNP and its oxides, as well as mechanisms such as load transfer, crack resistance, interface improvement, reduced porosity, and promotion of cement hydration.

This chapter primarily introduces the mechanical performances of CBSs and their mechanical sensing capabilities. The mechanical performance of CBSs was first presented in terms of the influence of conductive fillers of different dimensions and geometries. An overview was given of the impact of 0D, 1D, and 2D carbon nanomaterials, as well as macro-fillers, on the mechanical strengths of cement-based materials. Subsequently, the dynamic properties and elastic modulus of the materials used in CBSs were introduced. The third section outlines the mechanical sensing performance of CBSs, encompassing their classical responses in traditional mechanical tests such as compression, bending, and splitting tensile experiments. The emphasis was placed on highlighting the self-sensing capabilities of CBSs under the influence of impact energy. Furthermore, an in-depth analysis was conducted to explore the changes in the microstructure of cement-based materials and establish a correlation between resistance variations and microstructural changes.

DOI: 10.1201/9781032663685-5

Finally, the section concludes by summarizing the effects of factors like load frequency, amplitude, and duration on the self-sensing performance of CBSs.

5.2 MECHANICAL AND MICROSTRUCTURAL PROPERTIES

5.2.1 0D nanomaterials

NCB exhibits excellent conductivity. Incorporating it into CBSs can enhance the conductivity and self-sensing ability significantly. However, due to the high surface activity and water adsorption capacity of NCB, its inclusion can affect the flowability of the cementitious matrix. Therefore, it is often necessary to increase the w/b ratio of the matrix or employ additives to achieve satisfactory conductivity while incorporating NCB. Its influences on the compressive strength of CBSs are shown in Fig. 5.1 [3]. The CBSs filled with 0.1 wt.% and 0.5 wt.% NCB were provided with excellent compressive strength of both higher than 50 MPa and similar to the plain cement paste (PCP) of 55 MPa. It can be considered that a small amount of NCB particles could fill up the micropores in the composite and improve the compressive strength, while negativities were also accompanied by larger w/b ratios. However, considerable strength decreases occurred when the NCB content was larger than 1.0 wt.% to the weight of the binder, where the reduction rate reached 32.7%, 56.4%, and 72.7% for the content of 1.0 wt.%, 2.0 wt.%, and 4.0 wt.% NCB-filled composites, with only 37 MPa, 24 MPa, and 15 MPa remained, respectively. The reasons for their differences mainly came from the special physical characteristics of NCB, which possessed excellent water absorption ability. Therefore, the NCB particles had the capacity to absorb water content or attach to the cement surface when being mixed with a cement mixture, which affected the cement hydration process. Fortunately, this could be partially solved by adding more water according to the amount of NCB in the composite. But on the other hand, NCB particles with high surface energy

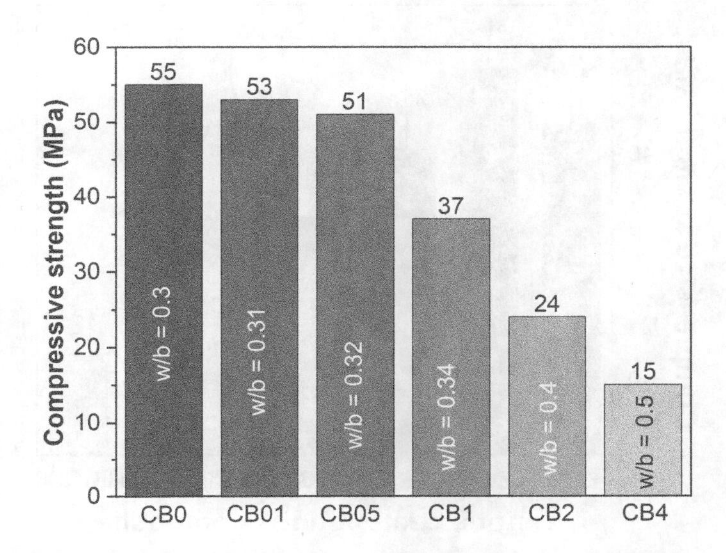

Figure 5.1 Compressive strength of NCB-filled CBS with multiple contents of CB [3].

were extremely inclined to absorb together and form agglomerations and clusters, which showed no resistance to the loading forces compared to the hardened hydration products.

5.2.2 1D nanomaterials

CNT and CNF have been widely studied for their potential to enhance the mechanical strength of concrete due to their exceptional mechanical properties and unique structural characteristics [4–6]. The specific enhancement of these nanomaterials on the mechanical properties of CBSs is identical to their effects on traditional cementitious materials, which can be found in various references [7–9]. Different from the previous studies, the compressive strengths of CBSs containing well-dispersed CNT or undispersed layer-distributed CNT are introduced in this section. The compressive strengths of the PCP, 0.5% well-dispersed CNT-reinforced CBS (UDCC), and the CBSs with various CNT layers (LDCC) are displayed in Fig. 5.2 [10]. Because of the anisotropic behavior of LDCC, it should be noted that the loading direction is vertical to the plane with CNT agglomerations and the same for the piezoresistive test. It shows that the uniformly distributed CNT reinforced the cementitious composites and increased the compressive strength of PCP from 44 MPa to 51 MPa. However, for the cementitious composites with CNT without dispersion, the compressive strength of the LDCC gradually decreased with the increase of CNT layers. The compressive strength was still acceptable for the LDCC with one layer of undispersed CNT (LDCC1), reaching approximately 40 MPa. However, it significantly decreased when the CNT layers increased ≥2, and the compressive strengths of only 31 MPa and 13 MPa remained for the LDCC with 2 (LDCC2) and 6 layers of CNT (LDCC6), respectively. Previous studies have drawn contradictory results on the enhanced/weakened compressive strength of CNT-engineered cementitious composites [11, 12]. Nevertheless, both of them reach the consensus that the CNT dispersion had a

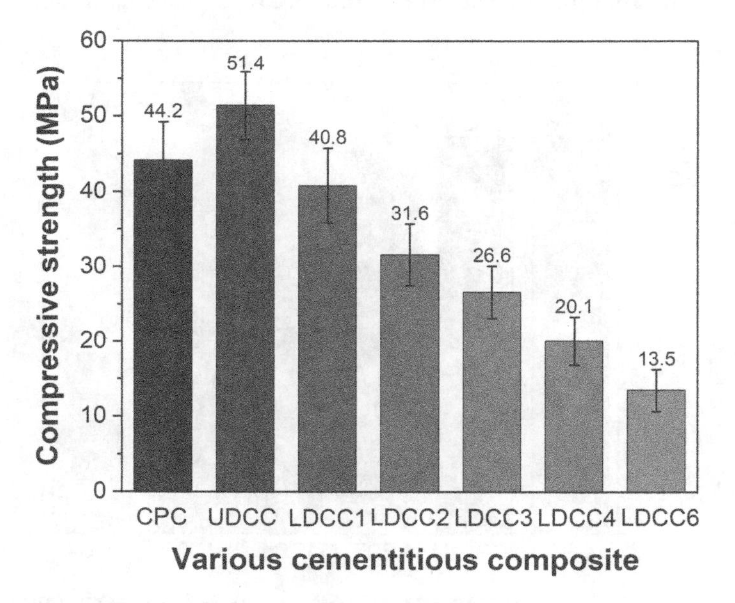

Figure 5.2 Compressive strength of PCP and the CBS with dispersed CNT or with layer-distributed CNT [10].

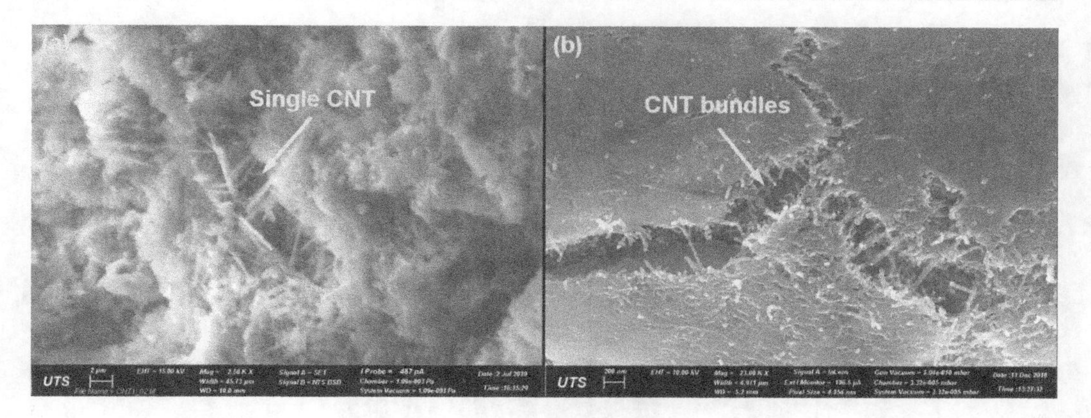

Figure 5.3 Microstructural morphology of CNT in cracks or gaps in different specimens (a) UDCC and (b) LDCC [10].

firm relationship to the mechanical properties. The well-dispersed CNT possessing filling and bridging effect is beneficial for the compressive strength, while the CNT agglomerations and the inappropriate surfactant can greatly affect the density and mechanical strengths of cementitious composites.

The microstructures of the UDCC and LDCC are depicted in Fig. 5.3 to explicate the compressive strength increase of UDCC and the strength reduction in LDCC. It could be found that the CNT in UDCC is straight and separated, while the CNT in LDCC is curved and intertwined. For the cracks in UDCC, it was observed that the well-dispersed CNT having the fiber-bridging effect could prohibit the further deformation of cracks. Hence, in comparison to the PCP, the compressive strength of UDCC was greatly improved. In terms of the LDCC, in the transition zone between a CNT layer and cement matrix, a proportion of CNT bundles were found possessing bridging effect as aforementioned. However, the reinforcement effect was much weakened because of the twined CNT. Not only the CNT bundles possessed weaker cohesion to cement matrix than single CNT, but also led to more porous structures surrounding the agglomerations, thus reducing the density and compressive strength of composites. In addition, the above discussions have proved the stress concentration in the pores induced by larger sized CNT agglomerations. More layers of CNT presented more CNT agglomerations in a larger size, thus generating severe stress concentrations and more defects for the LDCC with multiple layers of CNT and directly causing worse compressive strengths.

5.2.3 2D nanomaterials

The compressive and flexural strengths of CBSs with GNP and GP are illustrated in Fig. 5.4a and b [13]. For the composites filled with GNP, both compressive and flexural strengths were first increased and then decreased with the increase of GNP content. The optimal content of GNP reaching 1.0% could achieve the cementitious composites with the best mechanical properties, whose compressive and flexural strengths reached 53.6 MPa and 3.6 MPa, respectively. The strengths slightly decreased as the GNP content reached 2%, but was still larger than that of PCP. Until the concentration of GNP reached 3%, the negativity started to emerge with both lower compressive and flexural strengths to the PCP. Different from the GNP-filled cementitious composite, it seems that the GP

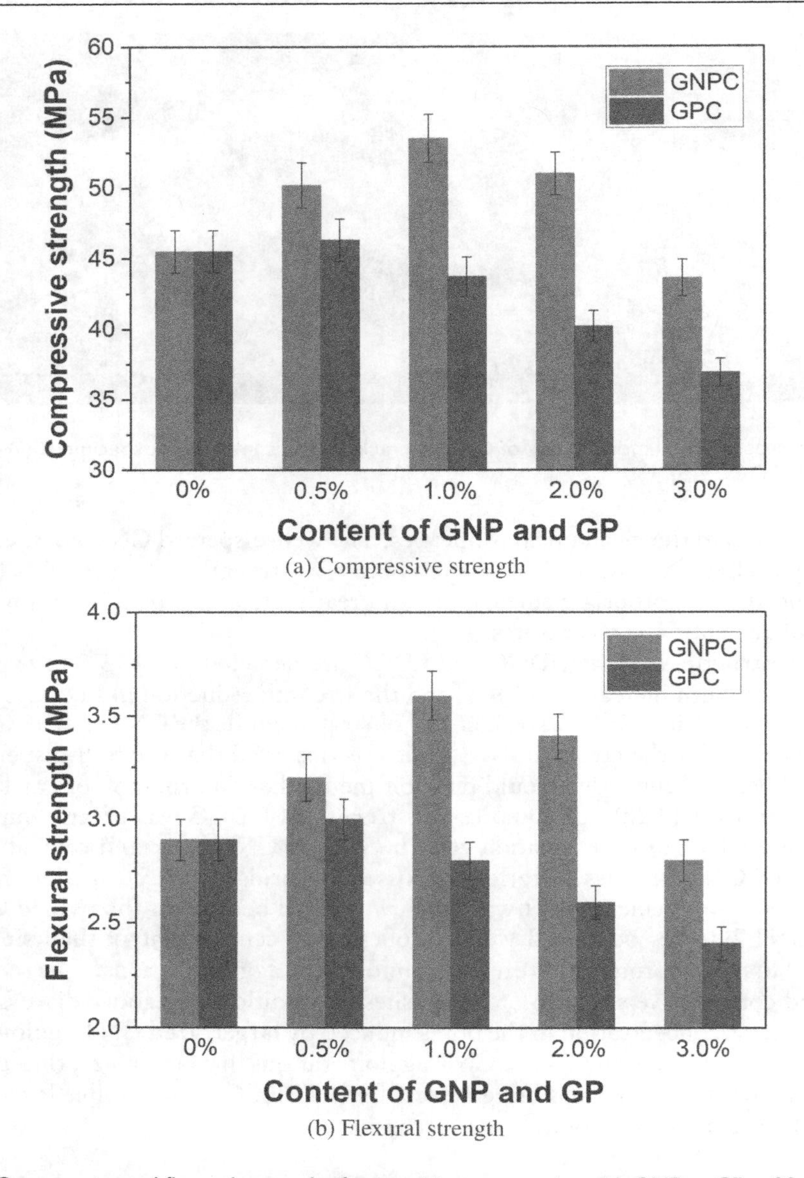

(a) Compressive strength

(b) Flexural strength

Figure 5.4 Compressive and flexural strength of cementitious composite with GNP or GP at 28-day age [13].

made nearly no improvements on the mechanical properties of cementitious composites. The compressive and flexural strengths only slightly increased in the case of a small dose of GP of 0.5%, and then the strengths dramatically declined with the increase of GP. For the GPC3, the reduction rates for the compressive and flexural strength peaked at 18.9% and 17.2%, respectively.

Fig. 5.5 shows the microstructures of CBSs containing 0.5% GNP, attempting to explain the reinforcement and reduction mechanism of the mechanical properties. It was seen that the thickness of GNP is in the nanoscale, which not only has high strength

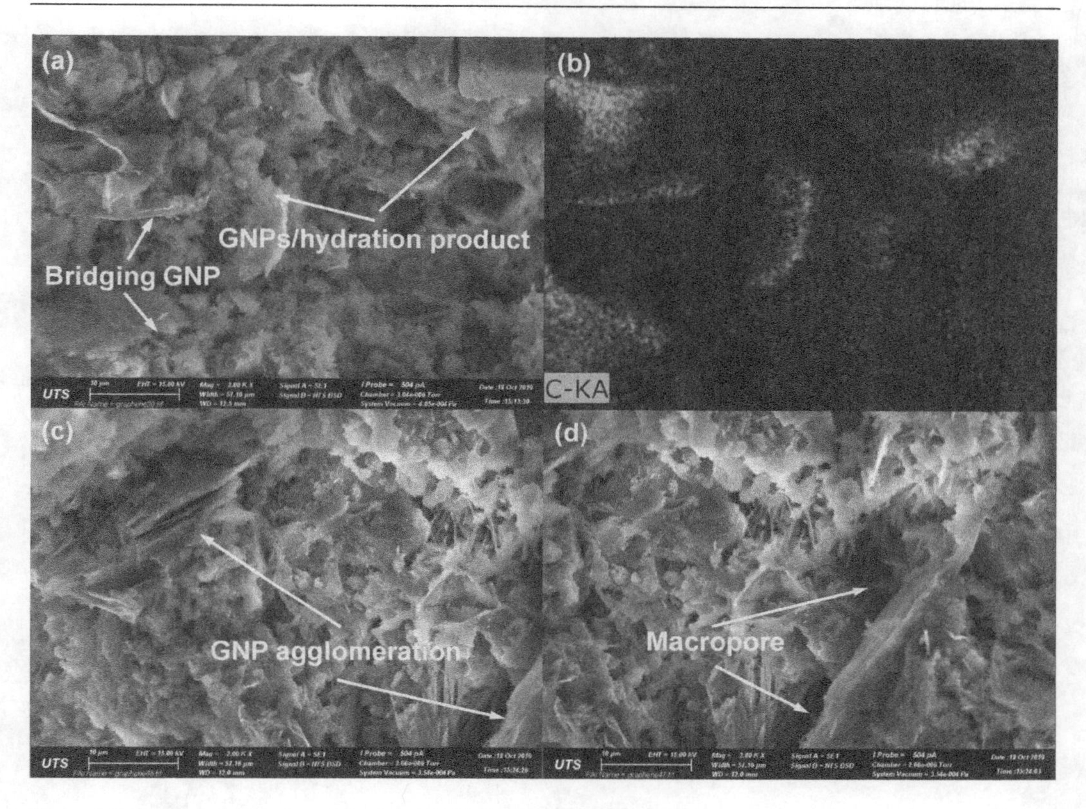

Figure 5.5 Microstructural morphology of GNP-reinforced CBS [13].

and modulus but also possesses an extremely high specific surface area. Therefore, it was clearly observed that the GNP provided a large space in the cementitious composites, so the hydration products generation could adhere to the surface, as illustrated in Fig. 5.5a. The EDX results on carbon element were applied to distinguish the GNP from the cement hydration products, as shown in Fig. 5.5b. In addition to the nucleation effect, GNP as the reinforcements in the cementitious composite also played a bridging effect to bond the cement matrix because of the structures of nanosheets. The reinforcement effect relates to the specific concentration of conductive fillers. It seems that cement hydration is stimulated with the increase of GNP. As for the decreased compressive strength for the composites with 3% GNP, the reason is mainly due to the poor distribution of GNP in the composites. Fig. 5.5c and d illustrates the microstructures of CBSs with 3.0% GNP, where significant GNP agglomerations and the induced micro-defects could be found. Since all the GNP solutions were prepared under identical circumstances, the higher concentration of GNP means the worse dispersion of GNP solutions and more GNP agglomerations. The GNP agglomeration without any bearing capacity is prone to lower the density of cementitious composites and make loose microstructures. Furthermore, the macropores around the GNP agglomerations could be observed.

In terms of the composites with a small GP dose, the large sheets of GP which are not involved in the hydration process have worse cohesion to cement matrix. As shown in

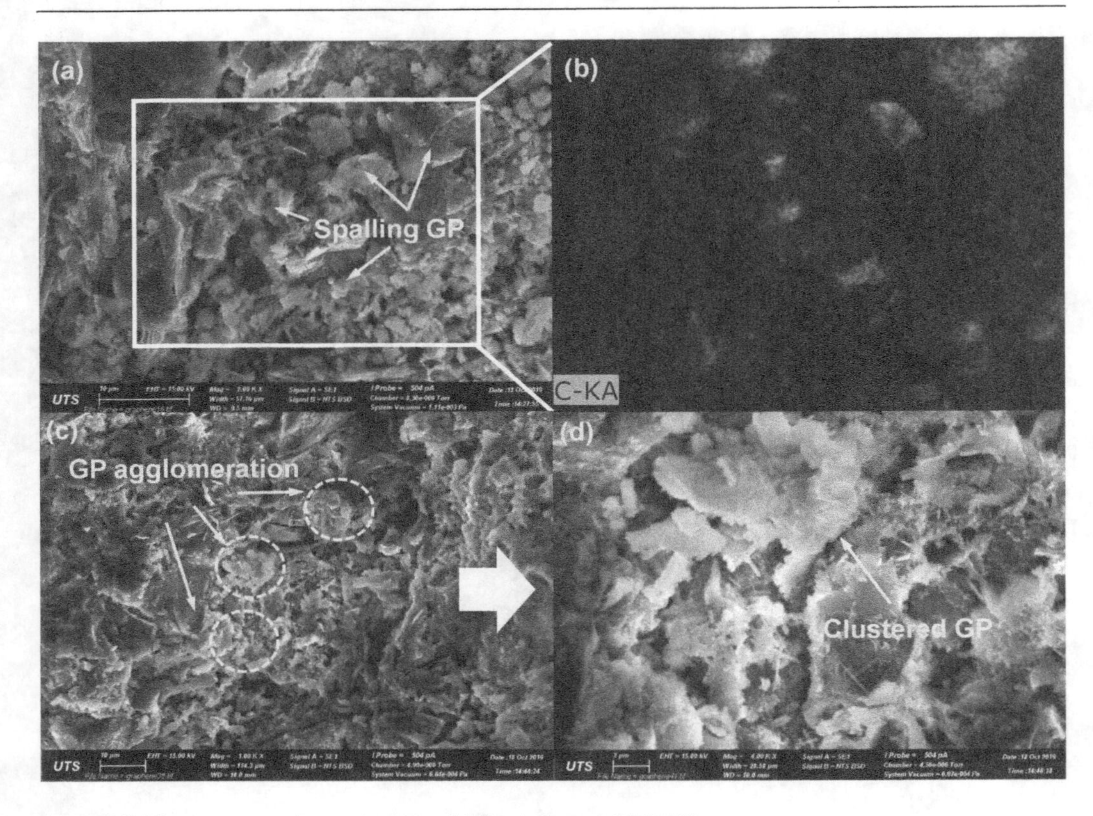

Figure 5.6 Microstructural morphology of GP-reinforced CBS [13].

Fig. 5.6a and b, it was observed that the GP in the CBS tended to spall from the cement matrix. In addition to the worse cohesion between GP and cement matrix, as there exists van der Waals force between sheets in GP, the small forces make the GP easily slip between layers [14]. When the GP filled CBSs under external forces, the slipped layers will greatly damage the microstructures and cause decreased mechanical properties. Furthermore, there existed similar agglomeration problems between GP when the dosage of the GP is high in the cementitious composites. Fig. 5.6c and d shows the micromorphology of the 3.0% GP-reinforced CBSs, where the GP agglomerations could be clearly observed. Different from the GNP which agglomerated layer by layer, it was observed that the GP clustered much randomly. This is probably due to the smaller surface activity for the GP than that of GNP. Overall, in comparison to the microstructures of the GNPC and GPC, an excellent connection between the cement matrix and GNP and a poor cohesion between the GP and cement matrix can be found, which should be responsible for their mechanical property differences.

In summary, the mechanism for the improvement of the compressive performance of CBSs could be primarily related to the following reasons: the nucleation effect, the large specific surface area of GNP, and the bridging effect. Fig. 5.7 shows the compressive strength of cementitious composites with different GOs and GNPs content in different curing periods. Baomin and Shuang [15] carried out experiments to examine the influence on compressive strength of cementitious composites with 0.03 wt.%, 0.06 wt.%,

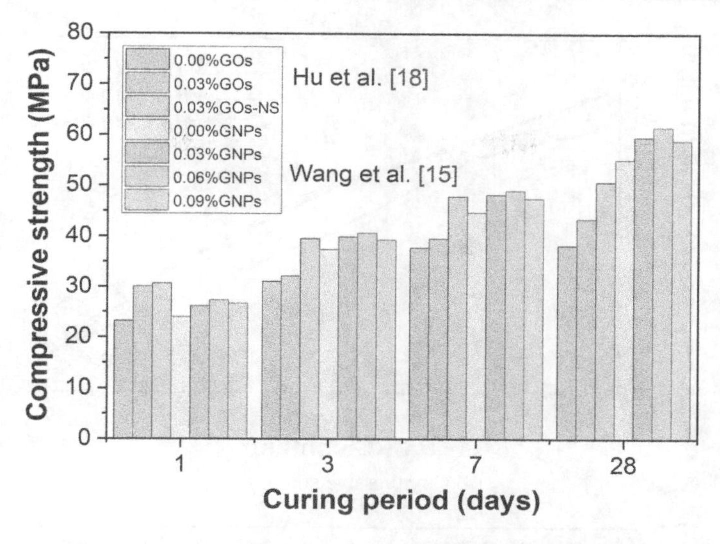

Figure 5.7 Compressive strength of graphene-based cementitious composites under different curing periods.

and 0.09 wt.% GNP. They stated that the cementitious composites with 0.06 wt.% GNP presented the largest compressive strength. Besides, the increasing addition of GNP could cause a slight improvement in the compressive strength. The results for the greatest addition content of GNP reported by Baomin and Shuang are also remarkably similar to that examined by other researchers [16–18]. However, a somewhat different finding was stated by Du et al. [19], who mentioned that the largest mechanical properties of the concrete were achieved with 1.0 wt.% of GNPs.

5.2.4 Other macro-fillers

5.2.4.1 Conductive rubber crumb (CRC)

The compressive strength of the CRC-filled cement mortar at w/b ratios of 0.40, 0.42, and 0.45 is depicted in Fig. 5.8a [20]. The plain cement mortar without any rubber fillers is used as a control group to study the effect of CRC on the compressive strength of cement mortar. It could be seen that the compressive strength of plain cement mortar decreased with the increase of w/b ratios. As for the CRC-filled cement mortar, the compressive strength of the cement mortar showed a monotone decrease with the increase of CRC content. In the case of the cement mortar filled with 40% CRC, the ultimate compressive strengths decreased to lower than 20 MPa regardless of the w/b ratios. Furthermore, when the CRC content is higher than 20%, the cement mortar at the w/b ratio of 0.42 started to possess the highest compressive strength compared to that of the cement mortar at the w/b ratios of 0.40 and 0.45.

Fig. 5.8b displays the reduction rate of compressive strength for the CRC-filled cement mortar. It could be observed that the cement mortar at the w/b ratios of 0.40 and 0.45 went through slightly larger compressive strength reduction than the cement mortar at the w/b ratio of 0.42. It indicated that the content of CRC in the cement mortar had less adverse effect on the cement mortar at the w/b ratio of 0.42. Despite the characteristics

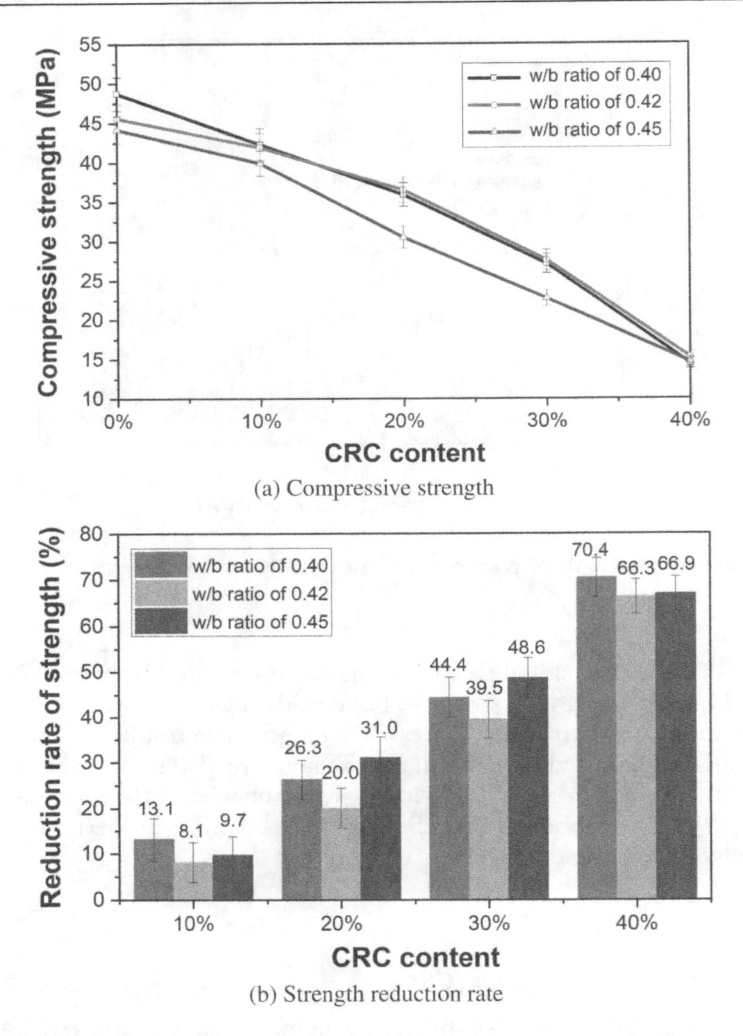

(a) Compressive strength

(b) Strength reduction rate

Figure 5.8 Compressive strength and the reduction rate for the CRC-filled cement mortar at various w/b ratios [20].

of poor water absorption for CRC, it has been reported that the added rubber crumbs will cause higher interparticle friction and influence the workability of cementitious composites [21]. Therefore, the dense structures of composites might be influenced when the workability is weakened. In other words, it demonstrated that both the defects caused by the weakened workability and the added CRC were responsible for the swift compressive strength reduction for the cement mortar at the w/b ratio of 0.40. As for the cement mortar at the w/b ratio of 0.45, it was believed that the compressive strength reduction was mainly due to the combined actions by the excessive water content and the increasing CRC. Overall, it could be deduced that the cement mortar at the w/b ratio of 0.42 was more suitable to be filled with CRC, in which the compressive strength was slightly higher than the counterparts at w/b ratio of 0.40 or 0.45 when filled with the CRC content was

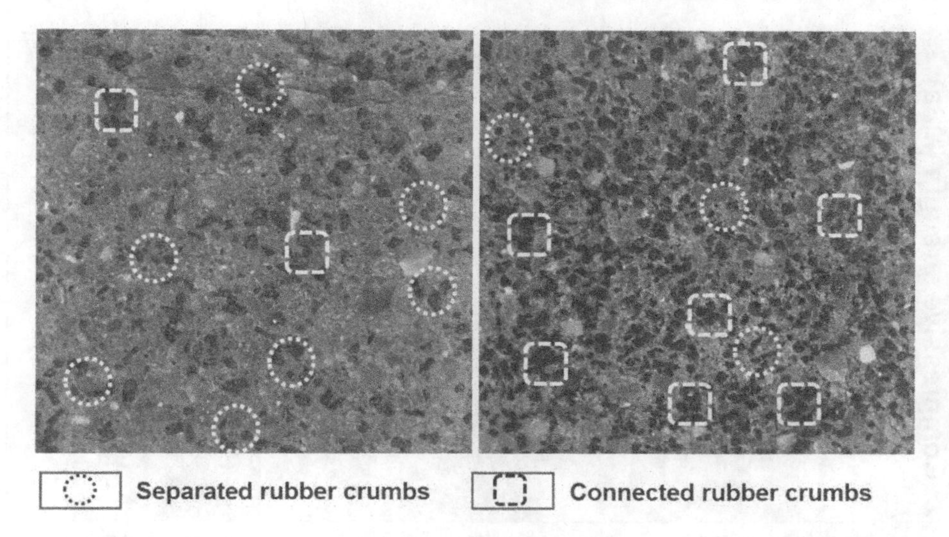

Figure 5.9 Cross-sectional morphology of rubberized cement mortar with (a) 20% CRC and (b) 40% CRC [20].

larger than 10%. The reason might be due to the relatively better workability than the composites at w/b ratio of 0.40.

To explain the reduced compressive strength, Fig. 5.9 shows the cross-sectional morphology of rubberized cement mortar filled with different contents of rubber crumbs. Because of the rare discrepancies among composites at various w/b ratios, only the composites at the w/b ratio of 0.42 were selected to illustrate their macro-morphology. Generally, it could be seen that the rubber crumbs were evenly distributed in the cement matrix for both composites with 20% and 40% CRC. In addition, a greater number of separated rubber crumbs could be observed in the rubberized mortar with 20% CRC, while a larger amount of connected rubber crumbs was found in the mortar with 40% CRC. These rubber crumbs possessing low strength and elastic modulus, especially for the connected crumbs which nearly have no cohesion in the boundaries, can easily produce a deformation under the external forces. As a result, the stress concentration can be caused in the boundary of the cement matrix to rubber crumbs, which leads to the generated cracks along the rubber crumbs. Overall, the aforementioned reduced compressive strength was exactly sourced from the substitution of cement matrix and fine aggregate to the rubber crumbs. In particular, the boundaries between the cement matrix and rubber crumbs, sand and rubber crumbs, and especially the interfaces among rubber crumbs (connected rubber crumbs) can generate high stress concentration and damage the mechanical properties of composites.

5.2.4.2 Conductive rubber fiber

Compressive strengths of rubber/CBS with various rubber fiber contents and w/b ratios are represented in Fig. 5.10 [22]. The compressive strengths of PCP (without rubber) were the largest, reaching 56 MPa, 45 MPa, and 33 MPa for the w/b ratios of 0.34, 0.38, and 0.42, respectively. Overall, it is found that the compressive strength gradually decreased with increase of rubber fibers and w/b ratios. The composite with 20 rubber fibers experienced

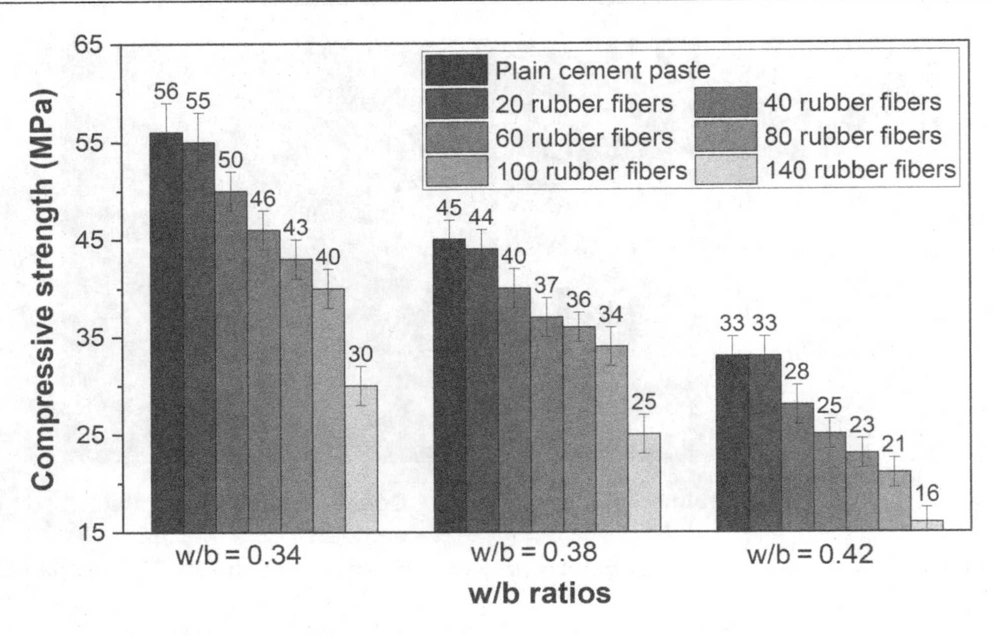

Figure 5.10 Compressive strength of rubber–cement composites with different contents of rubber fibers and w/b ratios [22].

slight strength reduction and the strength was very close to that of original composites; the strength of composites embedded with 80 rubber fibers fell to 43 MPa, 36 MPa, and 23 MPa by the decreasing rate of 23.2%, 20%, and 30.3%, respectively. For the composite with 140 rubber fibers, the reduction rate boomed to 46.4%, 44.4%, and 51.5% and the compressive strength left as 30 MPa, 25 MPa, and 16 MPa, respectively. In addition, it could be observed that the negativities on compressive strength were more with higher w/b ratios rather than rubber fibers. For example, the composite with 100 rubber fibers at the w/b ratio of 0.34 had higher compressive strength than that of PCP at the w/b ratio of 0.42. The reason might be due to the heterogeneity of rubber–cement composite, in which the rubber fibers were mainly concentrated in two layers and the central part, thus the other compact cement matrix could still bear the external forces. Furthermore, to explore the strength reduction with increase of rubber content, the interfaces between conductive rubber fibers and cement matrix were studied.

The SEM images on the cross section of rubber–cement matrix boundaries with different magnifications are shown in Fig. 5.11. It must be noted that the two phases were easy to distinguish because of their huge discrepancy on electrical conductivity. Hence, a general distribution of one conductive rubber fiber in cementitious composite could be observed in Fig. 5.11a, where the light color area means conductive rubber, and the slightly darker color area represents cement matrix. In their boundaries, two kinds of regions could be seen, with one named smooth boundaries for its good connections between rubber and cement matrix and another called disturbed boundaries with gaps and cracks. In addition to the lower mechanical properties of rubber fibers, it implied that the decreased compressive strength of rubber–cement composite might also be attributed to the gaps and cracks. Moreover, a large proportion of smooth boundaries and only a small disturbed region in the interface explained why the composite with 20 rubber

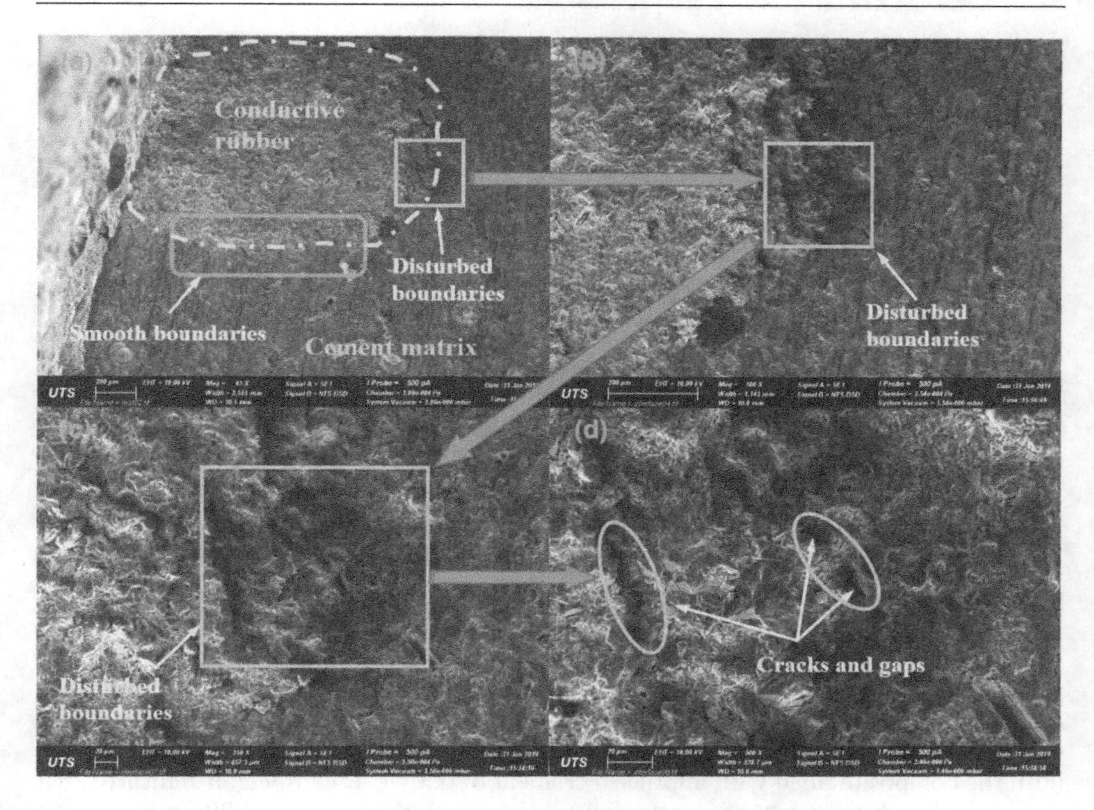

Figure 5.11 SEM images on the boundaries of rubber fiber to cement matrix [22].

fibers had nearly no influences on compressive strength. To have a better view of the disturbed boundaries, more images with higher magnification from 100× to 500× are displayed in Fig. 5.11b–d. Although the disturbed boundaries containing gaps and cracks are relatively limited in the boundaries of one fibrous rubber and cement matrix, it can be deduced that the composites with high content of rubber fibers could possess larger disturbed boundaries and more gaps and cracks. Therefore, for the composite with 140 rubber fibers, which was halved on compressive strength, was mainly owing to the long gaps and connected cracks in their boundaries.

5.2.4.3 PP fiber

Fig. 5.12 illustrates the compressive strength of the NCB-filled CBSs reinforced with different contents of PP fibers [23]. The strength reduction rate is also plotted to show the strength tendency with increase of PP fibers. It could be observed that the compressive strength of NCB-filled cementitious composite decreased with the increase of PP fibers, from 54.6 MPa for the composite without PP fibers to 51.4 MPa for the composite filled with 0.4 wt.% PP fibers. The slight reduction on the compressive strength could be explained by the lower elastic modulus of PP fibers, in comparison to that of hardened cement matrix. The addition of PP fibers decreased the elastic modulus of whole cementitious composites. In addition, cement hydration also made a difference on the mechanical

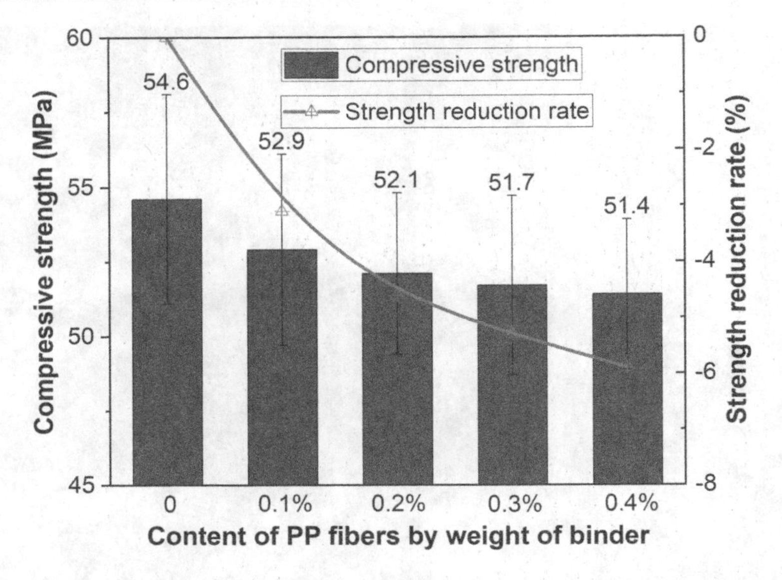

Figure 5.12 Compressive strength of NCB/cementitious composite with various contents of PP fiber [23].

properties, which will be discussed in detail in the next section. Particularly, 3% strength reduction occurred for the composite filled with 0.1 wt.% PP fibers compared to that of only NCB-filled cementitious composite, while another 3% strength reduction occurred until the composite filled with another increment of 0.3 wt.% PP fibers. It indicated that the strength reduction rate gradually slowed down with the increase of PP fibers.

Fig. 5.13 presents the flexural strength of the NCB-filled composite reinforced with various contents of PP fibers. For the composite without PP fibers, the flexural strength

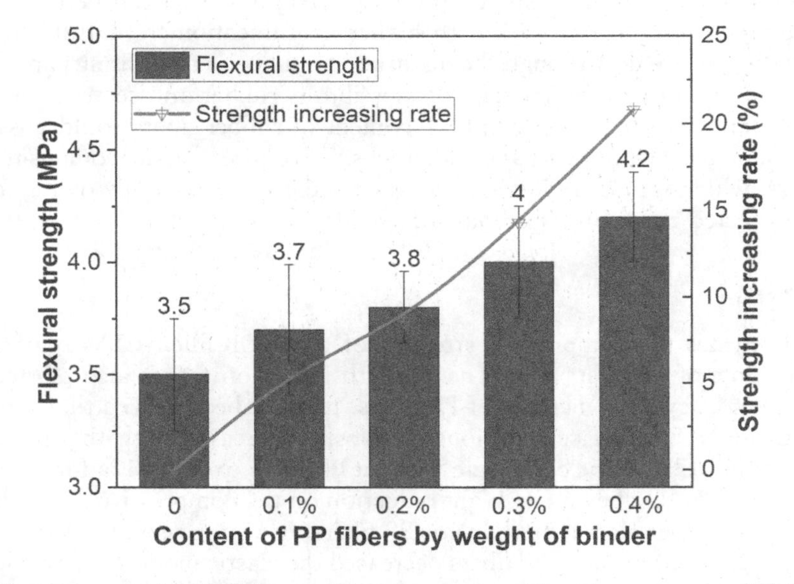

Figure 5.13 Flexural strength of NCB/cementitious composite with various contents of PP fiber [23].

reached 3.5 MPa. However, the strength significantly increased to 4.2 MPa when the added PP fibers reached 0.4 wt.%, indicating that the flexural strength of composite increased with the increase of PP fibers. In addition, the results show that the strength-increasing rate was almost linear to the content of PP fibers. The improved flexural strength was mainly due to the excellent tensile properties of PP fibers in the composite, which could release the concentrated stress and limit the generation of cracks. Moreover, the fibers crossing the pores and cracks had the capacity to prohibit their further propagation and improve the deformability of the cementitious composite.

As the compressive strength of composite decreased at a decreasing reduction rate with the increase of PP fibers, Fig. 5.14 shows the effect of NCB on the cement hydration in 72 h, including the rate of hydration heat and the accumulative heat of hydration [23]. In this study, the concentration of NCB in the composite is 0.5 wt.% to the weight of binder. Hence, the hydration test involving the composite filled with 0%, 0.5 wt.%, and 1.0 wt.% NCB can successfully elucidate the effect of NCB in the cementitious composites. Generally, it could be observed that the second peak gradually emerged with the increase of NCB, which indicated the delaying effect of NCB nanoparticles on the cement hydration acceleration stage. One reason is the dilution effect of NCB in the composite, which reduced the concentration of compositions involved in the cement hydration procedures [24]. Another reason is probably due to the high surface energy of NCB nanoparticles, which can absorb and enclose the water particles to hinder their contact to cement [25]. In addition, the second peak of composite even higher than the first peak representing the hydration was greatly improved at the acceleration stage, as confirmed by the cumulative heat of hydration in Fig. 5.14b. Moreover, it was found that both the first and second peaks gradually shifted to left, which means the decreased duration to reach the hydration peak with the increase of NCB content. Overall, for the cement hydration alone, it can be deduced that the NCB nanoparticles have the capacity to stimulate the cement hydration process, which are beneficial to the mechanical properties of cementitious composite.

Fig. 5.15 illustrates the microstructures of NCB-filled cementitious composite, and the distributions of NCB nanoparticles in the cement matrix and the PP fibers. Because of the nanosized properties of NCB and the filling effect, it could be observed that the NCB nanoparticles can fill some capillary pores and cracks in the cement matrix. This is consistent to the proposal that the mechanical properties of NCB-filled cementitious composite are slightly better than the plain cementitious composite when the NCB dosage is small [26]. On the other hand, NCB nanoparticles having high surface energy were easily absorbed with each other and form agglomerations, as illustrated in Fig. 5.15b. Since the NCB agglomerations had no resistance to external forces, the cementitious composite filled with worse-dispersed NCB nanoparticles would greatly decrease their mechanical performances. Overall, the comprehensive influences by NCB nanoparticles are determined by the combined effects of the cement hydration, NCB-filling effect, and agglomeration. The addition of PP fibers can decrease the NCB concentrations in the cement matrix and reduce the agglomerations, which might be another reason responsible for the improved flexural strength. As shown in Fig. 5.15c and d, the microstructural morphology is slightly different for the NCB nanoparticles on the PP fibers from those in the cement matrix. It could be seen that the NCB and the cement hydration products were integrated well on the surface of PP fibers, from which it could be deduced that the PP fibers actually had good cohesion with cement matrix in the cementitious composite, hence the flexural strength of the composite could be remarkably improved.

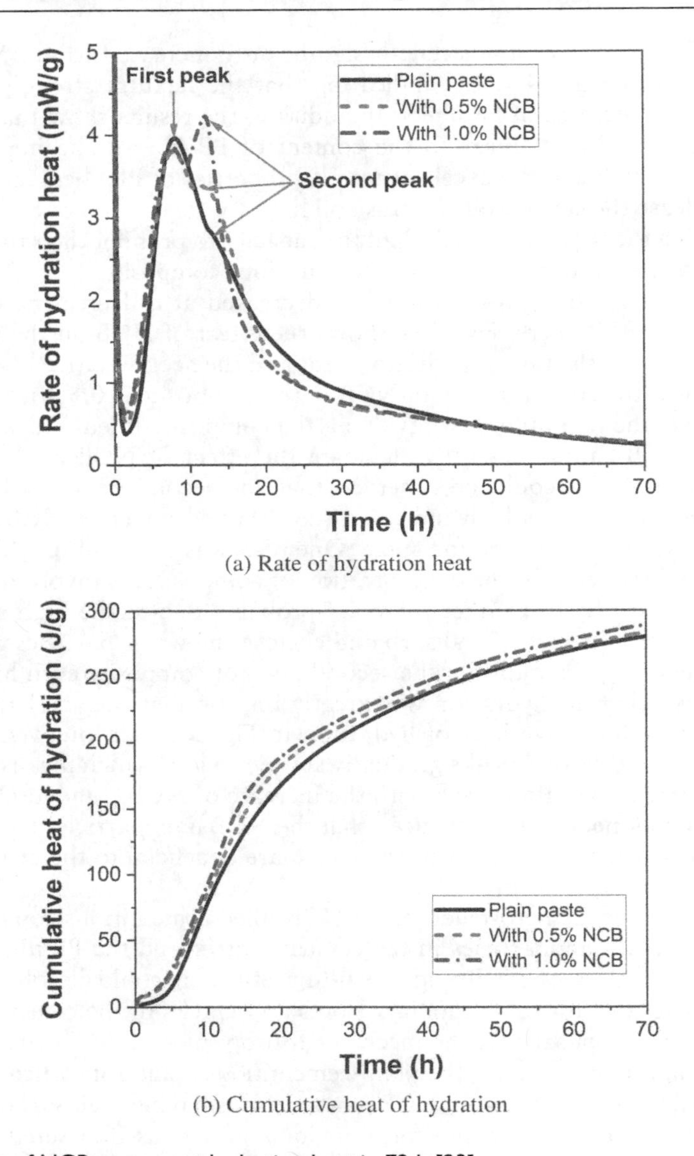

(a) Rate of hydration heat

(b) Cumulative heat of hydration

Figure 5.14 Effect of NCB on cement hydration heat in 72 h [23].

5.2.4.4 Conductive waste glass

Fig. 5.16 illustrates the compressive strength of cement mortar modified with different contents of CNTs-coated waste glass after 28 days curing [27]. The compressive strength of plain cement mortar reached 40.5 MPa. It was found that the compressive strength first increased and then decreased with the growth amount of waste glass. The highest strength was obtained for the cement mortar modified with 50% waste glass, reaching approximately 46.2 MPa. Similarly, the cement mortar with 25% waste glass had a significant enhancement compared to the plain cement mortar and reached a slightly lower value of 45.8 MPa. The mortar with 100% waste glass only showed slightly higher

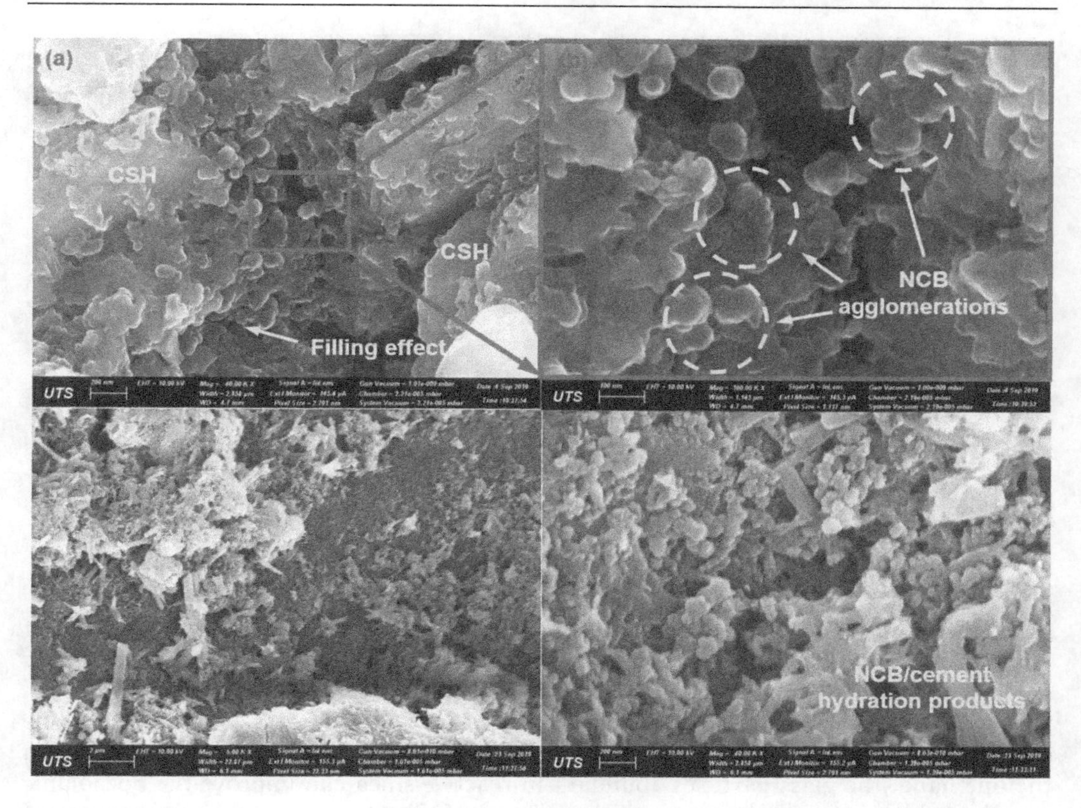

Figure 5.15 Microstructural morphology of NCB in the cement matrix and on the surface of PP fiber.

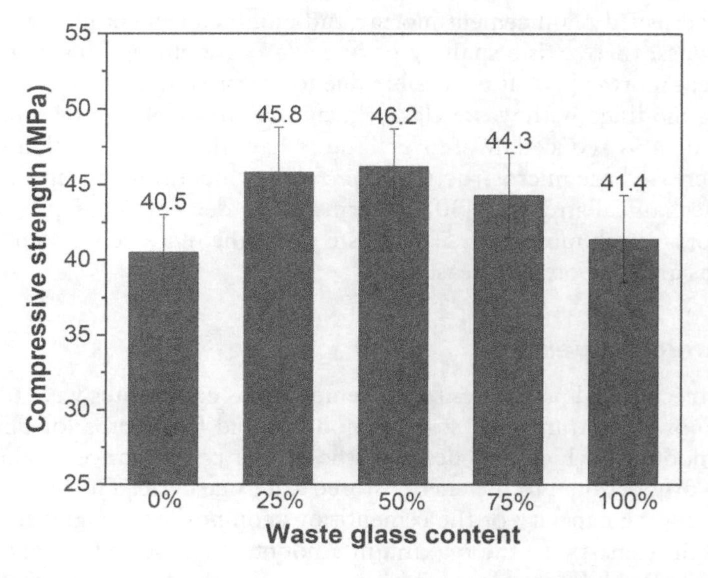

Figure 5.16 Compressive strength of modified cement mortar with multiple contents of CNTs-coated waste glass [27].

Figure 5.17 Optical images of plain cement mortar and modified mortar modified with CNT-coated waste glass [27].

compressive strength than that of plain cement mortar. Overall, with the contents of 25%, 50%, 75%, and 100% waste glass, the increasing rates of compressive strength reached 13.1%, 14.1%, 9.4%, and 2.2%, respectively.

Several potential reasons responsible for the increased compressive strength can be inferred based on the previous studies and the following macrostructure observations. On the one hand, the glass particles abundant in reactive silica can improve the pozzolanic reaction which is critical to strength development [28]. On the other hand, because of the irregular shape and sharp corners of crushed waste glass, it possesses better cohesion to cement matrix compared to that of natural sand [29]. As shown in Fig. 5.17a–c, for the identically crushed plain cement mortar and modified cement mortar with 50% and 100% waste glass, there exists spalling of fine aggregate during crushing, especially for the plain cement mortar [27]. It is possibly due to the broken fine sand. However, for the cement mortar modified with waste glass, the number of exfoliated or broken fine aggregate was considerably reduced. Moreover, studies have demonstrated that the addition of waste glass increased the microstructure fineness of cementitious composites, which also benefits the mechanical strengths [30]. In terms of the decreased compressive strength of the cement mortar with more than 50% waste glass, the increased amount of CNTs and agglomerations are the potential reasons.

5.2.5 Dynamic properties

The dynamic mechanical properties of the cementitious composites were usually assessed by the two following parameters: storage modulus and loss factor (or damping factor). The storage modulus, which can describe the elastic performance of the cementitious composites, is proportional to the energy stored after one cycle. The loss factor can characterize the damping capacity of the cementitious composites recognized as the ratio of the dissipated heat energy to the maximum amount of the stored energy [31]. Fig. 5.18 shows the storage modulus and loss factor versus temperature of cementitious composites with different GOs content in different curing periods [32]. The loss factor and storage

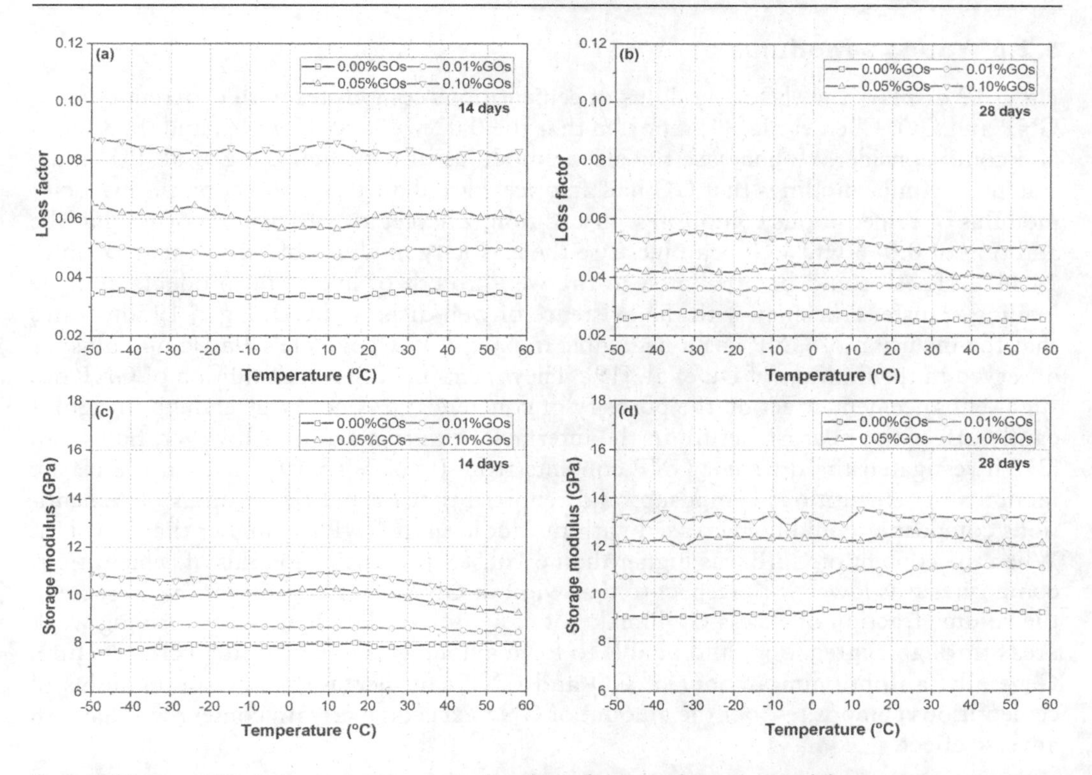

Figure 5.18 Dynamic mechanical analyzer results of loss factor and storage modulus versus temperature: (a) and (b) loss factor in 14 and 28 days, respectively; (c) and (d) storage modulus in 14 and 28 days, respectively [32].

modulus of the cementitious composites increased with increasing GO content regardless of the temperature. The maximum value increased in the storage modulus of 52.3% was examined with the added amount of 0.10 wt.% GO. The possible reason for why the addition of GO can enhance the damping capability is primarily related to the larger amount of increasing inner interfaces and uneven stress distributions under an outside force in the cementitious composites. In addition, Long et al. [33] also illustrated that the loss factor of cementitious composites also increased with the addition of GO. The maximum value increase of loss factor was 125% in contrast to the cementitious composites without GO content. They stated that the addition of GO had a positive effect on the loss factor of cementitious composites. As mentioned above, GO additions have a positive effect on the dynamic mechanical properties of cementitious composites. The primary reason for the increased loss factor of the cementitious composites could be related to the increasing internal friction and porosity of the cementitious composites. The increasing of the storage modulus with the addition of GO is significantly related to the differences between the elastic modulus of GO and the cementitious composites; it is also related to the chemical interactions between GO and the hydration products. However, there is no related report about the dynamic mechanical properties of cementitious composites with GNP. Further research can provide scope for validating the effects on the dynamic mechanical properties of GNP-filled CBSs.

5.2.6 Young's modulus

Fig. 5.19 displays the elastic modulus of cementitious composites with different content GNP and GO. Chen et al. [34] reported that the 0.02 wt.%, 0.05 wt.%, and 0.08 wt.% GO could considerably increase the elastic modulus of concrete. Long et al. [35] illustrated the similar findings that GO had a noticeable enhancement effect on the elasticity modulus of cementitious composites. In addition, Zhao et al. [36] also reported that the addition of 0.066 wt.% GO can increase the elasticity modulus of cement matrix up to 24.14%. These could be attributed to the well-growth of hydration products and the loading-transfer efficiency with the existence of well-dispersed GO. Fig. 5.19 illustrates that the inclusion of GNP has a negligible impact on the concrete's elastic modulus, as observed in the findings of Du et al. [19]. They mentioned that the addition of GNP has slight enhancement effect on the porosity of concrete and is not large enough to significantly affect the microstructure in the interfacial transition zone. However, Sun et al. [37] investigated the different GNP contents from 0 vol.% to 10 vol.% on the elastic modulus of cementitious composites and reported that the elastic modulus of cementitious composites could be increased with the addition of GNP not higher than 2 vol.%. When the content of GNP was higher than 6 vol.%, the elastic modulus of cementitious composites was lower with respect to that without GNP. They stated that the increase in the volume fraction of excess GNP makes it easier for GNP to aggregate, forming weak areas through aggregation, and unable to form a homogeneous hydrate microstructure. Generally, an optimum amount of GO and GNP can increase the elastic modulus of cementitious composites, but the amount of GNP exceeding certain content will have an adverse effect.

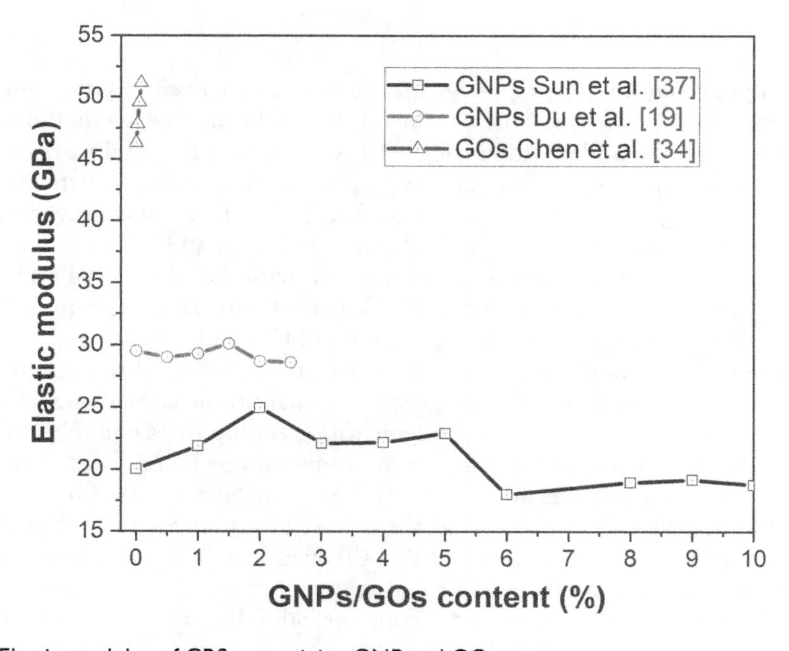

Figure 5.19 Elastic modulus of CBSs containing GNP and GO.

5.3 MECHANICAL SENSING PERFORMANCE

5.3.1 Typical compressive sensing

Fig. 5.20a–c depicts the piezoresistivity of PP fibers–reinforced CBSs filled with different dosages of NCB in both undried and dried states under the cyclic compression of 10 MPa [38]. Four loading–unloading compressive cycles and their synchronous response of FCR are plotted. Overall, both moisture and dry specimens exhibited characteristic piezoresistivity, with decreased resistivity during loading and increased resistivity during unloading. It is observed that the moisture specimens experience a degree of polarization effect, which is confirmed by a detectable time drift of FCR toward positive value. It is caused by the movement of dissolved ions in pore solution, which induce an opposite electrical field toward the applied DC electrical field [39]. Although moisture specimens are more conductive than dried specimens, more sensitive piezoresistivity toward stress with enlarged FCR values are observed for all groups after drying treatment, which is consistent with previous studies and can be partially ascribed to the increased concentration and conductivity of pore solution after drying treatment [40, 41]. In addition, it is observed that

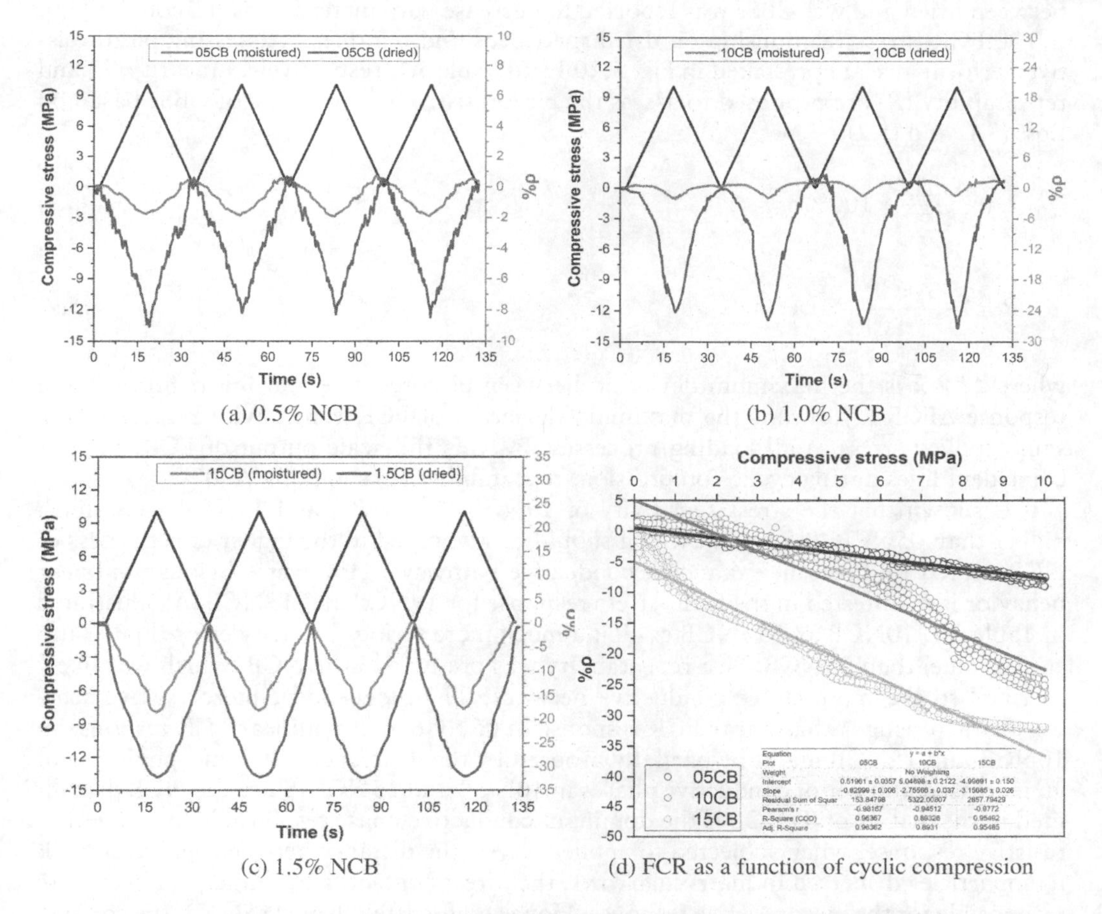

Figure 5.20 Piezoresistive responses of PP fibers–filled CBSs with different dosages of NCB under cyclic compression [38].

Table 5.1 Piezoresistivity of NCB/PP fibers–filled CBSs under cyclic compression [38]

	Group	Parameters of piezoresistive performance		
		Stress sensitivity (%/MPa)	*Linearity (%)*	*Repeatability (%)*
05NCB	Specimen-1	0.83	13.8	79.1
	Specimen-2	1.06	7.8	87.9
	Specimen-3	0.69	9.6	85.4
10NCB	Specimen-1	2.76	33.1	68.1
	Specimen-2	1.93	36.4	66.4
	Specimen-3	2.42	27.2	75.8
15NCB	Specimen-1	3.16	19.3	90.9
	Specimen-2	2.87	16.9	92.1
	Specimen-3	3.44	21.4	88.7

15NCB (1.5 wt.%) presents less variability in maximum FCR between the moisture and dried specimen, which is consistent with the previous study that the disparity of FCR between dried and wet CBSs was reported to decrease with increasing CNT content [42].

The FCR–stress relationship on dried specimens and parameters assessing piezoresistive performance are presented in Fig. 5.20d and Table 5.1, respectively. Linearity (L) and repeatability (R) are proposed to assess the piezoresistive performance of CBSs based on Eqs. (5.1) and (5.2):

$$L = \frac{\Delta \max}{\mathrm{FCR}_f} \times 100\% \tag{5.1}$$

$$R = 1 - \frac{\Delta FCR_{\max-i}}{\mathrm{FCR}_f} \times 100\% \tag{5.2}$$

where ΔMax is the maximum deviation between piezoresistive and linear fitting lines; response $\Delta\mathrm{FCR}_{\max-i}$ means the maximum deviation of FCR values with respect to the same applied stress in all loading processes. FCR_f is full-scale output of FCR. For the coincident lines during cyclic compression, repeatability (L) should be 100%.

It is shown that the stress sensitivity of 10NCB (1.0 wt.%) and 15NCB were much higher than 05NCB (0.5 wt.%), which should be attributed to the higher effectiveness of NCB-coated PP fibers in enhancing conductive pathways. However, obvious nonlinear behavior is manifested in the FCR–stress response for 10NCB and 15NCB. As illustrated in Table 5.1, 10NCB and 15NCB exhibit a much more serious linearity error (higher linearity value) than 05NCB. Best repeatability was presented in 15NCB, which was likely ascribed to the more stable conductive network. The independent piezoresistive sensing characteristic is that linear FCR response in 05NCB and nonlinear FCR response in 10NCB and 15NCB might be partially aroused by the difference in the morphologies of an increased number of conductive phases in 10NCB and 15NCB. Since the NCB distributed in the matrix of 05NCB is the dominant conductive phase responsible for the piezoresistive response, when subjected to applied stress, the distance between spherical NCB nanoparticles dispersed in matrix narrows, the direct contact, and tunneling effect were responsible for the piezoresistive response. However, for 10NCB and 15NCB, the conductive phases extended from the NCB distributed in matrix to both the NCB distributed in matrix and the conductive PP fibers effectively coated by NCB. Not only were the direct contact and tunneling conduction between fiber and NCB particles and between fiber and fiber responsible for piezoresistive response, but also the slight pull-in and pullout of NCB effectively coated fibers during loading–unloading significantly altered the fiber–matrix contact resistance and contributed to the piezoresistive response [43]. In addition, the possibility of the NCB agglomeration increases with the NCB content, which further complicates the piezoresistive mechanism. Hence, it is inferred that 10NCB and 15NCB are accompanied by more complicated piezoresistive mechanisms due to the more complex conductive phases.

5.3.2 Typical flexural sensing

The flexural stress, strain, and FCR–time curves for the PP fiber–reinforced CBSs with different amounts of NCB particles are presented in Fig. 5.21 [38]. Under the notched bending tests, it is observed that all specimens fail to demonstrate any nonlinear ascending and stationary stages before peak stress, or nonlinear descending stage after peak (black curves). Flexural stress abruptly drops to almost zero and a sudden increment in resistivity is observed once reaching the initial cracking strength. The specimens suddenly failed with a swift formation of a single macrocrack emanating from the notch tip, showing a brittle failure pattern. Hence, 9 mm PP fibers fail to achieve toughening enhancement and post-crack resisting effect on NCB-filled cementitious composites under flexure. The brittle failure should be attributed to the brittleness of the cement paste and the ineffective bonding between fibers and surrounding cement matrix so that no progressive pullout of fibers is achieved. The ineffective bonding is sourced from the hydrophobic nature of either PP fibers or NCB nanoparticles absorbed on PP fibers. Since the specimens suddenly failed with the initiation of a single macrocrack, FCR and CMOD are only valid up to the maximum stress. The FCR is evidently altered with the increase of stress and CMOD prior to the sudden brittle failure. However, both positive FCR increasing and negative FCR decreasing trends can be captured from parallel specimens within the same group. This is reasonable because the specific electrode configuration adopted in this study measured the entire cross section of the beam, not the regional compressive zone or tensile zone as in some previous research [44]. For the compressive region above the neutral axis, the micro-voids are compressed, and conductive phases tend to be closer and contact, leading to decreased FCR; contrarily, the microcracks and micro-voids are gradually induced and conductive phases become disconnected in the tensile zone below the neutral axis, leading to increased FCR. The overall FCR variation trend should be determined by conductive phases in which the zone is dominated. The dominance of either compressive or tensile zone is correlated to the distribution variance of conductive phases, including NCB nanoparticles and conductive PP fibers with effective coating efficiency between compressive and tensile zone. In addition, the NCB agglomeration states and the orientation of conductive fibers also raise uncertainties and affect the dominance of the region in the beam, so further quantitative research in this regard is required [45, 46].

Fig. 5.22a–c plots the comparative curves of FCR as a function of flexural stress and CMOD up to maximum stress, and both the parallel specimens and their averaged data are plotted. For the sake of assessing stress-sensing ability despite the specimens showing increased or decreased FCR trends, all data scatters pertain to specimens with decreased variation trends are symmetrically plotted about the horizontal axis, through which the decreased FCR and the increased FCR trends uniformly transformed to the

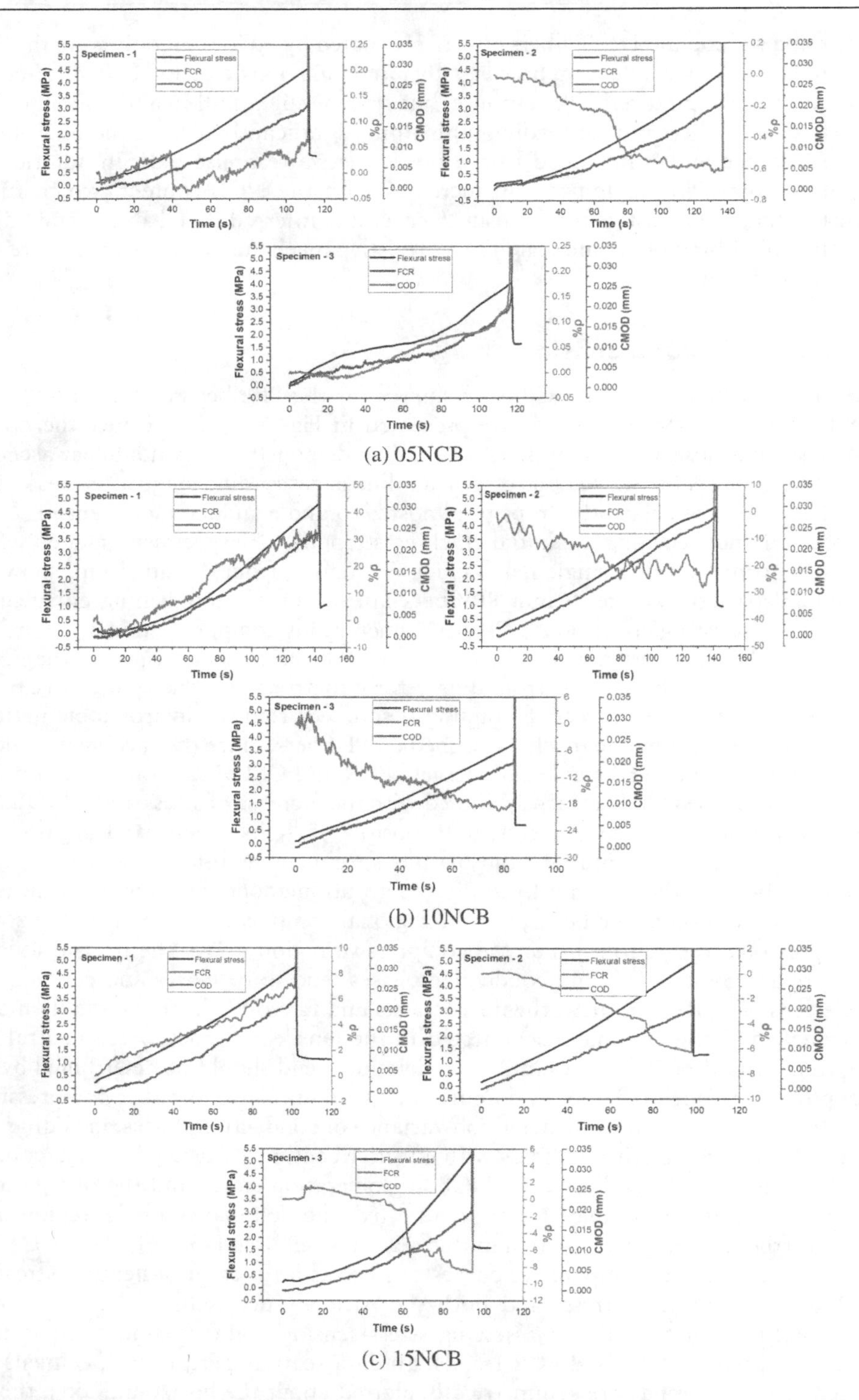

Figure 5.21 Relationship between flexural stress, CMOD, and FCR–time curves of 05NCB, 10NCB, and 15NCB [38].

same FCR trends. It is noted that the FCR magnitude differs very significantly among different groups. Very limited FCR irregularly altered with the increased flexural stress, proving that 05NCB fails to sense the flexural stress. Differently, 10NCB shows the highest FCR values with the maximum FCR value reaching approximately 30%, while a nonlinear FCR response to some extent is observed. In comparison to 10NCB, a strong linear correlation is observed for 15NCB, although the maximum FCR value has been compensated to approximately 8%.

The effective sensing capacity to sense the flexural stress and CMOD in the pre-cracking stage for NCB/PP fibers-filled CBSs, especially for 10NCB and 15NCB, is very promising. The ineffective sensing property was often manifested in the pre-crack stage for CBSs under flexure or tension, although sensitive FCR behaviors were shown in post-crack stage [47]. For example, the maximum fractional change in impedance (FCI) was negligible for all testing groups of self-sensing concrete containing carbon black, steel fiber, and carbon fiber either with single-crack or multiple-cracks characteristic under the notched bending test, while the FCI steeply ascended to the value comparative to the present study after first cracking. Besides, a maximum FCR of only 1.43% in the pre-peak stage was detected for reactive powder concrete (RPC) typed CBSs under flexure [48]. Likewise, Teomete [49] and Demircilioğlu et al. [50] investigated the self-sensing behaviors of steel fiber–reinforced concrete and brass fiber–reinforced concrete under the notched bending test, respectively; similar results that the FCR oscillated with negligible positive and negative values before the crack initiation were attained. In comparison to damage sensing in the post-crack stage, the sensing capacity in the pre-crack stage should be more valuable since it provides the early warning of material failure.

Based on linear and nonlinear characteristics, the averaged value of 15NCB and 10NCB was fitted by linear equation and exponential equation, respectively, to simulate the FCR–flexural stress and FCR–CMOD relationships. The fitting functions are given in Fig. 5.22. It is worth noting that although the FCR–flexural stress and FCR–CMOD relationships for the identical group are graphically similar, there is a disparity to some extent between them. The disparity is sourced from the presence of the nonlinear pre-crack stage in flexural stress–CMOD response, which follows the initial linear elastic stage. The nonlinear pre-crack stage can be observed from Fig. 5.23, and 15% deviation from linearity in the flexural stress–CMOD curve is referred as the onset of the nonlinear pre-peak stage [51]. In this stage, the nanocracks are formed and start to coalesce into microcracks. For 10NCB and 15NCB, the considerably large CMOD sensitivity (defined as the fractional change in resistivity per unit crack mouth opening displacement) can be attained by the small value of CMOD in the pre-crack stage, which is highly preferable for self-sensing cementitious sensors.

The representative strain maps measured from DIC for all specimens are shown in Fig. 5.24. As depicted in Fig. 5.24a, the limited strain localization surrounding the notch tip can only be captured in the very late stage of nonlinear prepack stage ahead of the brittle failure. The nonlinear pre-peak stage is relatively short. In consistent with the brittle failure, the cracking process, including strain localization, microcrack initiation, and macrocrack formation, showed a very swift manner for all specimens under flexure. The formation of microcrack (strain localization) is not evident until the short nonlinear stage ahead of the failure. Since the formation of microcracks is often associated with a sudden or an apparent change in slope of the FCR curve [52], the relatively long elastic linear stage and short nonlinear stage ahead of the brittle failure endowed 10NCB and 15NCB with a smooth/linear FCR response in conjunction with sensitive FCR values

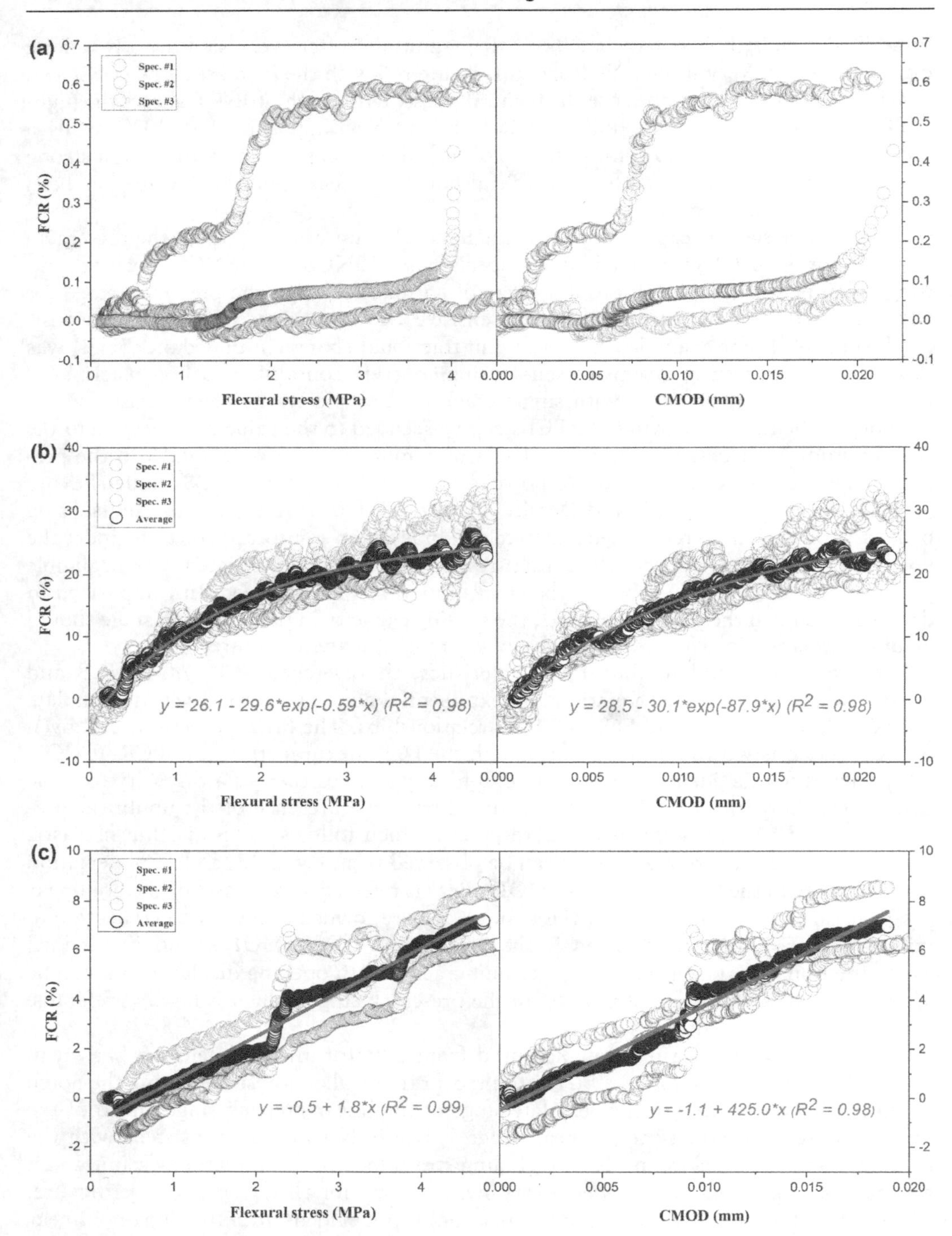

Figure 5.22 FCR-flexural stress and FCR-CMOD correlations in pre-peak stage for (a) 05NCB, (b) 10NCB, and (c) 15NCB [38].

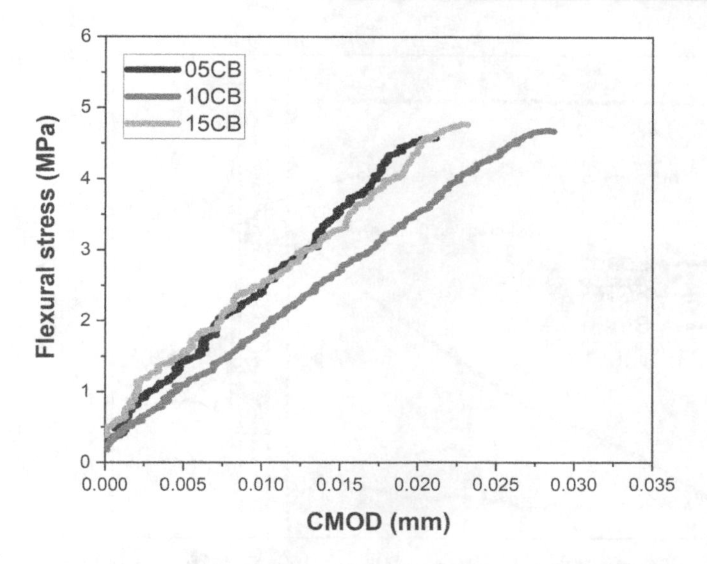

Figure 5.23 Representative flexural stress–CMOD curve in the pre-peak stage [38].

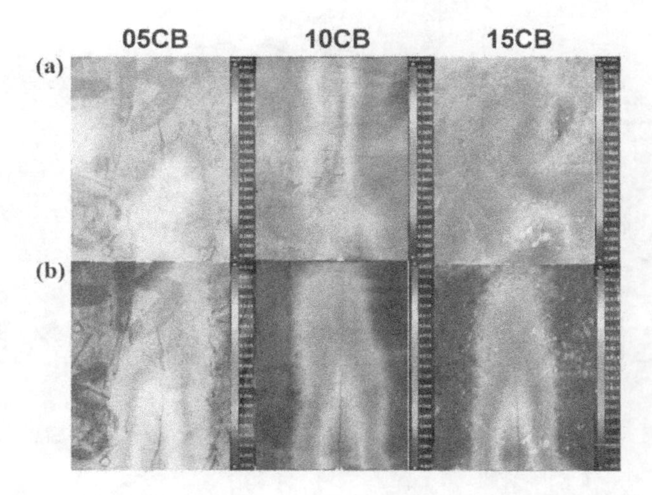

Figure 5.24 Horizontal strain E_{xx} at (a) nonlinear pre-peak stage before sudden failure and (b) after sudden failure [38].

across the nearly whole pre-peak stage. As illustrated in Fig. 5.24b, a nearly upright macrocrack emanating from the notch tip swiftly propagate toward top side after peak stress so that the cross section that the electrons transit greatly decreases, resulting in a sudden material failure and an increase in the FCR.

5.3.3 Typical split tensile sensing

The tensile load, tensile strain, and FCR curves for 05NCB, 10NCB, and 15NCB are compared with the reference value of time, as plotted in Fig. 5.25a–c [38]. Different stages

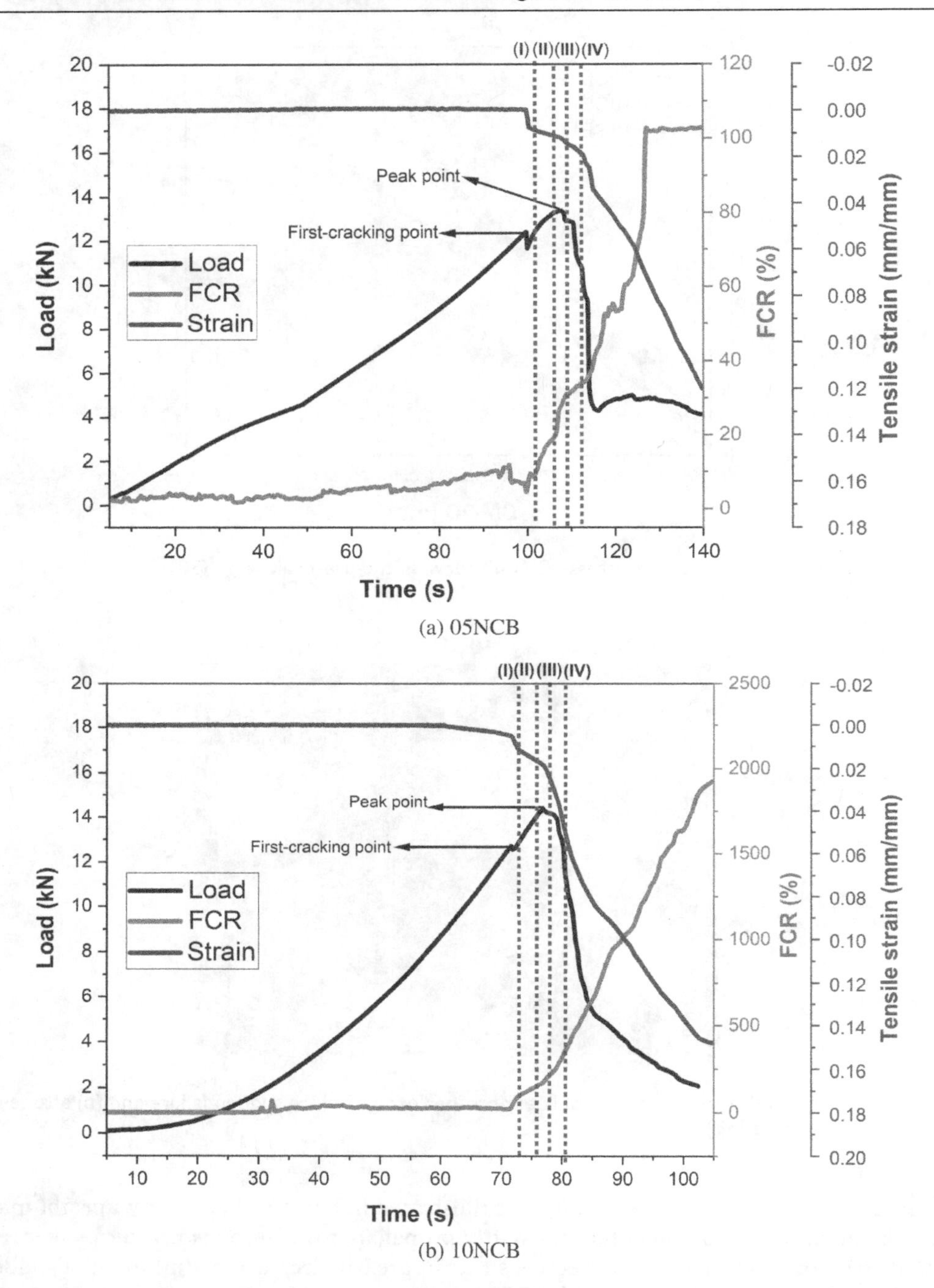

Figure 5.25 Relationship between the tensile load, tensile strain, and FCR–time curves for the CBSs containing 0.5%, 1.0%, and 1.5% NCB [38]. *(Continued)*

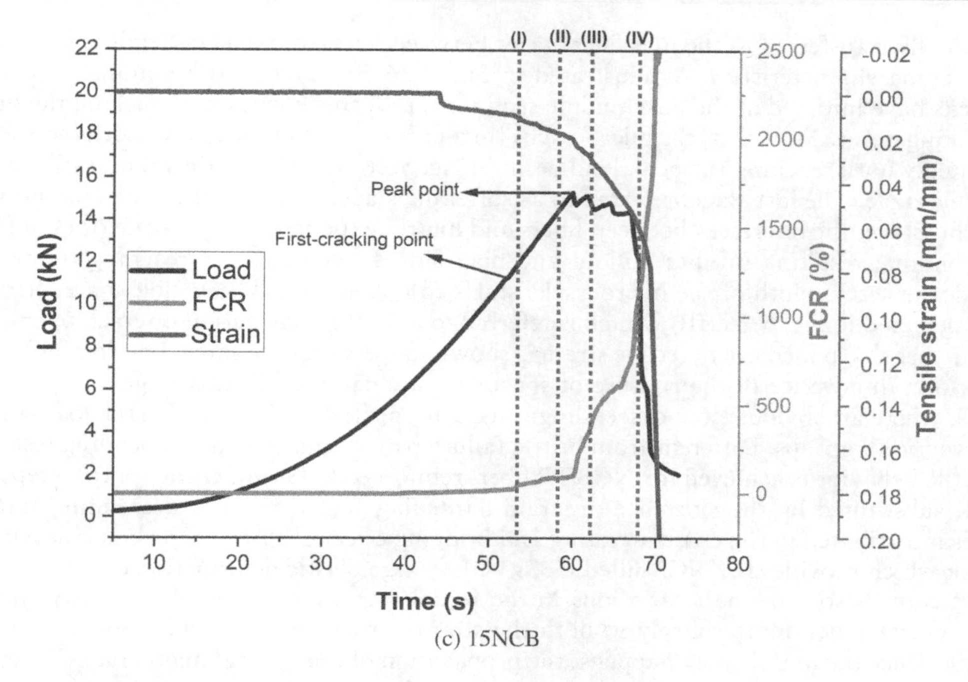

Figure 5.25 (Continued)

can be determined based on the degree of increasing steepness of the strain and their demarcation points corresponding to inflexion points of the increasing steepness. Before the initial crack forms, there is an initial stage which covers the linear elastic stage (from zero point to first cracking point). At this stage, tensile load grows sharply while the tensile strain increases very slightly. It is observed that the FCR of 05NCB has some small downward variations during the initial stage, which is likely attributed to the effect of local compressive strain developed at the interface of wooden strips and the specimen. As shown in Fig. 5.25a, there is an initial ascending of slope at the beginning of the load–time curve, which is attributed to the fact that the stiffness of wooden strips is much lower than the counterpart of specimens [53]. The applied stroke at the beginning of loading causes the wooden strips to deform first, thus the surface roughness of the wooden strips is gradually removed while the contact stiffness between strips and specimen gradually decreased. After the initial ascending of slope, the share of stroke taken by specimen significantly increased until the wooden strips are completely compacted. Since the fully compacted wooden strips were contained to use for the sequential tests for 10NCB and 15NCB, the initial increase of slope in the load–time curve is not observed in the tests of 10NCB and 15NCB. The FCR of 10NCB and 15NCB have increased to a small extent until the end of the initial stage, which are hardly read in Fig. 5.25b and c due to the large scale of the vertical axis. The poor self-sensing capacity manifested in the initial stage under splitting tensile test for the NCB/PP fiber–reinforced CBS is consistent with previous investigations on other types of CBSs [54].

The second stage is the strain-hardening stage from the first cracking point to the peak load point. The first crack occurs where tensile strain increases more steeply from this point. The tensile stress at the crack interface is mainly sustained by PP fibers and is gradually transferred to the interfacial faces between fibers and matrix, leading to strain hardening characteristics. As illustrated in Fig. 5.26 (I), vertical and nonlinear microcracks have initiated at the near-middle sections for all specimens once reaching the first cracking point. After that, the microcracks further proceed, but the crack width increases limitedly until reaching the peak load point in Fig. 5.26 (II). The strain value is still valid in this stage. The last stage, referred to as softening stage, begins at the peak load point. In this stage, the interfaces between fibers and matrix experience larger scale or even full debonding, resulting in fiber pullout and fiber slip. Therefore, once reaching the peak load, the crack width of the microcracks starts to increase quickly. Visible macrocracks are depicted in Fig. 5.26 (III), which is referred to as crack localization or crack opening. A further sharp increase in tensile strain is shown in the softening stage. Besides, multiple cracking in advance of the rupture of specimens are captured in this stage in Fig. 5.26 (IV), where an obvious slow-descending stage after peak load displayed in the load–time curve for all groups. Different from brittle failure pattern under notched bending tests, a ductile behavior is achieved for NCB/PP fiber–reinforced CBSs under the splitting tensile test, substituted by the existence of strain-hardening stage and slow-descending stage. This is attributed to the crack arresting and bridging effect of PP fibers on crack development which provides the NCB-filled CBSs with a large plastic deformation capacity.

In comparison to small variations in the initial stage, the electrical resistivity starts to increase much more sharply in both the strain-hardening stage and strain-softening stage. Once the initial crack happens, the propagation of the vertical microcracks reduces the cross section that allows electrons to transfer by damaging the electrical pathways, thus resistivity increases significantly. It is observed that the increased FCR of 05NCB

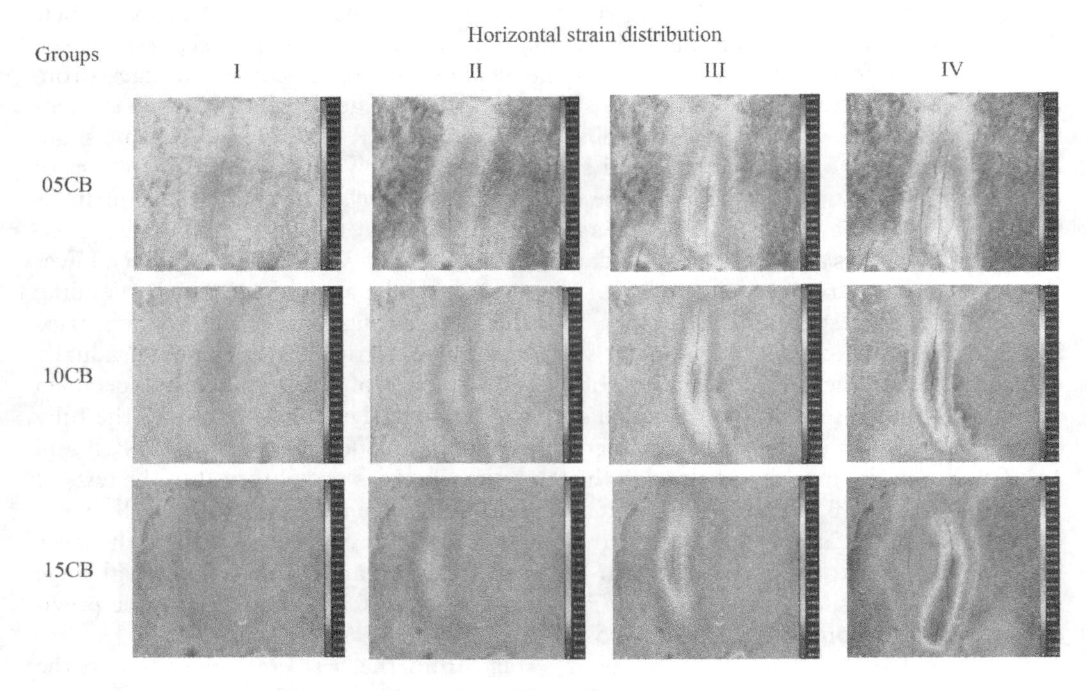

Figure 5.26 Horizontal strain E_{xx} (I) after initial cracking point, (II) before peak point, (III) in slow-ascending stage, and (IV) in rapid softening stage [38].

approaches 100% after failure, whereas this value exceeds the range of 2,000% for 10NCB and 15NCB. This is consistent with the results of microscopic characterization and static resistivity test that the PP fiber with ineffective coating efficiency is typically observed in 05NCB, while the PP fiber with effective NCB coating efficiency is commonly observed in 10NCB and 15NCB. It is revealed that a much greater number of continuous conductive passages exists inside 10NCB and 15NCB and the importance of NCB coating effectiveness on ameliorating the electrical conductivity of the composites. Fig. 5.27a–c plots the FCR–tensile strain relationships of all specimens in the strain-hardening stage where strain values are still valid. Interestingly, a strong linear correlation between FCR and strain in the strain-hardening stage is shown for all groups. The gauge factors in the strain hardening stage for different groups are summarized in Fig. 5.27d. The highest gauge factor of approximately 100 is presented in 10NCB, while the gauge factor of 05NCB is comparable to the counterpart of 15NCB, which is almost half of the 10NCB. The strong linear FCR–tensile strain correlation and the excellent gauge factor higher than 48 reveals the excellent self-damage sensing capacity of NCB/PP fiber–reinforced CBSs under the complex stress state of splitting tensile test.

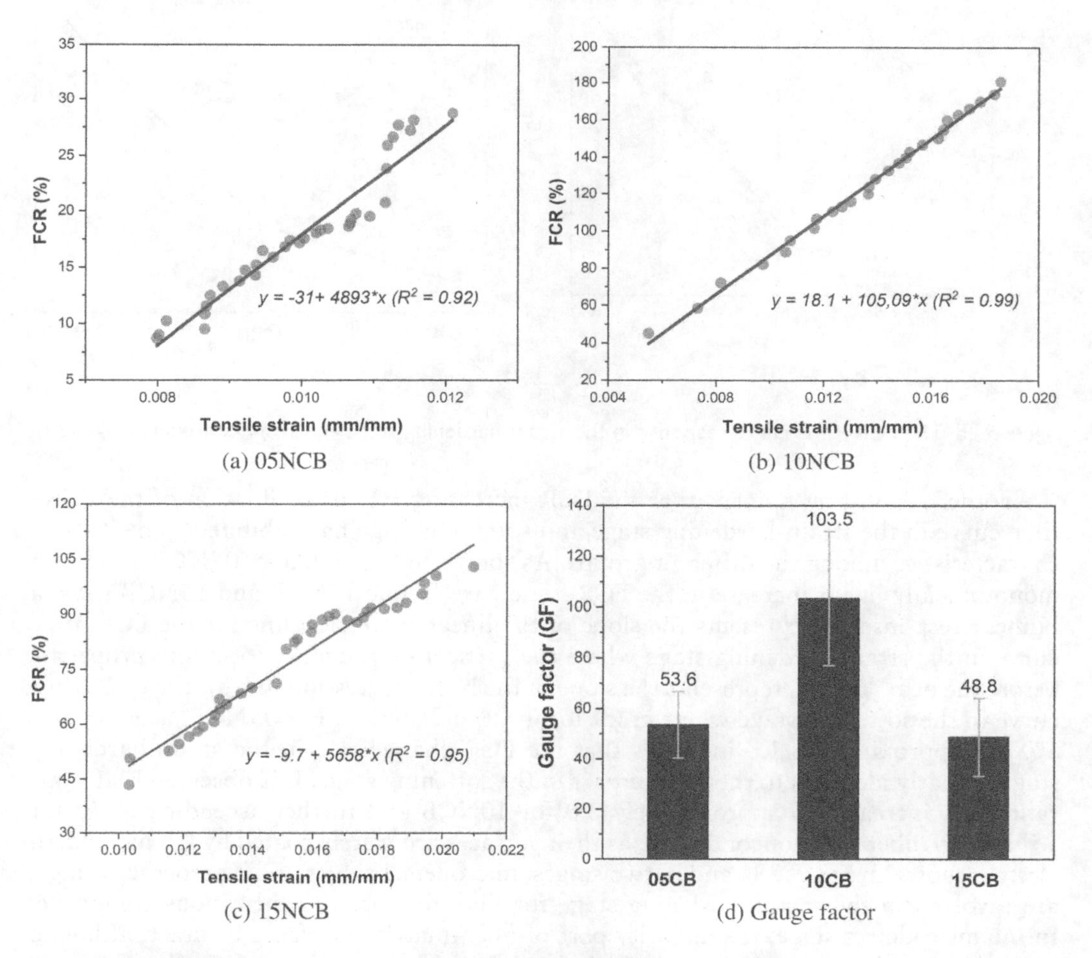

Figure 5.27 FCR–tensile strain correlation in the strain-hardening stage [38].

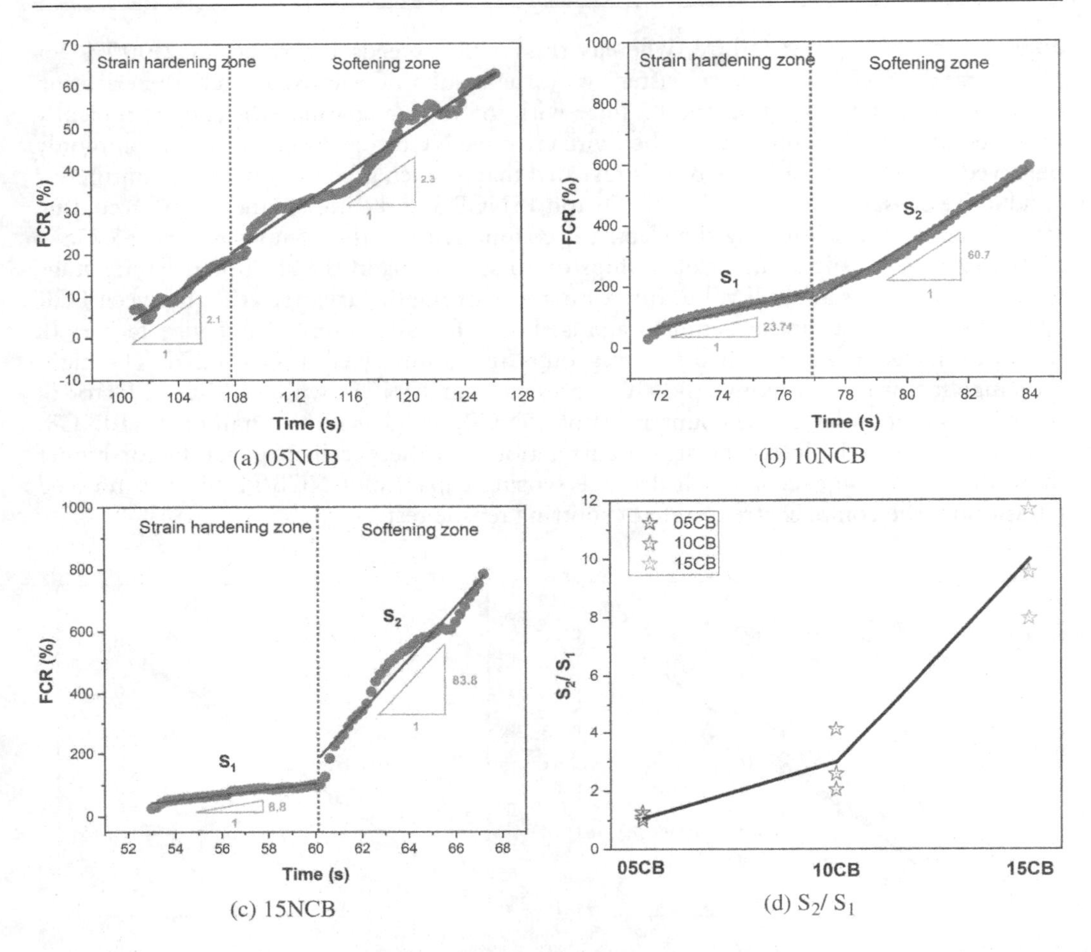

Figure 5.28 The FCR–time curve response in the strain-hardening stage and strain-softening stage [38].

Another interesting point is that the FCR increasing rate or inclination of the FCR–time curve in the strain-hardening stage and softening stage has exhibited some distinct characteristics among the different groups. As shown in Fig. 5.28a–c, 05NCB exhibits a monotonically linear increase in the FCR–time curve while 10NCB and 15NCB show a bilinear response. S_1 represents the slope of the linear regression line for the FCR–time curve in the strain-hardening stage where the vertical microcracks form and propagate before the peak load. S_2 represents the slope of the linear regression line for the FCR–time curve in the softening stage where crack localization happens. For 05NCB, a slope ratio S_2/S_1 of approximately 1.0 indicates that the FCR change rate in the strain-hardening stage is nearly identical to the counterpart in the softening stage. It is observed that slope ratio S_2/S_1 increases to approximately 3.0 for 10NCB and further ascending to 10 for 15NCB. PP fibers are nonconductive as their surfaces are merely coated by CB nanoparticles. As shown in Fig. 5.28a and b, two stages, microdefect stages and macrocracks stage, are involved in the strain-hardening stage for fiber-reinforced cementitious composite. In the microdefect stage, the majority part of fiber–matrix interfaces is intact, although multiple microdefects are initiated, which should correspond to the perfectly bonded state of fiber–matrix interfaces. Microdefects would extend into microcracks when approaching a saturated level where microcracks are initiated and propagate in the frontal process zone ahead of the crack tip. In the microcrack stage, the coalescence of multiple microdefects propagates into the microcracks, and the state of fiber–matrix interfaces converts to partially intact, with fiber–matrix debonding and matrix cracking approximately balanced [55]. In the strain-hardening stage, microsized PP fibers are capable of efficiently bridging and arresting the propagation of microcracks, thereby delaying the initiation of crack localization [56]. Since the PP fibers with ineffective NCB coating efficiency are nonconductive, the electrical current cannot flow through the fibers that bridge the macrocracks. Hence, in 05NCB, the electrons cannot pass through the microcracks area but only through weak conductive pathways composed by the distributed NCB in a noncracked cementitious matrix. Since the vertical macrocracks propagate swiftly with the cracking area expanding to the majority of the cross section of the specimen in the strain-hardening stage, as confirmed by the DIC results in Fig. 5.26 (II), the conductive pathways inside the 05NCB are severely broken immediately after the initial crack. As a result, the damage speed of conductive pathways in the strain-hardening stage where the microcracks and partial debonding occur between fiber–matrix interfaces is almost identical to the counterpart in the softening stage where large-scale and full debonding happens, consequently there is negligible difference in FCR increasing rate between the strain-hardening stage and the strain-softening stage for 05NCB and a slope ratio S_2/S_1 of approximately 1.0 is shown.

However, the major PP fibers in 10NCB and 15NCB are conductive due to the effective NCB coating efficiency. Therefore, in the strain-hardening stage, the bridging conductive fibers with the matrix–fiber interfaces states of almost intact or partially intact can still act as the conductive pathways for electrical current to flow through, as depicted in Fig. 5.29b, c, e, and f. The tensile softening is associated with the stage where large-scale and even full debonding of the matrix–fiber interfaces occur, with the bonding area and embedment length of the fibers being significantly decreased and shortened, respectively [57]. During the tensile-softening stage, characterized by the merging of microcracks and the initiation of localized macrocracks, the interfaces between the matrix and fibers undergo significant debonding, which can extend to a substantial or complete detachment. As a result, the conductive pathways are seriously damaged, as shown in Fig. 5.29h and i. This is why the damage speed of conductive pathways in the strain-hardening stage is slower than the counterpart in the softening stage. Thus, S_1 is smaller than S_2 for 10NCB and 15NCB. In comparison to 10NCB, more numbers of dispersed NCB nanoparticles and their small agglomerations existed in the fiber–matrix interface for 15NCB. As shown in Fig. 5.29f, the NCB particles and their small agglomerations broadly distributed in the fiber–matrix interface promote the more continuous and stronger conductive pathways for conductive fibers that are experiencing partial debonding by providing more contact points, which greatly compensate the damage degree and slow down the damage speed of the conductive pathways due to the debonding of the fiber–matrix interface.

5.3.4 Typical impact sensing

5.3.4.1 Impact regime

The impact load was generated by a self-made pendulum impact device. A solid metal ball of 80 g weight and 27 mm diameter was fastened by a wireline which was hooked

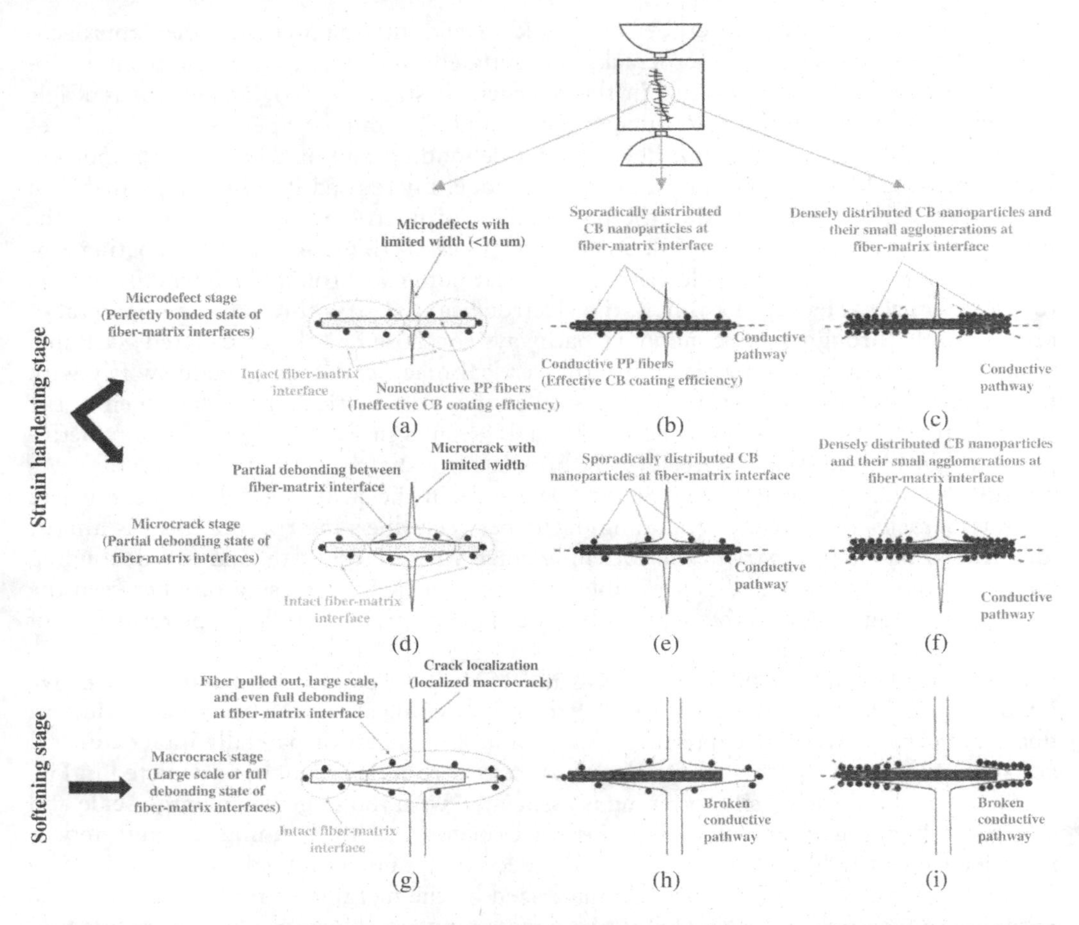

Figure 5.29 Damage mechanism on conductive pathways for (a) 05NCB (microdefect stage); (b) 10NCB (strain microdefect stage); (c) 15NCB (microdefect stage); (d) 05NCB (microcrack stage); (e) 10NCB (microcrack stage stage); (f) 15NCB (microcrack stage stage); (g) 05NCB (macrocrack stage); (h) 10NCB (macrocrack stage stage); and (i) 15NCB (macrocrack stage stage) [38].

on an angle scale plate. The CBS was fixed in the pedestal, and the right-end edge of the specimen was aligned with the wire when the ball reached the bottom. Therefore, the impact direction is perpendicular to the embedded electrodes to achieve the maximum variation of the electrical signal. Also, the altitude of the specimen was adjusted so that the solid ball impacted at the center of the right face of the cubic specimen. Different impact energy was generated by multiple pendulum degrees of 30°, 60°, and 90°, corresponding to the impact energy of 1.317×10^{-3} J/cm^3, 2.281×10^{-3} J/cm^3, and 2.634×10^{-3} J/cm^3, respectively. Each specimen was tested by two impact cycles which consisted of 60× impact in each cycle. Two replicates of cyclic compressive test containing three stress magnitudes were performed before/after each impact treatment to investigate the piezoresistive behaviors of 1% and 2% GNP-filled cement mortars (1GNPCM and 2GNPCM, respectively). The details of impact regimes and experimental procedures are illustrated in Fig. 5.30 [58].

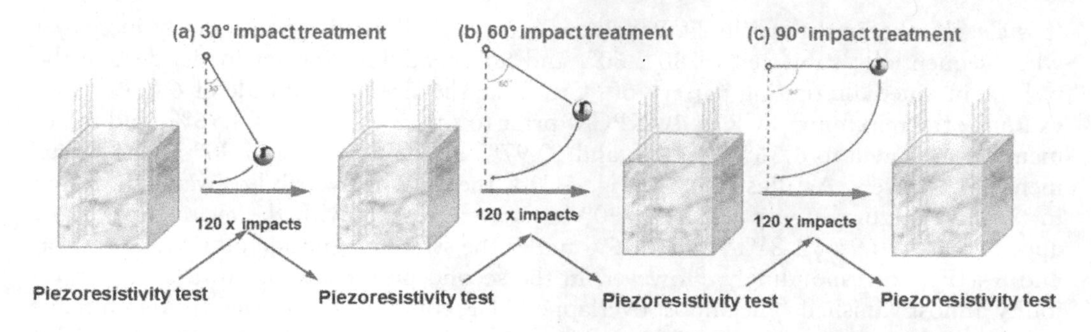

Figure 5.30 Schematic diagram and procedure of the impact load applied to GNP/CBSs [58].

5.3.4.2 Piezoresistive performance of 1GNPCM

Fig. 5.31a– d shows the piezoresistive results of 1GNPCM before and after 30°, 60°, and 90° impact loads, in which FCR(I) and FCR(II) are the results from the first and second measurements, respectively. The FCR changing patterns for 1GNPCM were similar before/after impact load. The resistivity smoothly decreased during loading and increased during unloading and exhibited an excellent signal-to-noise ratio with no perturbations.

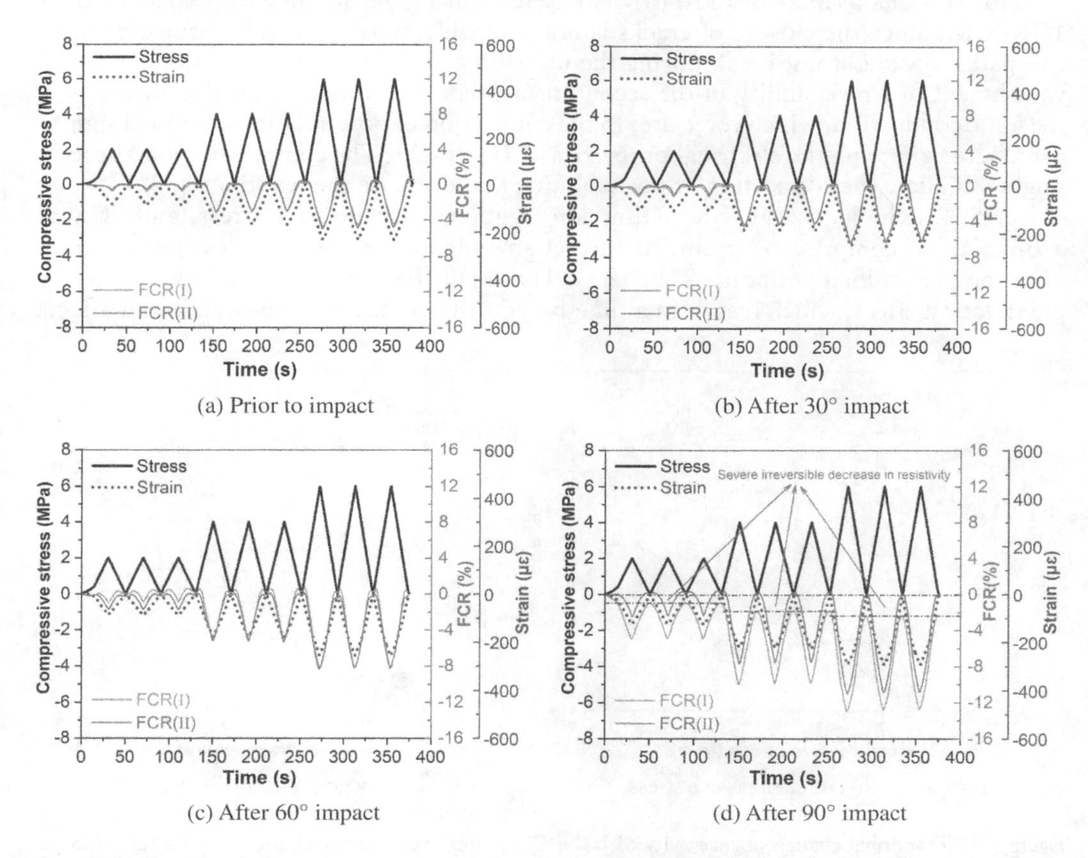

(a) Prior to impact
(b) After 30° impact
(c) After 60° impact
(d) After 90° impact

Figure 5.31 Fractional change of resistivity of 1GNPCM with strain/stress under cyclic compression [58].

It is clearly observed that the FCR value is incrementally and monotonously increased when sequentially subjected to 30°, 60°, and 90° pendulum impact loads, despite the well-maintained alternating pattern of FCR. Take the stress magnitude of 6 MPa as an example; the maximum FCR of 1GNPCM prior to the impact load is 4.96%, and incrementally ascended to 6.36%, 8.31%, and 10.97% after 30°, 60°, and 90° impact treatment, respectively. As illustrated in Fig. 5.31d, the FCR irreversibility was manifested in the first piezoresistivity test after 90° impact treatment, with the average values of approximately 0.9%, 1.81%, and 2.36% under the stress magnitude of 2 MPa, 4 MPa, and 6 MPa, correspondingly. However, in the second piezoresistivity test, the irreversibility almost vanished. The almost overlapped FCR contours are derived from two replicated measurements prior to the 90° impact treatment and the unaffected piezoresistivity by irreversibility after 90° impact treatment, suggesting the incrementally enlarged piezoresistivity stemming from the impact load is veritable and reproducible.

The irreversible decrease in resistivity might be related to the healing or closure of the cracks. Under compressive load, the removal of voids and weak matrix that separate the disconnected GNP would ameliorate the generation of more contacts points that form the conductive passages, thereby decreasing the resistance. The relatively severe irreversibility manifested after the 90° impact treatment implies a large quantity of newly generated cracks induced by the impact with the highest energy, which showed a good agreement with the continuous increase of resistivity baseline during the 60° and 90° impact load for 1GNPCM. Since the closure of cracks is not reversible, part of already formed conductive pathways might not break during the unloading, and this might be a reason for the vanishment of irreversibility in the second measurement. Moreover, as aforementioned, the ionic conduction, which correlates to the connection of pore solutions, is also a significant factor affecting the electrical properties of 1GNPCM. The compressive load process might stimulate the connection of pore solution and decrease the resistivity [59].

Fig. 5.32 exhibits the FCR as a function of applied compressive stress, and FCR as a function of compressive strain, to further investigate the piezoresistive performance after the impact load with different energy. The results have nullified the influence of the polarization effect, which results in a time-based drift of measured resistance and affects

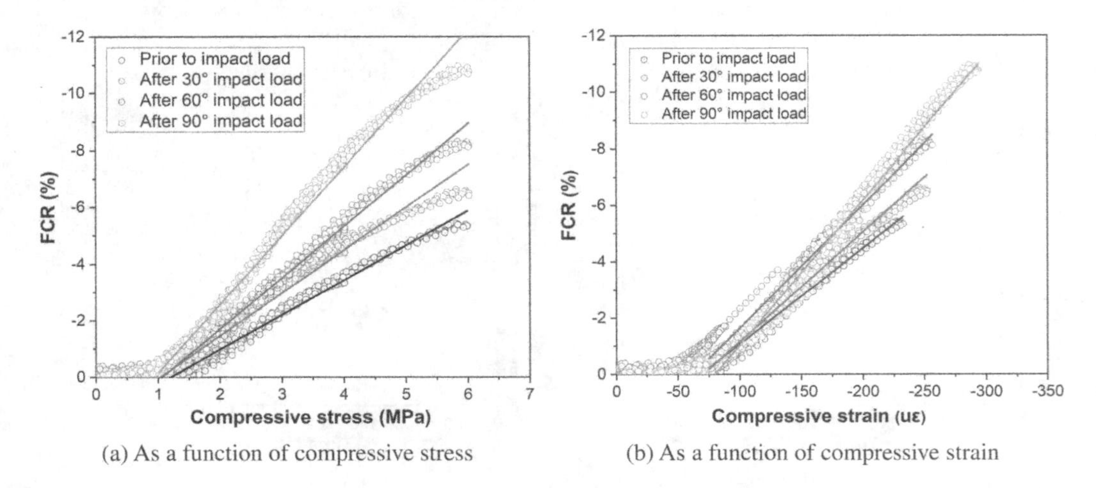

(a) As a function of compressive stress (b) As a function of compressive strain

Figure 5.32 Fractional change of resistivity of 1GNPCM under cyclic compression before/after different degrees of impact loads [58].

the precise characterization of piezoresistive behaviors. It is observed that the stress magnitude and loading rate have possessed negligible effect on the piezoresistive responses for the GNP-reinforced CBS since the piezoresistive behaviors under different stress magnitudes are almost identical. The small discreteness of whole scatter data points indicates excellent piezoresistive sensing stability and repeatability of 1GNPCM, which might be partially attributed to the excellent dispersion achieved by combined mechanical stirring and ultrasonication.

In addition to repeatability, the linearity and sensitivities are also the important parameters assessing the piezoresistive performance. The linearity is essentially correlated to the maximum offset between the piezoresistive data plots and their linear fitting curves, and can be calculated based on the following Eq. (5.3):

$$\text{Linearity} = \frac{\Delta\rho_{max}}{\Delta\rho_{fs}} \tag{5.3}$$

where $\Delta\rho_{max}$ is the maximum deviation between the scatters data points and their corresponding fitting curves and $\Delta\rho_{fs}$ is the full-scale output of FCR. The piezoresistive sensitivities, including stress sensitivity and strain sensitivity (also known as gauge factor), are essentially the slopes of the fitting curves between FCR and stress/strain. Table 5.2 summarizes the proposed parameters of the linear fitting curves for the linear portions of the relationship between FCR and stress/strain, excluding the static FCR varying sections under initial small stress/strain. Inconsistent with the increased maximum FCR values, the impact load could monotonously induce more sensitive piezoresistive behavior for 1GNPCM, confirmed by the incrementally increased sensitivities after impact loads. It seems that the increscent coefficient of the stress sensitivity is positively correlated to the impact energy, since the higher growth in stress sensitivity is achieved by the larger impact energy. However, the counterpart in strain sensitivity demonstrated is evidently compensated by the incremental growth of strain ability under equivalent stress after impact load, especially for a prominent ascending in strain value after 90° impact load. The conspicuous increase in strain after 90° impact load shows close agreement with the aforementioned severe irreversibility manifested after the 90° impact load. In addition,

Table 5.2 Evaluation of piezoresistivity of 1GNPCM after subjected to various impact loads [58]

	FCR as a function of stress			
Proposed parameters	*Prior to impacts*	*After 30° impacts*	*After 60° impacts*	*After 90° impacts*
Fitting equation	$y = -1.22x + 1.47$	$y = -1.51x + 1.57$	$y = -1.81x + 1.92$	$y = -2.41x + 2.26$
R^2	0.987	0.969	0.991	0.987
Stress sensitivity (%/MPa)	1.22	1.51	1.81	2.41
Linearity (%)	7.78	16.93	9.97	14.25
	FCR as a function of strain			
	Prior to impacts	*After 30° impacts*	*After 60° impacts*	*After 90° impacts*
Fitting equation	$y = 340.46x + 0.023$	$y = 388.87x + 0.027$	$y = 436.56x + 0.027$	$y = 513.93x + 0.041$
R^2	0.990	0.987	0.971	0.984
Strain sensitivity ($/\varepsilon$)	340.46	388.87	436.56	513.93
Linearity (%)	10.47	9.96	11.13	9.83

the results showed that impact loads seemed a failure to deteriorate the repeatability and linearity in both stress sensing and strain sensing, which is hardly explained since conductive phases were complicatedly rearranged after being subjected to impact loads. The competitive values of coefficient of determination (R^2) larger than 0.96 for all impact regimes suggests well-maintained repeatability for 1GNPCM when subjected to impact load. In particular, it seems that linearity in strain sensing is well-maintained and much more stable since there is a smaller variation in linearity after impact load. The slightly damaged linearity in stress sensing achieved after 30° and 60° impact load is ascribed to the slowdown of FCR changing rate under high compressive stress.

5.3.4.3 Piezoresistive performance of 2GNPCM

Fig. 5.33a–d displays the piezoresistive results of 2GNPCM, in which FCR(I) and FCR(II) are the results from the first and second measurements, respectively. The piezoresistive curves of 2GNPCM after the impact load are well-maintained, which keep nearly identical to the counterpart prior to impact load even after 90° impact load, with both the uninterrupted FCR alternating pattern and maximum FCR around 8%. The negligibly interrupted piezoresistive behavior of 2GNPCM suggests its well-maintained piezoresistive stability and reproducibility in resistance to impact load. Similarly, the irreversible

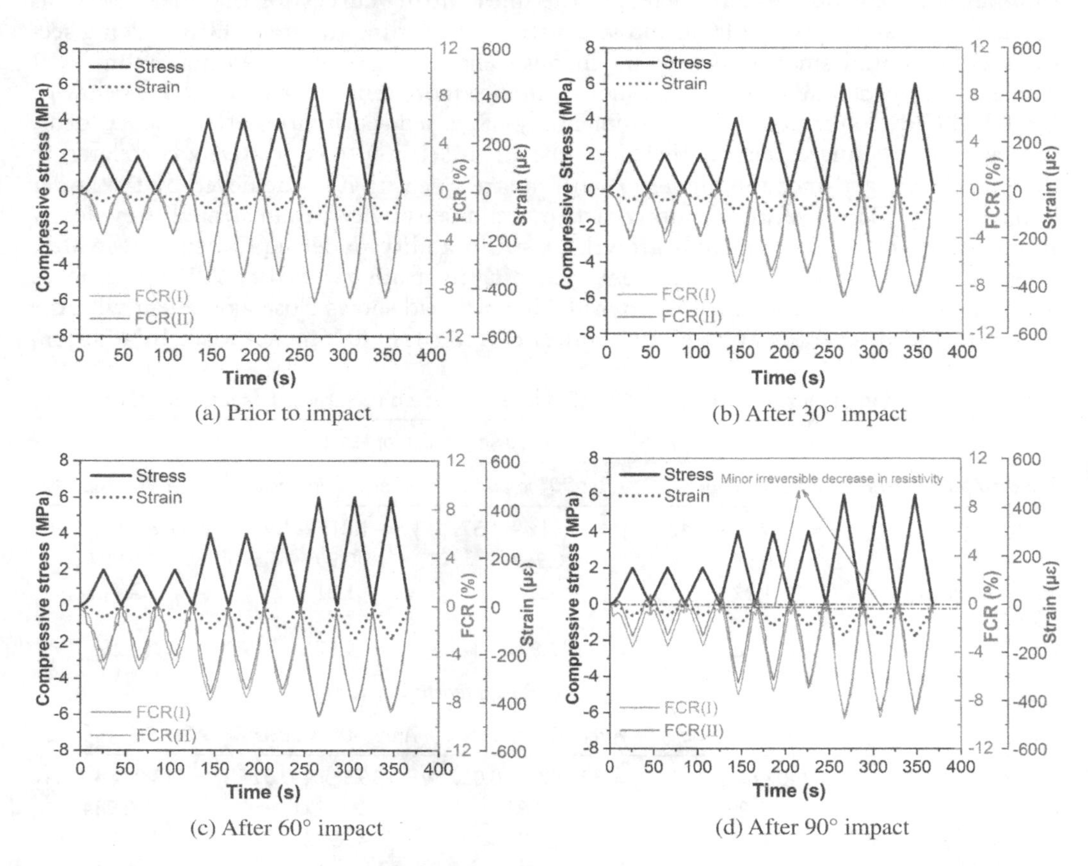

Figure 5.33 Fractional change of resistivity of 2GNPCM with strain/stress under cyclic compression [58].

decrease in resistivity also appeared in 2GNPCM after 90° impact load, while the irreversibility is much minor, which is one order of magnitude smaller than the counterpart that arose in 1GNPCM, showing a consistent correspondence with its well-maintained microstructures after impact load.

Fig. 5.34 shows the correlations between FCR and stress/strain of 2GNPCM. It is observed that data scatters that pertain to various impact regimes (different colors)

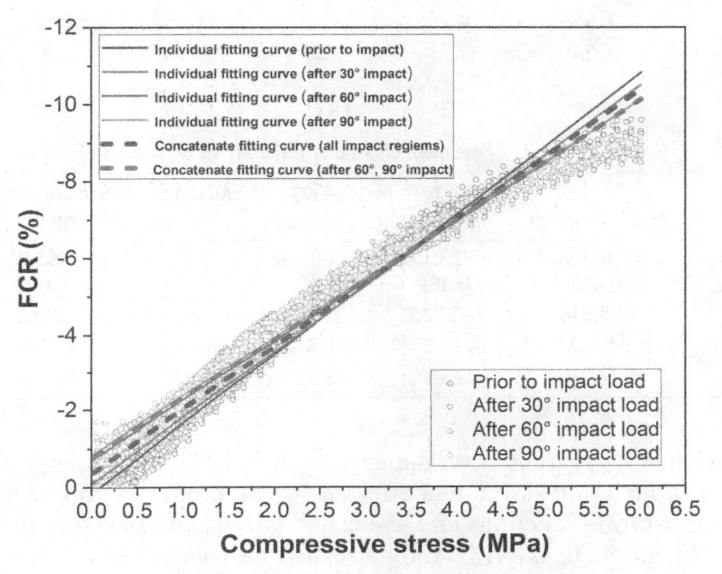

(a) As a function of compressive stress

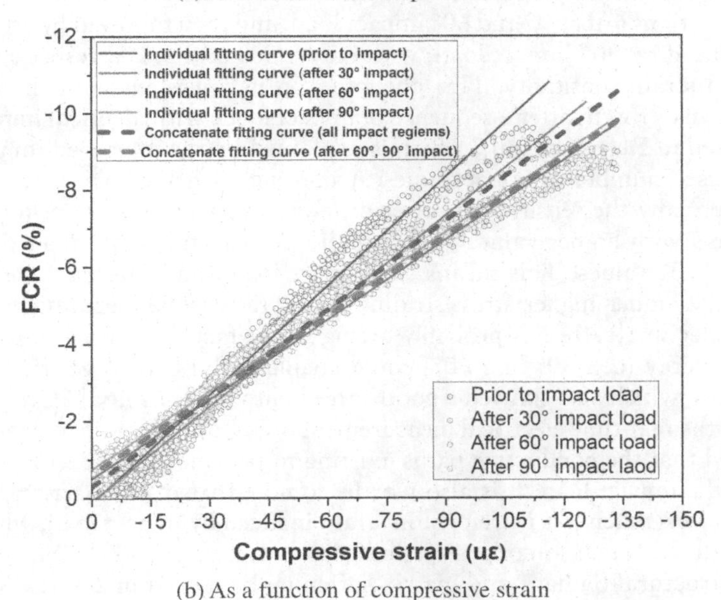

(b) As a function of compressive strain

Figure 5.34 Fractional change of resistivity of 2GNPCM under cyclic compression before/after various pendulum impact loads [58].

Table 5.3 Evaluation of piezoresistivity of 2GNPCM after subjected to various impact loads [58]

	Fractional change of resistivity as a function of stress					
Proposed parameters	*Prior to impacts*	*After 30° impacts*	*After 60° impacts*	*After 90° impacts*	*All impact regimes*	*After 60° and 90° impacts*
Fitting equation	$y = -1.83x + 0.17$	$y = -1.73x - 0.08$	$y = -1.57x - 0.69$	$y = -1.55x - 0.83$	$y = -1.67x - 0.36$	$y = -1.56x - 0.76$
R^2	0.986	0.965	0.965	0.975	0.963	0.970
Stress sensitivity (%/MPa)	1.83	1.73	1.57	1.55	1.67	1.56
Linearity (%)	16.67	19.39	18.16	16.18	19.80	18.01
	Fractional change of resistivity as a function of strain					
	Prior to impacts	*After 30° impacts*	*After 60° impacts*	*After 90° impacts*	*All impact regimes*	*After 60° and 90° impacts*
Fitting equation	$y = 940.55x + 0.001$	$y = 823.15x + 8.6E{-4}$	$y = 688.66x - 0.005$	$y = 683.31x - 0.007$	$y = 765.76 - 0.003$	$y = 685.23x - 0.006$
R^2	0.979	0.973	0.976	0.984	0.957	0.978
Strain sensitivity ($/\varepsilon$)	940.55	823.15	688.66	683.31	765.76	685.23
Linearity (%)	16.46	16.05	13.29	11.53	21.05	14.05

are closely matched or partially overlapped, which indicates that the piezoresistivity of 2GNPCM is well-maintained and less affected by the proposed impact load. Table 5.3 presents the linear fitting functions and associated parameters assessing the piezoresistive performance. In contrast to 1GNPCM, the stress/strain sensitivities of 2GNPCM progressively decrease after the impact load, while the reducing rate presents an asymptotic trend with rapid reductions after 30° and 60° impact load, and then followed by an imperceptible reduction induced by 90° impact load, with only 1.3% and 0.8%, respectively, for stress sensitivity and strain sensitivity. This phenomenon is consistent with the incrementally reduced resistivity growth after a sequence of impact loads with identical impact loads previously reported in Reference [60], which is explained by the improved impact resistance with an increased number of impact times. According to fitting curves, the reductions in the slope (essentially the sensitivities) after different impact loads are primarily triggered by data scatters with higher values under small stress/strain, rather than the descending of maximum FCR values. This means the impact load failure to alter the piezoresistivity of 2GNPCM under higher stress/strain. The more sensible piezoresistive response of 2GNPCM under small stress is probably owing to the fact that initial impact treatments with smaller energy (i.e., 30° and 60°) could enable 2GNPCM to get rid of some weak microstructures, which also showed a good agreement with the initial FCR baseline jumps manifested in the real-time electrical measurement of 30° and 60° impact loads. Therefore, it is envisioned that the conductive paths existing in pristine 2GNPCM have been nearly unaltered by the impact load. It is also worthy to note that in comparison to 1GNPCM, 2GNPCM exhibits much less increment in strain induced by impact load, which is in good agreement with the much minor irreversibility manifested in 2GNPCM, implying much minor microstructural damage and matrix softening happened in 2GNPCM when being subjected to impact load. Although the impact-induced strain increments slightly raise the discreteness of the FCR–strain curve and deteriorate the piezoresistive performance of 2GNPCM, its minor degree limits the negativities within the acceptable range.

In addition to fitting the scatter data pertaining to different impact regimes individually, a concatenate fitting of data scatters from all impact regimes is performed. The competitive coefficients of determination (R^2) of the fitting curves representing stress-sensing and strain-sensing are 0.963 and 0.957, respectively, which evidences a strong correlation between the FCR and cyclic compressive stress/strain. The strong correlation between the FCR–stress/strain relationships and fitting equations proves that 2GNPCM reserves a stable piezoresistivity with sufficient accuracy and repeatability after the proposed impact load. However, since the piezoresistivity is stabilized and tends to be unchanged after 60° impact load, stemming from the improved impact resistance as previously mentioned, it is reasonable to collectively fit the data scatters that pertain to both 60° and 90° impact regimes, and the fitted linear equations can precisely describe the piezoresistive response of 2GNPCM with reduced weak microstructures after being subjected to certain initial impact load. Since concrete structures are more frequently experienced impact loads such as live load during the service life, 2GNPCM is a desirable candidate as an impact-resistant CBS that can sustain stable piezoresistive behavior to monitor both stress and strain conditions prior to the exposure of certain amount/severity of impact loads.

5.3.4.4 Microstructural characterization

As shown in Fig. 5.35, the majority of GNP is in the form of single or small communities in 1GNPCM with fractured rims protruding from cement matrix that are likely to pull out or be detached. It is found that certain amounts of distinct and continuous crack paths pass through the mortar matrix. The presence of the crack branching, crack deflection, and tortuous crack paths have confirmed the energy absorption process due to the presence of the GNP [61]. Once the cracks propagate toward GNP, the sheet-like GNP could decelerate the crack development and deflect the crack in-plane or branch the cracks, either of which increased the length of the crack path and improved the energy absorption ability. The torn surface of GNP penetrated from the cement matrix and the detachment

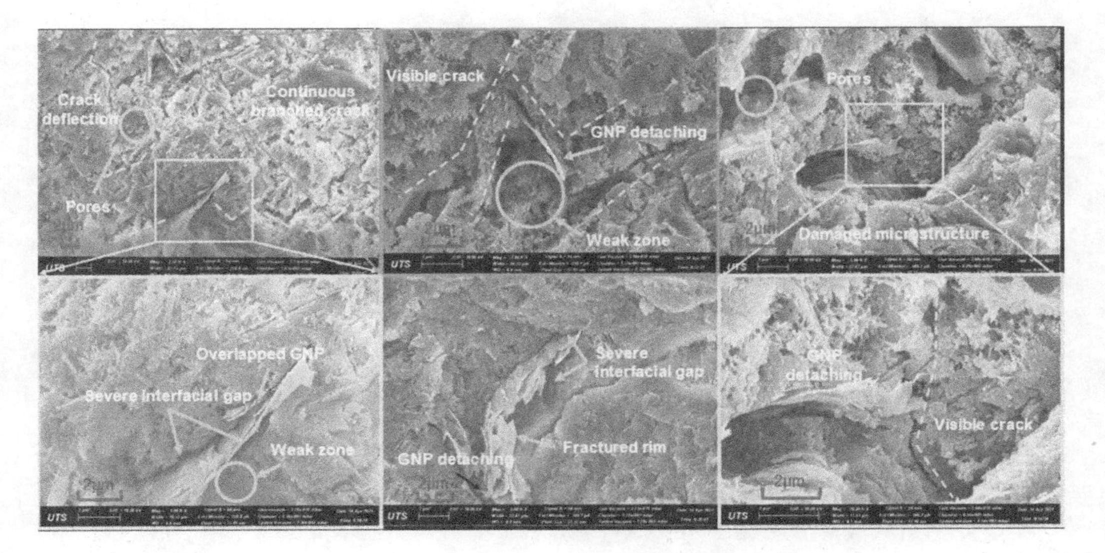

Figure 5.35 Microstructures of 1GNPCM after exposed to impact loads [58].

of GNP, which elucidate the resistance to the growth of cracks, also reveal the absorption process of energy with higher demand [62]. In addition, the impact energy was preferentially absorbed by GNP detaching from cement matrix instead of crack propagating. Fig. 5.35 captures the detachments of GNP particles with the microcracks reaching the proximity of the GNP. In other words, the matrix cracking is the primary dissipating form of the impact energy in cement paste happening after the GNP detachment.

Based on the above analysis, it can be concluded that although GNP particles in 1GNPCM played the positive reinforcing effect in absorbing impact energy, its quantity seems insufficient and could not successfully inhibit the crack initiation and propagation under the proposed impact regimes in this study. Still, a great deal of major cracks with branched secondary cracks are generated, passing through the fractured and detached GNP in the layouts of individuals or small communities, which is further substantiated by the obvious interfacial gaps between GNP and matrix. The presence of multiple cracks and voids elucidates the severe matrix deterioration of 1GNPCM.

By contrast, different microstructural morphologies are captured for 2GNPCM, as illustrated in Fig. 5.36. First, many clumped GNP clusters with no obvious wrinkles are entirely wrapped by cement hydrates, indicating strong bonding between the GNP and the surrounding matrix. Importantly, the fewer developed cracks and pores induced by impact loading are observed. Hence, it is inferred that microstructural damage resulting from impact loading is limited. In addition to some fractured rims shown in the curled GNP clusters which is an indicative of significant dissipation of impact energy, the crack bridging also demonstrates the strong energy absorption and impact resilience provided by the hauling effect of the graphene sheets, which significantly suppressed the potential shear stress and initiation of cracks [63]. The bridging GNP particles resist the extension of cracks, which involves excessive energy absorption due to the higher energy requirement for widening the cracks, leading to the pronounced impact-toughening behavior. Many bridging GNP particles existing in 2GNPCM, which are hardly captured in 1GNPCM, show a good agreement with the aforementioned minor localized damage in 2GNPCM, such as the generation of decreased FCR baseline during the impacts. The lesser visible

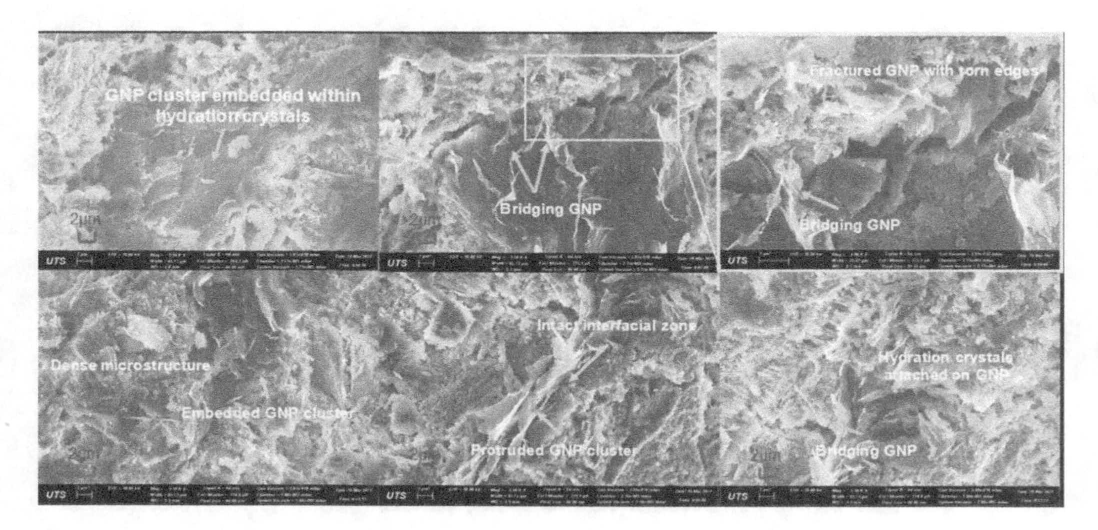

Figure 5.36 Microstructures of 2GNPCM after exposed to impact loads [58].

cracks, pores, and GNP detachments, and the stronger interfacial interaction between GNP clusters and matrix, have also evidenced a much minor microstructural deterioration for 2GNPCM in response to impact load.

The microstructural morphologies of 1GNPCM and 2GNPCM have proven enhancements of dynamic mechanical properties, fracture toughness, and energy absorption brought by the incorporation of GNP. However, different degrees of microstructural damage for 1GNPCM and 2GNPCM after identical impact loads suggest that higher energy absorption and impact resistance could be provided by a higher amount of GNP. It is known that lateral inertial stress and the induced confining pressure were generated under the dynamic impact load, the more prominent reinforcing effect brought by higher GNP content stemmed from a more integrated spatial GNP network more efficiently redistributed the stress and offset the inertial stress, which significantly limited the microstructural damage [64]. In addition, the featured structure of stacked two-dimensional nanosheets for GNP provided it superior stress transfer characteristic that enabled the prompt dislocation of the concentrated impact stress. A higher amount of GNP could improve the spatial homogeneity of the internal stress, thereby improving the energy absorption capacity of the cementitious composite. Prior to crack propagation, the interlaminar slip, the viscous friction between GNP and matrix, the GNP detachment, and structural failure of GNP itself can first release the strain energy liberated by matrix cracking, thereby significantly promoting the impact resistance of GNP-reinforced composite [65]. Meanwhile, the uneven distribution of stress aroused by the heterogeneity in elastic modulus among GNP and cement matrix could trigger a relative movement at the boundary surface of GNP, promoting the impact energy dissipation. A higher content of incorporated GNP could provide more internal boundary surfaces of GNP, thus a pronounced impact energy absorption should be achieved for 2GNPCM. Furthermore, since the featured netlike structure of GNP made it function as a template to optimize the crystallinity and morphology of the cement hydrations, 2GNPCM is expected to provide increasing number of functional templates [66]. It was also reported that GNP could modify the polymerization degree of calcium silicate hydrate (C-S-H) gel and possessed nucleation effect that promoted strong bonding between GNP and hydration products [67]. The above-mentioned reinforcing effects of GNP could explain the varying degree of microstructural damage for 1GNPCM and 2GNPCM after being subjected to the same impact load. The distinctly different microstructural morphologies indicate an inconsistent degree of microstructural deterioration between 1GNPCM and 2GNPCM after being subjected to the proposed impact treatments, a higher amount of GNP provided the composite with a higher energy absorption efficiency and impact resistance.

5.3.4.5 Mechanism discussion

To explain the variations of electrical and piezoresistive properties after impact load, five potential conductive passages of cementitious composites are presented in Fig. 5.37, namely, CP1 to CP5. CP1 refers to the conductive path constituted by interconnected GNP, while CP2 and CP3 represent the neighboring GNP separated by cement matrix and virgin voids/cavities, respectively. CR4 and CR5 refer to the disconnected liquid-filled pores and the conductive passage made up of connected GNP and liquid-filled pores. Therefore, the conductive mechanism of CP1, CP2, and CP3 are determined by the connection of conductive GNP; CP4 is associated with the ionic conduction, which relies on moisture content, volume fraction, and electrical conductivity and connectivity

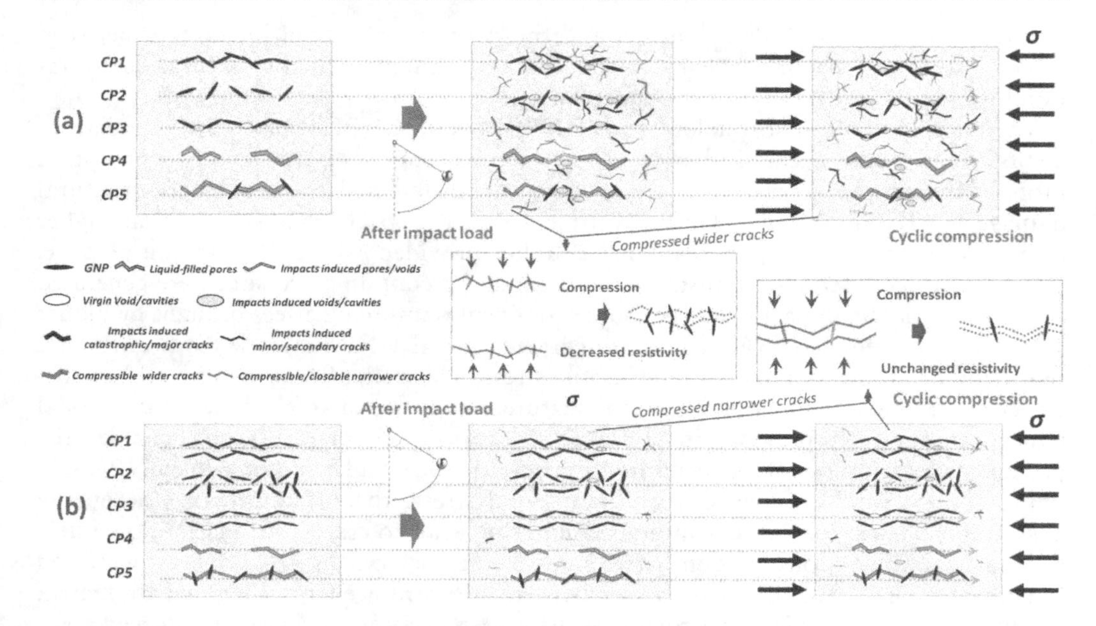

Figure 5.37 Mechanism of potential conductive passages in response to impact loads and cyclic compression for (a) 1GNPCM and (b) 2GNPCM [58].

of pore solutions; CP5 involves both electronic and ionic conduction. It should be noted that another potential conductive passage, which is formed by GNP connection stemming from the localized damage due to the slight movement of bridging GNP, is not considered herein, given this minor microstructural damage slightly influences electrical and piezoresistive properties.

According to SEM morphologies of 1GNPCM after impact load, severe deteriorations in microstructures, including continuous and branched cracks, voids, pores, and cavities, were induced by impact load, which are schematically illustrated in Fig. 5.37a. The impact-induced microstructural damage, which transforms the morphologies of the conductive passages, is believed to be the main influential factor varying the electrical and piezoresistive properties of GNP-reinforced cementitious composites. There are two possible transformed morphologies of CP1 after being subjected to the impact load. The first one is that the impact-induced voids or pores might break the connection of detached GNP particles and accordingly increase the potential electrical barrier between them. Although the disconnected GNPs are provided with higher electrical resistivity, the resistivity significantly decreases since the disconnected conductive phases could generate more conductive points and might be reconnected under the applied external load, thereby resulting in an enlarged FCR. The second case is that the impact-induced cracks slightly enlarge the originally connected GNP or make it slightly slip along the crack propagating direction, which significantly reduces the tightness degree and increases the contact resistance between GNP. Similarly, the contact between GNPs becomes closer, and the tightness degree swiftly increases once the compressive load has been applied, which causes the pronounced electron transition and reduced contact resistivity and contributes to a larger FCR. As for CP2, the impact-induced cracks and voids/cavities might develop and weaken the cement matrix intruded in the neighboring GNP, the matrix is compressed under the external load which reduces the distance between GNP and increases the possibility of the GNP connection. The CP3 is similar to the CP2, while the possibility and efficiency for the GNP coalescence in CP3 are expected to be more prominent since the original voids/cavities provide no resistance to the external force. In addition, disconnected pores (CP4) containing solutions might be either transversely or vertically connected by impact-induced pores or voids, and the filled water might become connected with enhanced ionic conduction under the compressive load. CP5 is similar to CP1, while the only difference is the conductive passages made up of connected liquid-filled pores and GNP instead of directly contacted GNP. It is worth noting that the enlarged strain capacity induced by impact load is due to the softer and looser matrix stemming from the microstructural deterioration promotes the five potential transformations of conductive passages, thus further intensifying their promoting effect in enlarging the FCR values and piezoresistive sensitivities.

As illustrated in Fig. 5.37b, the 2GNPCM showed less microstructural damages, including minor cracks, and less generated pores/voids possess very weak interference to the virgin conductive passages. In other words, the conductive passages are well-maintained and are hardly influenced by the impact load. It implies that a higher content of GNP provides 2GNPCM with more conductive passages, which formulate a stable conductive network against 1GNPCM according to the percolation theory. More conductive phases with a denser conductive network not only provide 2GNPCM with more stable electrical and piezoresistive properties prior to the impact load, but more significantly, further eliminating the interferences on the conductive passages due to the impact-induced microstructural damages. Moreover, the much smaller strain capacity of 2GNPCM which is less than half of the 1GNPCM, coupled with a less increased strain induced by impact load, also contribute to the stable electrical and piezoresistive properties of 2GNPCM.

As for the difference in the magnitude of irreversible resistivity in 1GNPCM and 2GNPCM, it can be explained by the different severity of the microstructural damage. Since the wider cracks have a higher possibility to be compressed and are associated with more detached GNP particles, more contacting opportunities of GNP particles are provided by wider cracks in comparison to the narrow cracks when being subjected to the external compression. Hence, it is deduced that the wider cracks/voids are associated with a more significant reduction in resistivity in comparison to the narrower cracks. The 1GNPCM is accompanied by severe microstructural damages with the continuous, branched, and wider cracks, which shows a good agreement with the facts of the severe irreversibility of approximately 2.36% for 1GNPCM and much smaller irreversibility of 0.43% for 2GNPCM under the cyclic compression of 6 MPa. Overall, the magnitude of irreversible decrease in resistivity might potentially be an indicative of the severity of microstructural damages in GNP/cementitious composites.

5.3.5 Loading style and amplitude

Loading style and amplitude affect the failure mechanism of CBSs and indirectly affect their electrical properties like resistance sensitivity and repeatability. Cyclic loading is one of the basic loading styles to evaluate repeatability and sensitivity of CBSs, as well as tiredness and service life. The divergence of mechanical and electrical performances of self-sensing concrete beams under the monotonic and cyclic flexural loading were investigated by Wang et al. [68]. It shows that there is reversible resistivity under cyclic loading with the amplitude below the ultimate load. Otherwise, the monotonous loading approaching

Table 5.4 Effects of loading cycles and amplitudes on the fractional changes of resistivity

Loading types	*Stress amplitude (MPa) or ratio to fracture stress*	*Number of cycles*	*Minimum $\Delta R/R_0$ (%)*	*Reference*
Compression	0.30	123	−0.003	Fu and Chung [69]
	0.50	217	−0.008	
	0.70	306	−0.020	
Tension	0.30	146	−0.006	
	0.50	252	−0.012	
	0.70	347	−0.020	
Compression	0.03	3	−2.90	Sun et al. [70]
	0.06		−6.20	
	0.12		−10.4	
	0.25		−15.6	
Compression	0.05	2	−0.050	Azhari and Banthia [71]
	0.09	3	−0.170	
	0.18	5	−0.310	
	0.27	5	−0.380	
	0.55	5	−0.500	
Flexure (compression zone)	44.1	3	−0.045	Wang and Chung [72]
	86.1		−0.077	
	129.7		−0.097	
	218.5		−0.183	

the ultimate load could cause accumulated electrical irreversibility and unneglectable errors. The amplitude of harmonic loading is normally lower than the fracture strength of cementitious composites, or the sensor will be damaged by the force that shows a sudden increase in electrical resistivity. Table 5.4 sums up the fractional changes of resistivity for CBSs under different amplitudes and cycles of harmonic loadings [69–72]. Under the compressive, tensile, or flexural loading, the average FCR at lower loading amplitude is less than that at higher amplitude when subjected to the same loading cycles. The increased amplitude can lead to closer conductors and smaller porosity. Meanwhile, the cement matrixes separating the nearby fibers could be damaged, increasing the possibility for adjacent fibers to touch one another to increase the fractional changes of resistivity. Similar conclusions were drawn by Galao et al. [73] who observed that the gauge factor of CBS increased in both the loading and unloading processes with cyclic load amplitude increasing from 3 to 9 kN. Fu [69] and Wang and Chung [72] reported experimental results with tensile tests, showing that their fractional changes of resistivity were negative and positive, respectively. It is reasonable for the electrical resistivity in the tension zone to increase because of the large cracks, reported by Wang and Chung [72]. However, in the tests by Fu and Chung [69], the decreased electrical resistance in the tension zone is mainly explained by the higher contact possibility of conductive fibers because of the cracks in cementitious solid, and the effect of the increased contact outweighs the influence resulting from the slightly larger cracks.

Fig. 5.38a–c illustrates the FCR as a function to compressive stress for the NCB-filled CBSs with various contents of PP fibers under various loading rates and amplitudes [23]. For all composites, the FCR (absolute value, the same below) increased with the increment of compressive stress and decreased when the stress returned to zero and performed good piezoresistivity. Linear fittings of these results were attached, showing that the relationship between FCR changes and compressive strength was almost linear. Moreover,

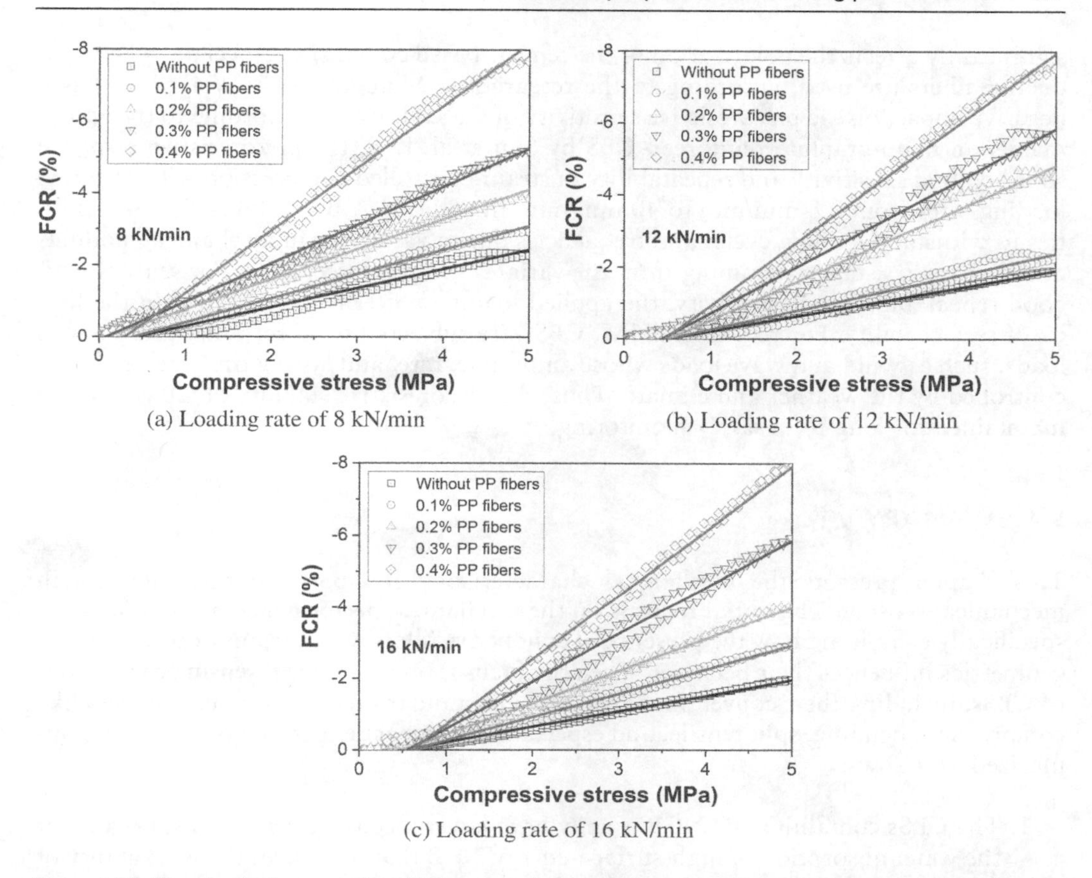

(a) Loading rate of 8 kN/min

(b) Loading rate of 12 kN/min

(c) Loading rate of 16 kN/min

Figure 5.38 Fractional changes of resistivity as a function of compressive stress under various loading rates and amplitudes [23].

the overlapped or very close curves in the loading and unloading processes indicated that the piezoresistivity of composite showed good repeatability. Therefore, it could be deduced that the NCB-filled CBSs with/without PP fibers were eligible to be a stress sensor, regardless of the loading rates. Generally, the tendency of FCR changes was similar under various loading rates, which demonstrated that the loading rates almost possess no impact on the piezoresistivity. Since the loading rates in reality are always unpredictable and multiple, the results lay the foundation for applying the NCB/PP fibers-filled CBSs to the real engineering project.

5.3.6 Loading cycles and frequency

As for the loading cycles and frequency, it seems that fatigue damage in each cycle is responsible for the irreversible electrical resistivity of sensor, illuminating the gradually increased fractional changes of resistivity under loading cycles, as can be seen from the test results reported by Fu and Chung [69]. Normally, with increasing loading cycles, the irreversibility increases for generation of permanent cracks and deformations. But if the loading cycles are within the threshold, the fractional changes of resistivity show periodicity with loadings until the cycles exceed the limit. Meanwhile, loading frequency significantly affects the polarization of the cement-based composite, especially when conductive fibers are used, according to the research by Materazzi et al. [74]. This has a negative impact on the piezoresistive sensitivity of CBS. However, according to the experiments on nano-graphite reinforced-CBS by Sun et al. [70], the piezoresistive response showed good sensitivity and repeatability in strain-controlled compression with different loading rates from 0.8 mm/min to 40 mm/min. In addition, CBSs exhibit differing abilities to withstand loading cycles and frequencies due to variations in mechanical modulus and damage resistance stemming from the variances in reinforcements. For sensors with good repeatability and sensitivity, the applied loading patterns must be within the limits of serviceability. For real-time SHM, CBSs are subjected to uncertainty and various loads, such as wind and wave loads whose amplitude, rate, and lasting time are randomly controlled by the weather and climate. Thus, the piezoresistive stability of CBSs must be maintained for long-term safety monitoring.

5.4 SUMMARY

This chapter presents the mechanical characteristics of CBSs and their abilities in mechanical sensing. The initial focus is on the mechanical performance of these sensors, specifically examining how the presence of conductive fillers with various dimensions and geometries influences their behavior. Then the focus is on mechanical sensing capabilities of CBSs, including their conventional reactions in standard mechanical assessments like compression, bending, split tensile, and especially the impact experiments. It can be summarized as follows:

1. The CBSs containing 0D NCB usually have lower mechanical strengths, because of the water absorption of high-surface-energy NCB that influences the workability of cementitious matrix. One-dimensional CNT can significantly enhance the mechanical performance of CBSs only under the condition of well-dispersed nanomaterials. The layer-distributed (undispersed) CNT decreases the mechanical strengths considerably. Two-dimensional GNP is more efficient in improving the mechanical properties of CBSs than 2D GP. Still, the excellent dispersion of GNP is a prerequisite here.
2. Compared to the nanofillers, the application of macro-fillers in CBSs does not need the dispersion process. The introduction of CRC can bring about the piezoresistive effect, but the increased efficiency is limited, and it significantly affects the mechanical characteristics of CBSs. Similar results can be found for the sensors with conductive rubber fibers. Despite having electrical insulating properties, the incorporation of PP fibers greatly enhances the electrical and mechanical characteristics of carbon black CBSs. This is due to the realignment of NCB along with PP fibers in CBSs. In addition, the PP fiber as a reinforcement can restrict the cracks propagation and increase the flexural strength of CBSs greatly. The waste glass with enclosed CNT also benefits the compressive strength of CBSs. Moreover, the 2D GNP and GO can enhance the dynamic mechanical properties and Young' modulus of CBSs.
3. In current studies, the sensitivity of CBSs is often assessed based on the rate of resistance change per unit strain under cyclic compressive loads. Through cyclic loading, it is also possible to evaluate the sensor's repeatability, recoverability, and other stability aspects. In addition, the linearity can be observed through cyclic

compressive sensing. In typical flexural sensing, different from the reduced electrical resistivity in compression, the resistivity gradually increases with flexural stress and then has a sharp increase at the failure point of CBSs. The sensing performance of CBSs subjected to split tensile has similar changing mode with that under flexural stress, but the FCR value under split tensile sensing is much larger than that under flexural sensing.

4. The real-time electrical response during impact showed good correspondence of electromechanical effect for GNP-filled CBSs under individual impact loads, the magnitudes of FCR corresponding to rapid drop portion reflected impact sensing capacity of GNP/cementitious composites. Based on the microstructural analysis, it was found that the microstructural deterioration induced by impact load was responsible for the variation of mechanical properties, electrical resistivity, and piezoresistive performances of CBSs. The microstructural damage and deterioration degraded the morphologies of the conductive passages.
5. The mechanical sensing performances of CBSs subjected to different loading styles, amplitude, cycle, and frequency are not yet very clear. The studies by authors demonstrated that the loading rates and amplitudes will not greatly affect the sensing behaviors, laying the foundation of applying the CBSs into the real engineering project where loading rates are always unpredictable and multiple.

REFERENCES

1. W. Dong, W. Li, L. Shen, D. Sheng, Piezoresistive behaviours of carbon black cement-based sensors with layer-distributed conductive rubber fibres, Materials & Design 182 (2019) 108012.
2. R.K. Abu Al-Rub, A.I. Ashour, B.M. Tyson, On the aspect ratio effect of multi-walled carbon nanotube reinforcements on the mechanical properties of cementitious nanocomposites, Construction and Building Materials 35 (2012) 647–655.
3. W. Dong, Multifunctional cement-based sensors with integrated piezoresistivity and hydrophobicity toward smart infrastructure, Thesis, University of Technology Sydney, 2021.
4. L. Silvestro, P.J.P. Gleize, Effect of carbon nanotubes on compressive, flexural and tensile strengths of Portland cement-based materials: A systematic literature review, Construction and Building Materials 264 (2020) 120237.
5. N. Yazdani, V. Mohanam, Carbon nano-tube and nano-fiber in cement mortar: Effect of dosage rate and water–cement ratio, International Journal of Material Science 4(2) (2014) 45.
6. M.S. Konsta-Gdoutos, C.A. Aza, Self sensing carbon nanotube (CNT) and nanofiber (CNF) cementitious composites for real time damage assessment in smart structures, Cement and Concrete Composites 53 (2014) 162–169.
7. T. Shi, Z. Li, J. Guo, H. Gong, C. Gu, Research progress on CNTs/CNFs-modified cement-based composites: A review, Construction and Building Materials 202 (2019) 290–307.
8. W. Dong, W. Li, Z. Tao, K. Wang, Piezoresistive properties of cement-based sensors: Review and perspective, Construction and Building Materials 203 (2019) 146–163.
9. R. Siddique, A. Mehta, Effect of carbon nanotubes on properties of cement mortars, Construction and Building Materials 50 (2014) 116–129.
10. W. Dong, W. Li, Z. Luo, Y. Guo, K. Wang, Effect of layer-distributed carbon nanotube (CNT) on mechanical and piezoresistive performance of intelligent cement-based sensor, Nanotechnology 31(50) (2020) 505503.
11. L. Lu, D. Ouyang, W. Xu, Mechanical properties and durability of ultra high strength concrete incorporating multi-walled carbon nanotubes, Materials 9(6) (2016) 419.
12. G.Y. Li, P.M. Wang, X. Zhao, Mechanical behavior and microstructure of cement composites incorporating surface-treated multi-walled carbon nanotubes, Carbon 43(6) (2005) 1239–1245.

13. W. Dong, W. Li, K. Wang, S.P. Shah, Physicochemical and piezoresistive properties of smart cementitious composites with graphene nanoplates and graphite plates, Construction and Building Materials 286 (2021) 122943.
14. M. Van Wijk, M. Dienwiebel, J. Frenken, A. Fasolino, Superlubric to stick-slip sliding of incommensurate graphene flakes on graphite, Physical Review B 88(23) (2013) 235423.
15. W. Baomin, D. Shuang, Effect and mechanism of graphene nanoplatelets on hydration reaction, mechanical properties and microstructure of cement composites, Construction and Building Materials 228 (2019) 116720.
16. V.D. Ho, C.-T. Ng, C.J. Coghlan, A. Goodwin, C. Mc Guckin, T. Ozbakkaloglu, D. Losic, Electrochemically produced graphene with ultra large particles enhances mechanical properties of Portland cement mortar, Construction and Building Materials 234 (2020) 117403.
17. R.A. e Silva, P. de Castro Guetti, M.S. da Luz, F. Rouxinol, R.V. Gelamo, Enhanced properties of cement mortars with multilayer graphene nanoparticles, Construction and Building Materials 149 (2017) 378–385.
18. M. Hu, J. Guo, P. Li, D. Chen, Y. Xu, Y. Feng, Y. Yu, H. Zhang, Effect of characteristics of chemical combined of graphene oxide–nanosilica nanocomposite fillers on properties of cement-based materials, Construction and Building Materials 225 (2019) 745–753.
19. H. Du, H.J. Gao, S. Dai Pang, Improvement in concrete resistance against water and chloride ingress by adding graphene nanoplatelet, Cement and Concrete Research 83 (2016) 114–123.
20. W. Dong, W. Li, K. Wang, K. Vessalas, S. Zhang, Mechanical strength and self-sensing capacity of smart cementitious composite containing conductive rubber crumbs, Journal of Intelligent Material Systems and Structures 31(10) (2020) 1325–1340.
21. E. Güneyisi, M. Gesoğlu, T. Özturan, Properties of rubberized concretes containing silica fume, Cement and Concrete Research 34(12) (2004) 2309–2317.
22. W. Dong, W. Li, K. Wang, Z. Luo, D. Sheng, Self-sensing capabilities of cement-based sensor with layer-distributed conductive rubber fibres, Sensors and Actuators A: Physical 301 (2020) 111763.
23. W. Dong, W. Li, K. Wang, Y. Guo, D. Sheng, S.P. Shah, Piezoresistivity enhancement of functional carbon black filled cement-based sensor using polypropylene fibre, Powder Technology 373 (2020) 184–194.
24. W. Li, X. Li, S.J. Chen, Y.M. Liu, W.H. Duan, S.P. Shah, Effects of graphene oxide on early-age hydration and electrical resistivity of Portland cement paste, Construction and Building Materials 136 (2017) 506–514.
25. Y. Dai, M. Sun, C. Liu, Z. Li, Electromagnetic wave absorbing characteristics of carbon black cement-based composites, Cement and Concrete Composites 32(7) (2010) 508–513.
26. S. Wen, D. Chung, Partial replacement of carbon fiber by carbon black in multifunctional cement–matrix composites, Carbon 45(3) (2007) 505–513.
27. W. Dong, Y. Guo, Z. Sun, Z. Tao, W. Li, Development of piezoresistive cement-based sensor using recycled waste glass cullets coated with carbon nanotubes, Journal of Cleaner Production 314 (2021) 127968.
28. M. Małek, W. Łasica, M. Jackowski, M. Kadela, Effect of waste glass addition as a replacement for fine aggregate on properties of mortar, Materials 13(14) (2020) 3189.
29. W. Her-Yung, A study of the engineering properties of waste LCD glass applied to controlled low strength materials concrete, Construction and Building Materials 23(6) (2009) 2127–2131.
30. K. Afshinnia, P.R. Rangaraju, Influence of fineness of ground recycled glass on mitigation of alkali–silica reaction in mortars, Construction and Building Materials 81 (2015) 257–267.
31. G. Foray-Thevenin, G. Vigier, R. Vassoille, G. Orange, Characterization of cement paste by dynamic mechanical thermo-analysis: Part I: Operative conditions, Materials Characterization 56(2) (2006) 129–137.
32. W.-J. Long, J.-J. Wei, F. Xing, K.H. Khayat, Enhanced dynamic mechanical properties of cement paste modified with graphene oxide nanosheets and its reinforcing mechanism, Cement and Concrete Composites 93 (2018) 127–139.
33. W.-J. Long, J.-J. Wei, H. Ma, F. Xing, Dynamic mechanical properties and microstructure of graphene oxide nanosheets reinforced cement composites, Nanomaterials 7(12) (2017) 407.
34. Z. Chen, Y. Xu, J. Hua, X. Wang, L. Huang, X. Zhou, Mechanical properties and shrinkage behavior of concrete-containing graphene-oxide nanosheets, Materials 13(3) (2020) 590.

35. W.-J. Long, Y.-C. Gu, B.-X. Xiao, Q.-M. Zhang, F. Xing, Micro-mechanical properties and multi-scaled pore structure of graphene oxide cement paste: Synergistic application of nanoindentation, X-ray computed tomography, and SEM-EDS analysis, Construction and Building Materials 179 (2018) 661–674.
36. L. Zhao, X. Guo, C. Ge, Q. Li, L. Guo, X. Shu, J. Liu, Mechanical behavior and toughening mechanism of polycarboxylate superplasticizer modified graphene oxide reinforced cement composites, Composites Part B: Engineering 113 (2017) 308–316.
37. S. Sun, S. Ding, B. Han, S. Dong, X. Yu, D. Zhou, J. Ou, Multi-layer graphene-engineered cementitious composites with multifunctionality/intelligence, Composites Part B: Engineering 129 (2017) 221–232.
38. Y. Guo, W. Li, W. Dong, Z. Luo, F. Qu, F. Yang, K. Wang, Self-sensing performance of cement-based sensor with carbon black and polypropylene fibre subjected to different loading conditions, Journal of Building Engineering 59 (2022) 105003.
39. S. Wen, D. Chung, Electric polarization in carbon fiber-reinforced cement, Cement and Concrete Research 31(1) (2001) 141–147.
40. H. Kim, I. Park, H.-K. Lee, Improved piezoresistive sensitivity and stability of CNT/cement mortar composites with low water–binder ratio, Composite Structures 116 (2014) 713–719.
41. L. Wang, F. Aslani, Mechanical properties, electrical resistivity and piezoresistivity of carbon fibre-based self-sensing cementitious composites, Ceramics International 47(6) (2021) 7864–7879.
42. D. Jang, H. Yoon, S. Farooq, H.-K. Lee, I. Nam, Influence of water ingress on the electrical properties and electromechanical sensing capabilities of CNT/cement composites, Journal of Building Engineering 42 (2021) 103065.
43. E. Teomete, The effect of temperature and moisture on electrical resistance, strain sensitivity and crack sensitivity of steel fiber reinforced smart cement composite, Smart Materials and Structures 25(7) (2016) 075024.
44. Y. Ding, G. Liu, A. Hussain, F. Pacheco-Torgal, Y. Zhang, Effect of steel fiber and carbon black on the self-sensing ability of concrete cracks under bending, Construction and Building Materials 207 (2019) 630–639.
45. D.-Y. Yoo, S. Kim, S.H. Lee, Self-sensing capability of ultra-high-performance concrete containing steel fibers and carbon nanotubes under tension, Sensors and Actuators A: Physical 276 (2018) 125–136.
46. J. Song, D.L. Nguyen, C. Manathamsombat, D.J. Kim, Effect of fiber volume content on electromechanical behavior of strain-hardening steel-fiber-reinforced cementitious composites, Journal of Composite Materials 49(29) (2015) 3621–3634.
47. S.-J. Lee, I. You, S. Kim, H.-O. Shin, D.-Y. Yoo, Self-sensing capacity of ultra-high-performance fiber-reinforced concrete containing conductive powders in tension, Cement and Concrete Composites 125 (2022) 104331.
48. S. Dong, B. Han, J. Ou, Z. Li, L. Han, X. Yu, Electrically conductive behaviors and mechanisms of short-cut super-fine stainless wire reinforced reactive powder concrete, Cement and Concrete Composites 72 (2016) 48–65.
49. E. Teomete, Measurement of crack length sensitivity and strain gage factor of carbon fiber reinforced cement matrix composites, Measurement 74 (2015) 21–30.
50. E. Demircilioğlu, E. Teomete, O.E. Ozbulut, Characterization of smart brass fiber reinforced concrete under various loading conditions, Construction and Building Materials 265 (2020) 120411.
51. N. Zohhadi, B. Koohbor, F. Matta, A. Kidane, Characterization of fracture behavior of multi-walled carbon nanotube reinforced cement paste using digital image correlation, fracture, fatigue, failure, and damage evolution, Volume 5: Proceedings of the 2014 Annual Conference on Experimental and Applied Mechanics, Springer, 2015, pp. 73–79.
52. F. Naeem, H.-K. Lee, H. Kim, I. Nam, Flexural stress and crack sensing capabilities of MWNT/cement composites, Composite Structures 175 (2017) 86–100.
53. E. Demircilioglu, E. Teomete, O.E. Ozbulut, Strain sensitivity of steel-fiber-reinforced industrial smart concrete, Journal of Intelligent Material Systems and Structures 31(1) (2020) 127–136.
54. E. Teomete, O.I. Kocyigit, Tensile strain sensitivity of steel fiber reinforced cement matrix composites tested by split tensile test, Construction and Building Materials 47 (2013) 962–968.

55. C. Bian, J.-Y. Wang, J.-Y. Guo, Damage mechanism of ultra-high performance fibre reinforced concrete at different stages of direct tensile test based on acoustic emission analysis, Construction and Building Materials 267 (2021) 120927.
56. P.K. Nelson, V.C. Li, T. Kamada, Fracture toughness of microfiber reinforced cement composites, Journal of Materials in Civil Engineering 14(5) (2002) 384–391.
57. B. Chun, D.-Y. Yoo, N. Banthia, Achieving slip-hardening behavior of sanded straight steel fibers in ultra-high-performance concrete, Cement and Concrete Composites 113 (2020) 103669.
58. W. Dong, W. Li, Y. Guo, K. Wang, D. Sheng, Mechanical properties and piezoresistive performances of intrinsic graphene nanoplate/cement-based sensors subjected to impact load, Construction and Building Materials 327 (2022) 126978.
59. Y. Wang, X. Zhao, Y. Zhao, Piezoresistivity of cement matrix composites incorporating multiwalled carbon nanotubes due to moisture variation, Advances in Civil Engineering 2020 (2020) 1–11.
60. W. Dong, W. Li, L. Shen, Z. Sun, D. Sheng, Piezoresistivity of smart carbon nanotubes (CNTs) reinforced cementitious composite under integrated cyclic compression and impact, Composite Structures 241 (2020) 112106.
61. M. Saafi, L. Tang, J. Fung, M. Rahman, J. Liggat, Enhanced properties of graphene/fly ash geopolymeric composite cement, Cement and Concrete Research 67 (2015) 292–299.
62. M. Tabatabaei, A.D. Taleghani, N. Alem, Nanoengineering of cement using graphite platelets to refine inherent microstructural defects, Composites Part B: Engineering 202 (2020) 108277.
63. N. Ranjbar, M. Mehrali, M. Mehrali, U.J. Alengaram, M.Z. Jumaat, Graphene nanoplatelet-fly ash based geopolymer composites, Cement and Concrete Research 76 (2015) 222–231.
64. X. Li, A.H. Korayem, C. Li, Y. Liu, H. He, J.G. Sanjayan, W.H. Duan, Incorporation of graphene oxide and silica fume into cement paste: A study of dispersion and compressive strength, Construction and Building Materials 123 (2016) 327–335.
65. D. Dimov, I. Amit, O. Gorrie, M.D. Barnes, N.J. Townsend, A.I. Neves, F. Withers, S. Russo, M.F. Craciun, Ultrahigh performance nanoengineered graphene–concrete composites for multifunctional applications, Advanced Functional Materials 28(23) (2018) 1705183.
66. Y. Zhao, Y. Liu, T. Shi, Y. Gu, B. Zheng, K. Zhang, J. Xu, Y. Fu, S. Shi, Study of mechanical properties and early-stage deformation properties of graphene-modified cement-based materials, Construction and Building Materials 257 (2020) 119498.
67. L. Zhao, X. Guo, Y. Liu, Y. Zhao, Z. Chen, Y. Zhang, L. Guo, X. Shu, J. Liu, Hydration kinetics, pore structure, 3D network calcium silicate hydrate, and mechanical behavior of graphene oxide reinforced cement composites, Construction and Building Materials 190 (2018) 150–163.
68. W. Wang, H. Dai, S. Wu, Mechanical behavior and electrical property of CFRC-strengthened RC beams under fatigue and monotonic loading, Materials Science and Engineering: A 479(1–2) (2008) 191–196.
69. X. Fu, D.D.L. Chung, Self-monitoring of fatigue damage in carbon fiber reinforced cement, Cement and Concrete Research 26(1) (1996) 15–20.
70. S. Sun, B. Han, S. Jiang, X. Yu, Y. Wang, H. Li, J. Ou, Nano graphite platelets-enabled piezoresistive cementitious composites for structural health monitoring, Construction and Building Materials 136 (2017) 314–328.
71. F. Azhari, N. Banthia, Cement-based sensors with carbon fibers and carbon nanotubes for piezoresistive sensing, Cement and Concrete Composites 34(7) (2012) 866–873.
72. S. Wang, D. Chung, Self-sensing of flexural strain and damage in carbon fiber polymer-matrix composite by electrical resistance measurement, Carbon 44(13) (2006) 2739–2751.
73. O. Galao, F.J. Baeza, E. Zornoza, P. Garcés, Strain and damage sensing properties on multifunctional cement composites with CNF admixture, Cement and Concrete Composites 46 (2014) 90–98.
74. A.L. Materazzi, F. Ubertini, A. D'Alessandro, Carbon nanotube cement-based transducers for dynamic sensing of strain, Cement and Concrete Composites 37 (2013) 2–11.

Chapter 6

Durability of CBSs and impact of various environments on sensory performance

6.1 INTRODUCTION

Considering that cement-based sensors (CBSs) are composed of a blend of ordinary concrete and conductive fillers, their durability is remarkably excellent compared to the conventional metal or polymer-based sensors, and their lifespan might even be as long as that of concrete structures. Normal concrete exposed to a comprehensive environment is susceptible to various physical and chemical degradation processes, and even with improved mechanical and durable properties for the nano-modified CBSs, their electrical and piezoresistive performances could be altered because of the multiple interferences from the environments [1, 2]. For instance, in the case of concrete in typical environments, temperature and humidity fluctuations caused by weather and seasonal factors can result in changes in the electrical performance of CBSs. In some special environments, such as hydraulic concrete, various invaded aqueous solutions and free ions can also lead to variations in electrical performance, let alone the alterations in the electrical and piezoresistive performance of CBSs under extreme conditions, such as high temperatures or highly corrosive environments. Therefore, the durability investigation of CBSs is crucial for their long-term secure application in the field of civil engineering, even though there is very limited research regarding the effects of various environmental factors on the performance of CBSs at present [3, 4].

This chapter first introduces the durability performance of electrically conductive cementitious composites containing various conductive fillers ranging from chemical attacks, to transport properties, shrinkage deformation, and high-temperature performance. Some of the studies do not explicitly address the durability of CBSs; however, due to the impact of nanomodification, cement-based materials possess both good electrical conductivity and piezoresistivity potentials. Therefore, we also believe that these studies are equally applicable to CBSs. Afterward, the environmental sensing performance of CBSs, including temperature, water content, chloride penetration, sulfuric corrosion, and freeze-thaw cycles, are summarized. These include changes in the self-sensing performance of degraded CBSs in various environments, as well as their utilization in monitoring the environmental factors through these changes. Subsequently, a series of complex mechanisms and illustrations, after being summarized and categorized, are used to explain the specific reasons for this variation.

DOI: 10.1201/9781032663685-6

6.2 DURABILITY PERFORMANCE

6.2.1 Chemical attacks

Fig. 6.1 illustrates the surface appearance of GNP-filled cementitious composites that have been stored in 0%, 1%, 2%, and 3% H_2SO_4 after 90 and 180 days [5]. For the preconditioned control specimen cured for 28 days, a smooth surface devoid of surface defects is observed. After 90 days of storage, the specimens in the water remain unchanged and intact while its counterparts in H_2SO_4 are observed to have undergone significant change with additional products deposited on the specimen surfaces. It should be noted that the appearance of the specimens stored in 3% H_2SO_4 after 90 and 180 days appear different from each other. This is likely due to the replacement of the solution after the first 90 days and the removal of the detritus from the specimen surfaces. In addition, the number of deposited products has been observed to have increased with increased H_2SO_4 concentration. Based on previous studies, the deposited products have been reported to contain traces of elements relating to the presence of silica gel, gypsum, and ettringite [6, 7]. In addition to these deposited products, worn edges on the specimens have been observed with increased H_2SO_4 concentration. For the specimens stored in water, a smooth surface without deposited products is apparent after 180 days. In contrast, the number of deposited products formed specimen surfaces as a function of increasing acid concentration has intensified, especially for those specimens stored in 3% H_2SO_4. Due to the H_2SO_4 solution being replenished after the first 90 days, some deposited products may have been removed away, which may explain why relatively fewer products have been found on specimen surfaces after 180 days. Despite this occurrence, continual propagation of cracks throughout the edges of the specimens is observed for those specimens stored in 3% H_2SO_4. This feature has intensified and contributed to the spalling of the specimen edges. In summary, the rate of deterioration was found to increase with an increase in acid concentration and immersion duration. Conversely, the specimens stored in ≤2% H_2SO_4 exhibited relatively intact surfaces after 180 days of storage.

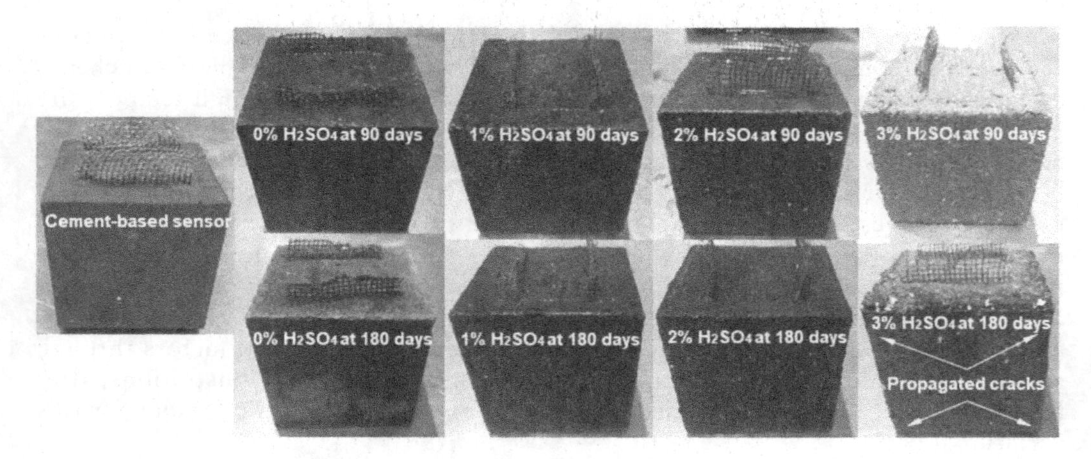

Figure 6.1 Surface appearance of GNP-filled cementitious composites stored in 0%, 1%, 2%, and 3% H_2SO_4 after 90 and 180 days [5].

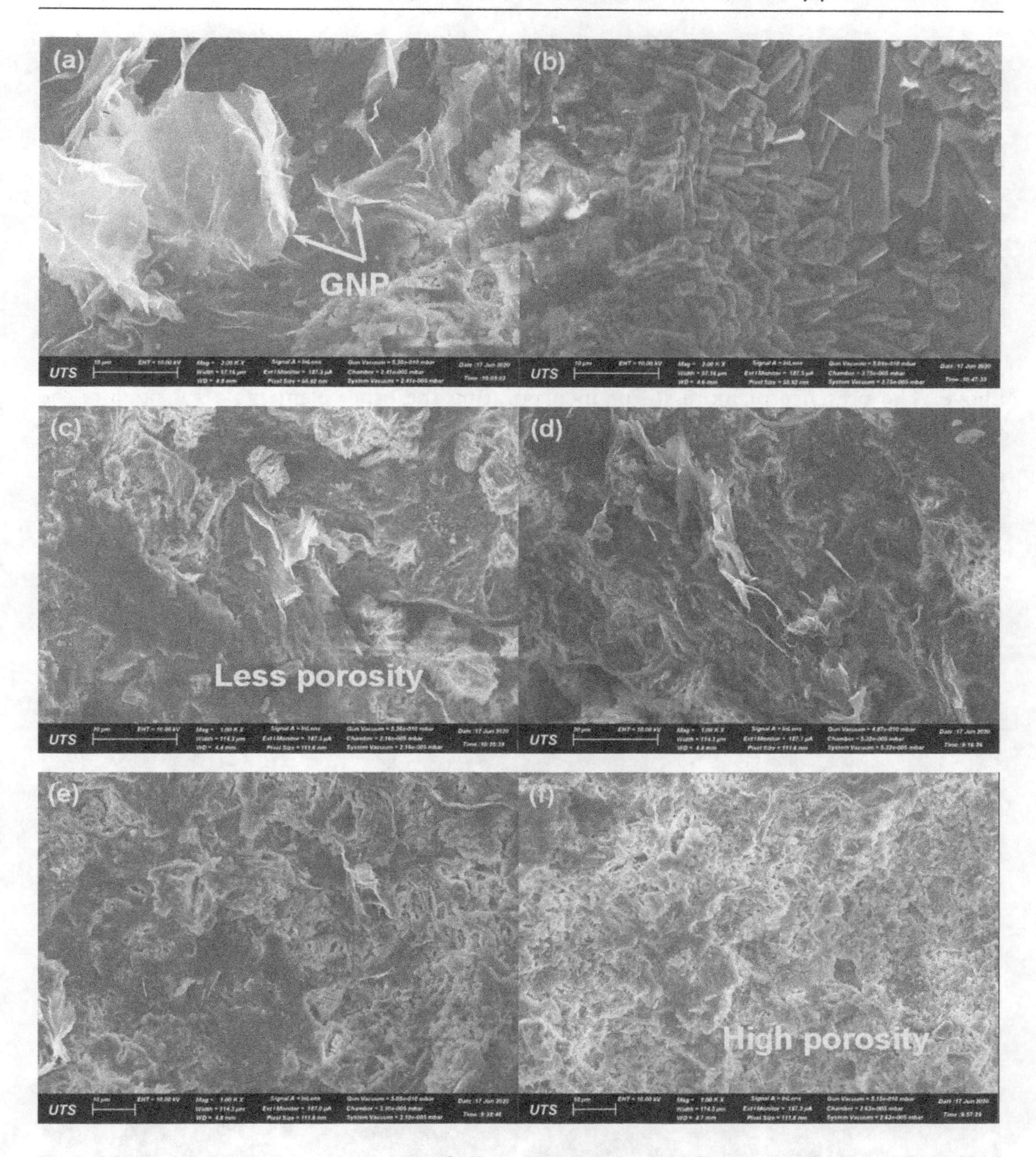

Figure 6.2 Microstructural morphology. (a) GNP in cement matrix. (b) Erosion products and GNP-filled cementitious composite subjected to (c) 0%; (d) 1%; (e) 2%, and (f) 3% H_2SO_4 solution [5].

Fig. 6.2 depicts the microstructure of the GNP and the distinct features of different products formed in the cement matrix collected from cementitious composites after 180 days storage in 0%, 1%, 2%, and 3% H_2SO_4. Fig. 6.2a and b illustrate the presence of GNP in silicate sheet-like layers and the inclusion of needle-like material found in the deposited product. This needle-like material is reminiscent of the formation of later-age ettringite [8]. The SEM images were taken from specimens exposed to 0%, 1%, 2%, and 3% H_2SO_4.

For the specimen exposed to water (Fig. 6.2c), a dense and impervious microstructure is observed. This feature is most likely due to the hydration of cement, contributing to the formation of more interlayered C–S–H products. In terms of the specimen exposed to 1% H_2SO_4 (Fig. 6.2d), a relatively dense microstructure is apparent; however, there also appears to be the formation of a sporadically generated pore structure. Further, in Fig. 6.2e, a higher volume of pores is apparent for the specimen exposed to 2% H_2SO_4. The highest volume of pores though is observed for the specimen exposed to 3% H_2SO_4 in Fig. 6.2f.

Fig. 6.3 illustrates the EDX data of the elements of Ca, Si, Al, Na, and S for cementitious composite specimens stored in 0%, 1%, 2%, and 3% H_2SO_4 after 180 days. The amount of sulfur found in specimens was observed to increase as a function of increasing sulfuric acid concentration. This result is consistent with the microstructural features reported above. The presence of more sulfur incorporating the other elements also suggests the formation of more sulfate-bearing hydration products such as gypsum and ettringite.

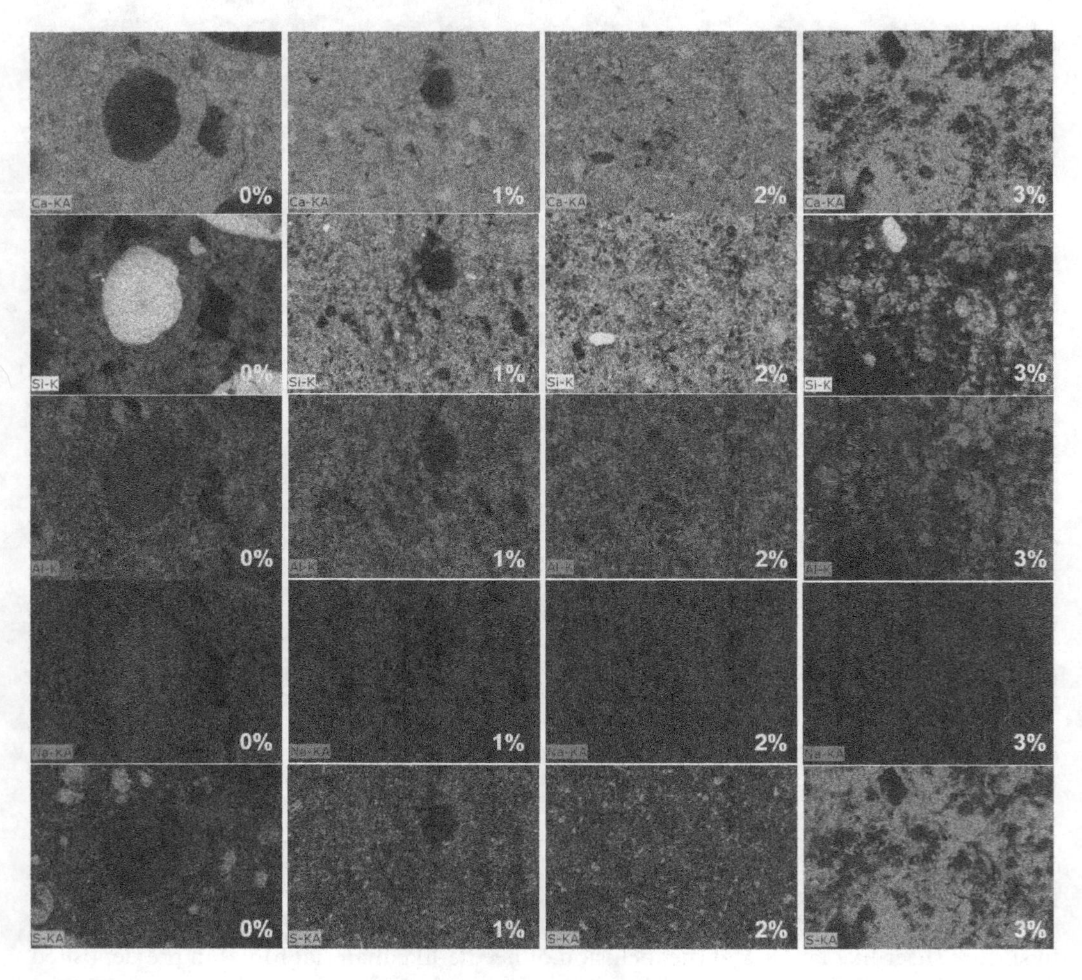

Figure 6.3 EDX analysis on elements of Ca, Si, Al, Na, and S in the GNP-filled cementitious composite subjected to 0%, 1%, 2%, and 3% H_2SO_4 solutions after 180 days [5].

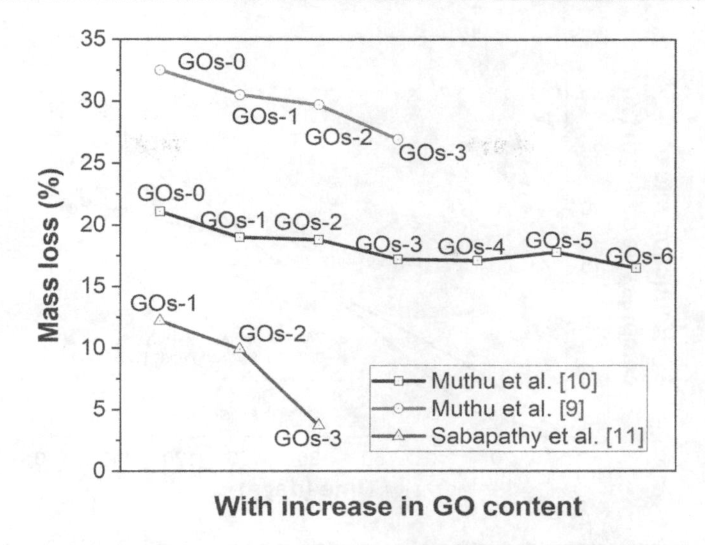

Figure 6.4 Mass loss of cementitious composites containing GOs after 28-day exposure to acid attacks.

Further, the appearance of these later-age hydration products can lead to expansion and cause distress in the cement matrix. This distress can also contribute to a higher pore volume and interconnectivity between these pores leading to higher permeability arising from the sulfuric acid attack.

Fig. 6.4 shows the losses in mass of cementitious composites containing different dosages of GOs after being subjected to 28 days of acid attack. The mass loss of cementitious composite has firm relationship with the diffusion of aggressive acid and their contact with hydrates. For all studies, the plain cementitious composites without GOs had highest degree of damage and the mass loss decreased with increasing GOs content, implying a positive effect of GOs on improving the acid resistance of cementitious composites [9–11]. The superiority can be attributed to the enhanced pore structures and fineness of porosity that restrict the entrance of aggressive acid. In addition, the studies were conducted under various types and contents of GOs, acids, and water to binder ratios, the enhancement of GOs on the acid resistance of cementitious composites would not be greatly influenced.

In addition to acid erosion, the chemical attacks consist of the reactions among nitrate and sulfate solutions with cementitious composites. Fig. 6.5 shows the reduction of compressive strength of cementitious composites incorporating multiple contents of GNPs exposure to ammonium nitrate and sulfate solutions. Based on the studies by Tong et al. [12], apart from the value at 90 days, the strength reduction rate of 0.1% GNPs-reinforced concrete was always lower than that of plain concrete. Similar results could be observed from the study by Sharma and Arora [13], who found decreased reduction rate of compressive strength for the cement mortars with 0.05% and 0.1% GNPs compared to the plain cement mortar. The cement mortar containing 0.05% GNPs performed better on both the compressive strength and sulfate resistance, indicating an optimal concentration of GNPs in the cementitious composites. In summary, it can be concluded that the suitable addition of 2D GBNs are able to enhance the resistance to chemical attack of cementitious composites.

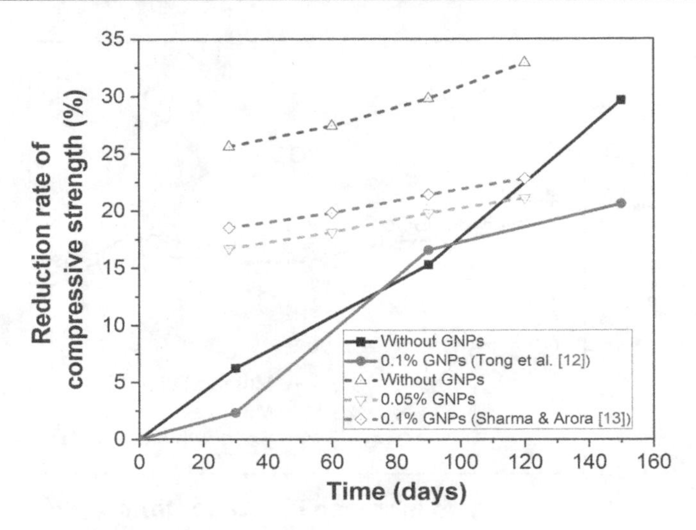

Figure 6.5 Compressive strength reduction of cementitious composites with GNPs exposed to ammonium nitrate and sulfate solutions.

6.2.2 Transport properties

6.2.2.1 Plain conductive filler

The chloride penetration depths of cementitious composites containing various concentrations of GBNs are displayed in Fig. 6.6. Du and Pang investigated the effect of multiple contents of GNPs ranging from 0% to 7.5% on the chloride penetration depth and found better erosion resistance to chloride for the cement mortar mixed with GNPs. The enhancement would not be increased with increasing GNPs content, but reach the smallest penetration depth when the GNPs content was 5% by weight of cement [14]. Later,

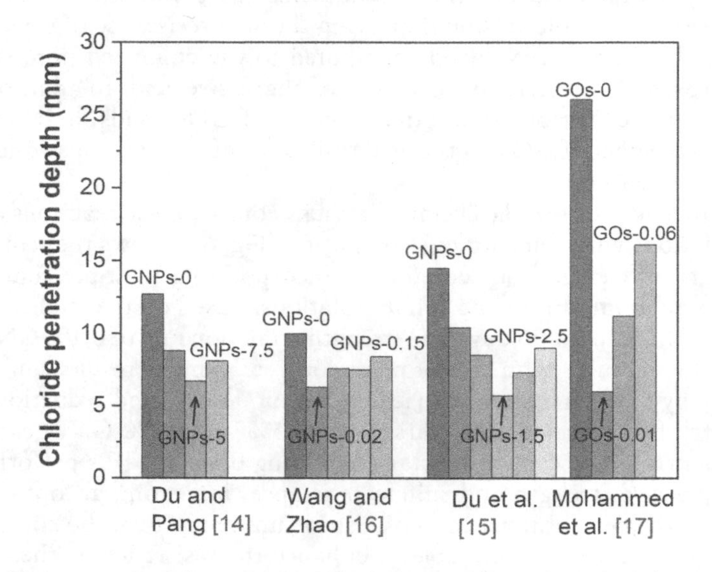

Figure 6.6 Chloride penetration of cementitious composites with graphene-based nanomaterials.

they applied lower dosage of GNPs ranging from 0% to 2.5% in cement concrete and found that the optimal concentration of GNPs to reduce the chloride penetration was 1.5% by weight of cement [15]. Wang and Zhao [16] found the lowest chloride penetration depth of cement paste filled with 0.02% GNPs. They attributed the improvement to the enhanced cement hydration, filling, barrier, and crack-arresting effects of GNPs. Similarly, Mohammed et al. [17] observed that 0.01% GOs could efficiently hinder the penetration of chloride ions, but the enhancement was weakened with increasing content of GOs to 0.06%. They proposed that a very low concentration of GOs could improve the resistance of chloride penetration of cement paste because of the layer structures of GOs and its barrier effect.

The sorptivity is one of the critical factors that influence the fluid transport properties of cementitious composites. Table 6.1 lists the initial and final sorptivity of cementitious composites incorporating various contents of graphene-based nanomaterials. Almost all of studies found decreased initial and final sorptivity of cementitious composites filled with GOs or GNPs compared to the plain cement matrix [18]. Based on the studies by Alharbi et al. [19], they observed decreased sorptivity for the cement paste with low content of 0.01% GOs. However, higher final sorptivity was found for the cement paste filled with 0.5% and 1.0% GNPs, indicating a negative effect of highly concentrated GOs

Table 6.1 Initial and final sorptivity of cementitious composites with graphene-based nanomaterials [18]

Specimens	*Initial sorptivity ($mm/s^{1/2}$)*	*Changing rate (%)*	*Secondary sorptivity ($mm/s^{1/2}$)*	*Changing rate (%)*	*Reference*
Plain paste	0.0183	—	0.0035	—	Alharbi et al. [19]
0.01% GOs	0.0147	−19.7	0.0028	−20.0	
0.05% GOs	0.0130	−29.0	0.0018	−48.6	
0.5% GOs	0.0170	−7.1	0.0040	+14.3	
1.0% GOs	0.0183	0	0.0042	+20.0	
Plain paste	0.00046	—	—	—	Mohammed et al. [17]
0.01% GOs	0.00060	+30.4	—	—	
0.03% GOs	0.00020	−56.5	—	—	
0.06% GOs	0.00040	−13.0	—	—	
Plain paste	0.0463	—	0.0056	—	Li et al. [20]
0.02% GOs	0.0426	−8.0	0.0036	−35.7	
0.04% GOs	0.0441	−4.8	0.0031	−44.6	
Plain concrete	0.0025	—	—	—	Matalkah and Soroushian [21]
0.1% GNPs	0.0019	−24	—	—	
0.2% GNPs	0.0016	−36	—	—	
0.3% GNPs	0.0019	−24	—	—	
Plain mortar	0.018.5	—	0.00074	—	Krystek et al. [22]
0.01% GNPs	0.0177	−4.3	0.00066	−10.8	
0.03% GNPs	0.017	−8.1	0.00057	−23	
0.05% GNPs	0.0141	−23.8	0.00056	−24.3	
0.075% GNPs	0.0165	−10.8	0.00067	−9.5	
0.1% GNPs	0.0184	−0.5	0.0007	−5.4	

on the cementitious composites. Apart from the cement paste filled with 0.01% GOs [17], all the cementitious composites reinforced with 2D GBNs showed decreased sorptivity when the filler content was lower than 0.5%. It implied that the negative effect was possibly due to the high surface energy of 2D GBNs that easily agglomerated to affect the density and compactness of cementitious composites. Overall, a small proportion addition of well-dispersed GBNs is supposed to enhance the transport properties of GBCCs. In addition, it should also be noted that the quality of the dispersion of GBNs should be guaranteed.

6.2.2.2 Conductive filler with additives

Fig. 6.7a and b show the water absorption and sorptivity of hardened CB/cementitious composites reinforced with different contents of crystalline waterproofing admixture (CWA) and silicone hydrophobic powder (SHP) after 28 days of curing [23]. The composites without additives had the highest rate of water absorption, and the coefficient I_t reached nearly 3.2 on the fourth day. Also, the initial and secondary sorptivity reached 5.5×10^{-3} and 3.2×10^{-3} mm/s$^{1/2}$, respectively. The composites with 1.0% CWA showed the second-highest rate of water absorption, reaching approximately 2.6 on the final measuring day. It possessed a very similar initial sorptivity of 5.6×10^{-3} mm/s$^{1/2}$ to that of plain CB/cementitious composites, while the secondary sorptivity was considerably decreased to 1.7×10^{-3} mm/s$^{1/2}$. It implies that the addition of CWA failed to improve the water impermeability and hydrophobicity of CB/cementitious composites immediately. However, with the entrance of water that could react with CWA, the formed crystals such as calcium carbonate and magnesium carbonate could fill the pores and gradually improve the water impermeability of cementitious composites [24].

The lowest water absorption occurred on the CB/cementitious composites containing 1.0% SHP, whose final water absorption arrived at less than 0.6. In comparison to the counterparts without additives, the addition of SHP significantly declined the water absorption of composites by five times. Moreover, the initial and secondary sorptivity reached the bottom to 7.0×10^{-4} and 6.0×10^{-4} mm/s$^{1/2}$, respectively. Similarly, the

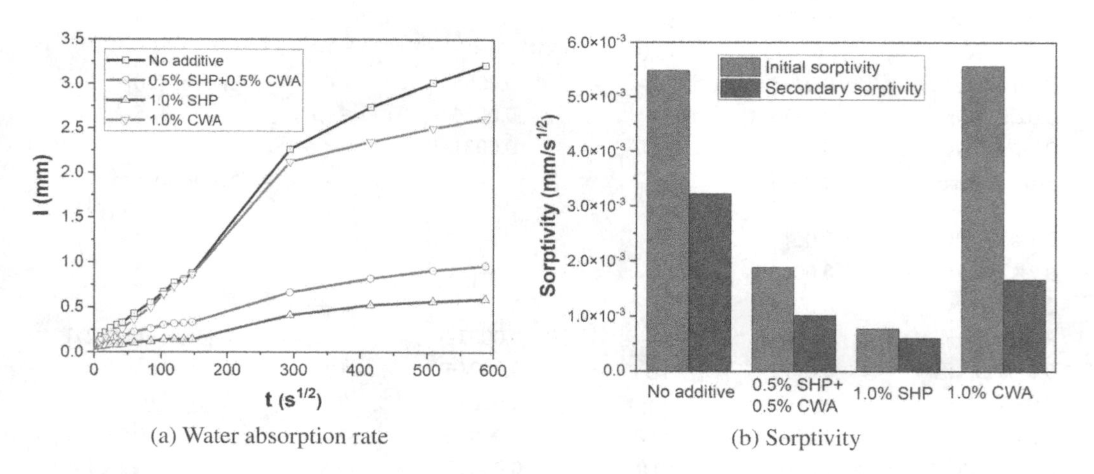

Figure 6.7 Water absorption rate and sorptivity of NCB/cementitious composites containing different amounts of SHP and CWA [23].

Figure 6.8 Chloride penetration depth of CB/cementitious composites with/without CWA and SHP [23].

composites with 0.5% SHP and 0.5% CWA showed an excellent waterproofing ability compared to the plain composites and those only with CWA. The results are consistent with our previous investigations, where we found considerably reduced permeability for the graphene-filled cementitious composites with 1.0% and 2.0% SHP. Generally, the surfaces and pores of cementitious composites are enclosed with hydrophobic silicone/resin particles, which lead to the improved hydrophobicity of composites to block water entrance [25, 26].

Fig. 6.8 shows the chloride penetration depth of CB/cementitious composites with/without CWA and SHP, and Table 6.2 displays the chloride migration coefficient and other parameters of CB/cementitious composites during non-steady-state migration test. In comparison to the composites without additives whose chloride migration coefficient D_{nssm} (measured based on NT BUILD 492) reached 9.03 ± 0.25 ($\times 10^{-12}$ m^2/s), it could be observed that the CB/cementitious composites with 1.0% CWA possessed the highest resistance to chloride penetration, and the migration coefficient reached the smallest value of 5.26 ± 0.06 ($\times 10^{-12}$ m^2/s). In addition, the composites with 0.5% CWA also showed a small chloride migration coefficient with a slightly higher value of 5.88 ± 0.11 ($\times 10^{-12}$ m^2/s). Although the addition of SHP could retard chloride penetration to some extent, it seemed that the efficiency was lower than that of CWA, and the migration coefficient of 6.87 ± 0.14 ($\times 10^{-12}$ m^2/s) was obtained during the non-steady-state migration test. There are several potential reasons for the better performance of CWA than SHP regarding chloride penetration resistance. First, the addition of CWA improves the microstructures through the formation of needle-shaped crystal neoplasms in pores which can disconnect the interconnected matrix pores and reduce the chloride penetration [27]. Second, the penetration of chloride into concrete is mainly through absorption and diffusion. Given the cementitious composites with SHP had lower water absorption than that of counterparts with CA, the lowest chloride migration coefficient indicated that CA had better performance to inhibit the diffusion of chloride than SHP in CB/cementitious composites. The electrical resistivity of composites with CWA was lower because of their higher concentrations of ions in pore solutions, which contributed to the poorer diffusion of chloride ions.

Table 6.2 Parameters of CB/cementitious composites during non-steady-state migration test [23]

Additives	Applied voltage (V) and time (h)	New initial current (mA)	Initial and final temperatures (°C)	Specimen depth (mm)	Chloride penetration depth (mm)	Migration coefficient D_{nssm} ($\times10$–$12\ m^2/s$)
0%	15 and 24	61.8*	21.2 and 23.02 (±0.2)	50.22	10.57 ± 0.24	9.03 ± 0.25
1.0% CWA	15 and 24	82.2	21.5 and 23.0 (±0.2)	49.66	6.76 ± 0.06	5.26 ± 0.06
1.0% SHP	15 and 24	69.9	20.8 and 23.0 (±0.2)	49.95	8.41 ± 0.15	6.87 ± 0.14
0.5% CWA + 0.5% SHP	15 and 24	76.1	21.0 and 23.0 (±0.2)	49.89	7.39 ± 0.12	5.88 ± 0.11

Note: The selection of testing voltage and time are based on scale range of initial current under 30V, and the new initial current means the current value after the application of the selective testing voltage should be within a reasonable range, which are detailed in NT BUILD 492. For example, the 61.8* means the current value under the applied voltage of 15V.

6.2.3 Shrinkage deformation

6.2.3.1 *Drying shrinkage*

The drying shrinkage test was commonly adopted to examine the effect of 2D GBNs on the cracking possibility of cementitious composites. Fig. 6.9 displays the relationship between the GNPs/GOs content and the micro-strain of cement paste drying shrinkage [18]. Zhao et al. [28] investigated the effect of 0.03, 0.06, and 0.09 wt.% GNPs on the cementitious composites with W/B of 0.3. They have pointed out that the proper amount of GNPs can inhibit the drying shrinkage of cementitious composites; when the dosage is 0.06%, the best suppression effect is obtained, and the suppression rate is 27.4%. When the amount of GNPs is less than 0.06%, the GNPs lamellar structure can store a certain amount of water in the gap between the sheets, and the internal curing effect can inhibit the shrinkage of the cement slurry. The microstructure system becomes denser, and the graphene sheet blocks the larger pores, dividing the microcapillary pores in the cement stone into nanoscale capillary pores, play a lapped role, and inhibit the shrinkage of cement paste. However, when the graphene content is too high, it causes agglomeration due to uneven dispersion in cement

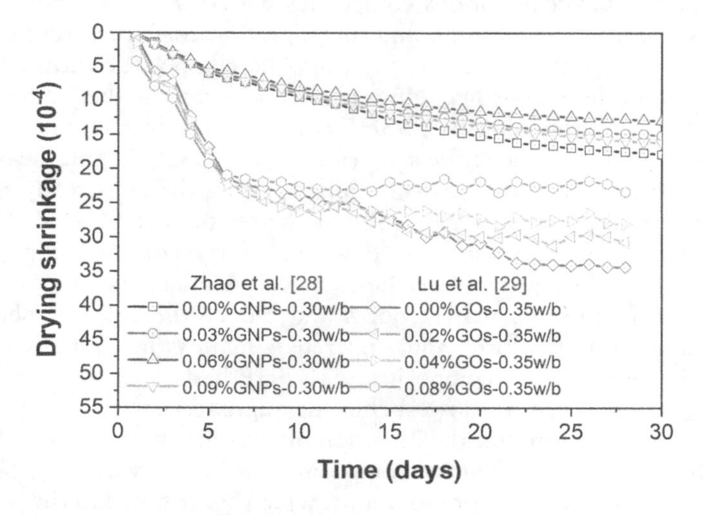

Figure 6.9 Effects of GNPs and GOs on drying shrinkage of cement paste.

matrix, which may weaken the drying shrinkage inhibition effect on the cement-based material. In addition, Lu et al. [29] examined the effect of 0.02%, 0.04%, and 0.08% GOs on the drying shrinkage of cementitious composites with a W/B of 0.35. It shows that the dry shrinkage rate of the cementitious composites added with GOs was slightly higher than that without GOs. The larger the amount of GOs added, the greater the dry shrinkage rate, which is primarily related to the nucleation effect of GOs on hydration reaction. The addition of GOs could accelerate the hydration of the cementitious composites, resulting in a decrease in the total absolute volume and an increase in early auto-shrinkage and chemical shrinkage. After that, due to the hydrophilicity of GOs and the generation of chemical bonds between GOs and cement, the GOs-based cementitious composites showed a lower drying shrinkage rate than that without GOs [30]. On the one hand, the drying shrinkage of the cementitious composites is possibly attributed to the capillary surface tension formed by the meniscus in the capillary pores. Since the amounts of GOs added had a significant effect on reducing the total porosity and the number of capillary pores of cementitious composites, it is understandable to reduce the drying shrinkage in 28 days. More importantly, the higher stiffness, densified, and compact microstructures of the cementitious composites caused by the addition of GOs could also limit the shrinkage. On the other hand, due to its high specific surface area, GOs could absorb water to wet its surface at an early stage. Due to its high specific surface area, GOs can absorb water to wet its surface at an early stage. As the curing age increases, the adsorbed water will be released from GOs into the cement matrix, and the adsorbed water could be released from GOs into the cementitious composites as the curing age increases [31]. The water storage and release from GOs can provide a self-curing effect to the hydration reaction at later ages, which is also beneficial for reducing the drying shrinkage. In summary, the proper amounts of GOs and GNPs are beneficial for inhibiting the drying shrinkage of cementitious composites at both early and later curing ages.

6.2.3.2 Autogenous shrinkage

Fig. 6.10 shows autogenous shrinkage of cement paste reinforced with different dosages of GNPs and GOs in the early curing age [18]. Generally, three stages could be found on the autogenous shrinkage of cementitious composites: the sharply decreased volume in

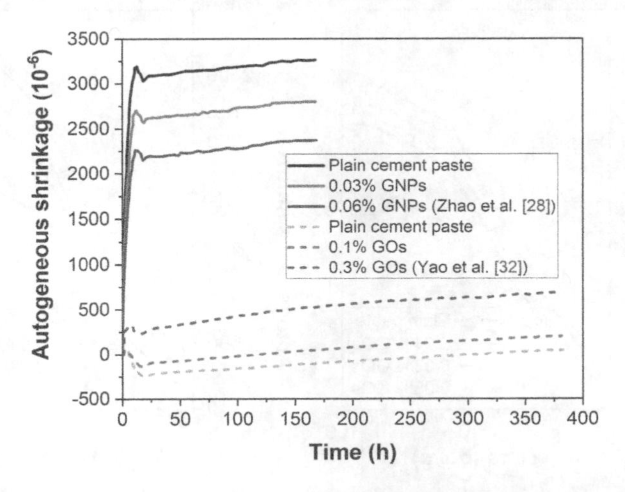

Figure 6.10 Autogenous shrinkage of cement pastes with GNPs and GOs at the early age.

the early stage; the expansion of specimen in a short period; and the linearly reduced volume with increasing curing age. The fast cement hydration in the early age could explain the sharp increase in shrinkage. As for the reduced shrinkage, it was mainly caused by the thermal expansion from cement hydration heat, which failed to be released in time. The reduced volume resulted from the joint effect of continuous cement hydration and reduced water content inside the cement paste. For the specimen with GNPs, the two turning points were 16 h and 20 h, respectively [28]. The GNPs helped to reduce the autogenous shrinkage of cement paste. However, the initially considerable autogenous shrinkage shortened to first 0–6 h for the cement paste with GOs, and the second turning point was promoted to 24 h. The GOs seemed to intensify the shrinkage of cement paste. The discrepancy possibly was due to the different water-to-binder ratios in the two studies. Also, it could be observed that smaller autogenous shrinkage of cement paste cast by Yao et al. [32] mainly contributed to the larger water-to-binder ratio.

Fig. 6.11 shows the autogenous shrinkage of alkali-activated slag cement mortar reinforced with different contents of GNPs and GOs from 1 day to 100 days [18]. Similarly, Yao et al. [32] observed an increased autogenous shrinkage of cement mortar filled with GNPs and GOs. Because the autogenous shrinkage of cement mortar was controlled by fine pores, they attributed the enhanced autogenous shrinkage adding GNPs and GOs [33]. This is consistent to previous studies, which proposed the enhanced pore fineness of cementitious composites with GNPs or GOs [34]. In general, the effect of GNPs was larger than that of GOs on the cement mortar with larger autogenous shrinkage, indicating a better performance of GNP to increase the pore fineness. As for the cementitious composites containing cellulose nanofibers, investigators found that cellulose nanofibers can not only slow down the autogenous shrinkage of cement slurry but also effectively suppress shrinkage cracking. The reduction of shrinkage is more pronounced in mortar than in cement paste because of the existence of aggregate [35].

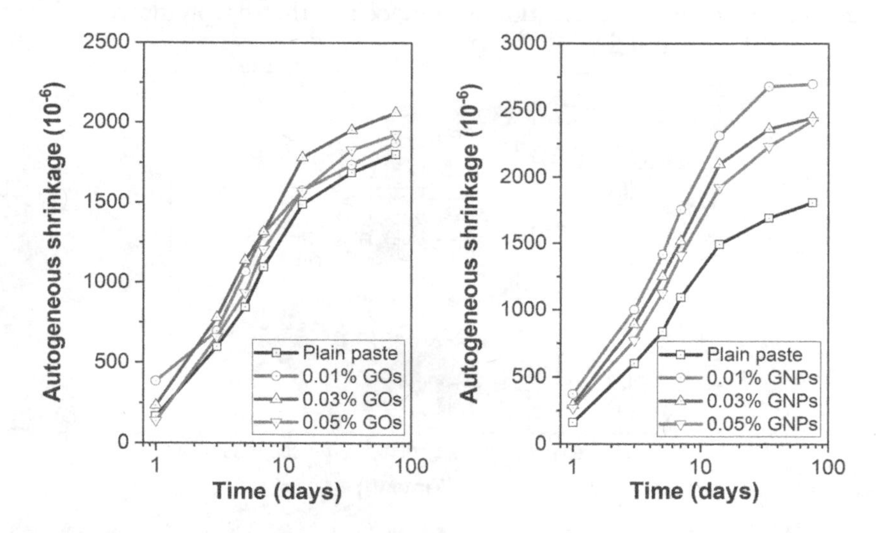

Figure 6.11 Autogenous shrinkage of cement paste filled with GNPs and GOs in long-term age [33].

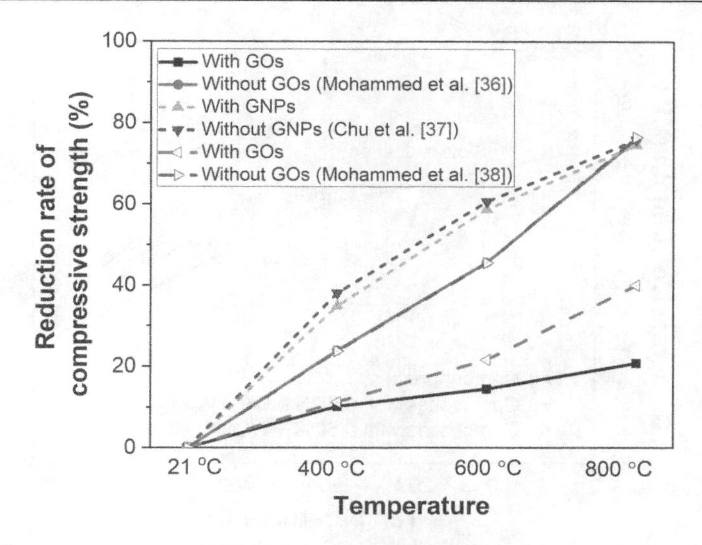

Figure 6.12 Reduction of compressive strength of cementitious composites with GNPs or GOs.

6.2.4 High-temperature performance

6.2.4.1 Mechanical properties

Fig. 6.12 displays decreasing rate of compressive strength of cementitious composites containing with or without GNPs and GOs after exposure to elevated temperatures of 400°C, 600°C, and 800°C [18]. It happens due to deterioration of hydrates and increase of porosity. It could be observed that the composites with graphene nanomaterials possessed lower reduction rate than the plain counterparts, indicating the enhanced resistance to high temperature of cementitious composites with graphene nanomaterials [36–38]. Previous studies also proposed that the nanomaterials could work as additional channel to release the heat toward outside, which could be one of the reasons for the better resistance to elevated temperature [39]. Moreover, given the modification of pore structures of cementitious composites with graphene nanomaterials, the improved fineness of gel porosity and reduced capillary pores could be another reason for the enhanced high-temperature resistance. In summary, cementitious composites incorporating 2D graphene-based nanomaterials are provided with enhanced resistance to elevated temperature.

Fig. 6.13 depicts the compressive strength of plain cement paste and MWCNT-reinforced cementitious composites with and without the treatment of elevated temperatures at 300°C and 600°C [39]. Significant strength enhancements were observed from the MWCNT-reinforced cementitious composites. The compressive strength was approximately 46 MPa for specimens with no MWCNT, 53 MPa for specimens with 0.25% MWCNT, and 52 MPa for specimens with 0.50% MWCNT. The slightly lower strength of specimens with 0.50% MWCNT, in comparison with that of specimens with 0.25% MWCNT, might be due to the poor dispersion of the MWCNT. After the elevated temperature of 300°C treatment, all the composites underwent a drop on the compressive strength, probably related to calcium silicate hydrate (C–S–H) and the $Ca(OH)_2$ decomposition, which quickly released a large amount of CaO, gas, and water and generated a large internal pressure, thus enlarging pore size and microcracks in the composites [40].

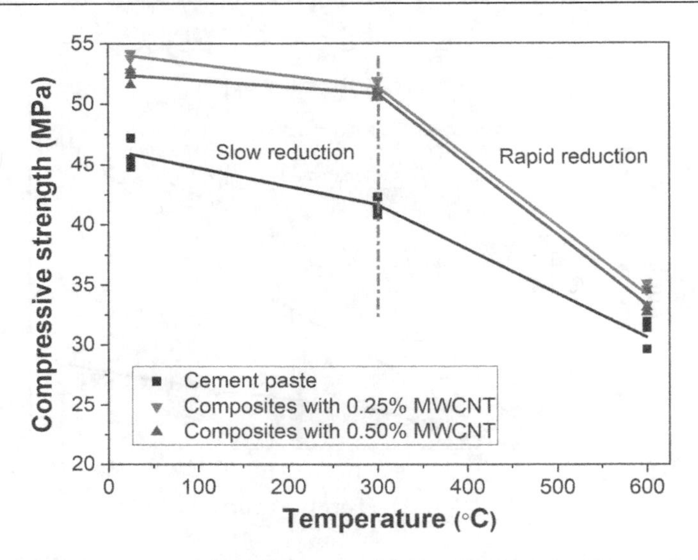

Figure 6.13 Compressive strength of cement paste and MWCNT/cementitious composites exposed to elevated temperature [39].

The plain cement paste went through a most severe reduction to nearly 42 MPa, while less reduction was observed for the composites with MWCNT, especially the composites reinforced with 0.50% MWCNT, whose compressive strength was well preserved to slightly over 51 MPa. Since the main components were not greatly decomposed at the elevated temperature of 300°C and the possibility for the further cement hydration process due to the water vapour, the compressive strength of cementitious composites could be maintained to same extent. In addition, the reason for the less strength reduction on the MWCNT-reinforced cementitious composites was owing to the heat and steam transmission effects by small fibers [41]. The more the contents of MWCNT inside the composites, the higher the efficiency for the transmission, and that is why the composites with 0.50% MWCNT suffered less negativity than the counterparts with 0.25% MWCNT on the strength reduction. It seems that the high-pressure steam could be easily transferred to the external through the hollow MWCNT. Moreover, the added fibers often increase porosity of cement matrix, and the additional pores might help to release pressure inside the composites, to avoid the generation of pores and cracks. For the composites subjected to temperature of 600°C, more rapid compressive strength reduction was observed, with 31 MPa for the composites with no MWCNT and nearly 33 MPa for the composites with 0.50% and 0.25% MWCNT. It demonstrates that the MWCNT failed to improve the compressive strength of cementitious composites. During this temperature range, results show that not only more C–S–H gel was decomposed but also MWCNTs are stable up to temperature of 400°C and start degrading after the initial combustion of amorphous carbon [42].

To further understand the effects of elevated temperature treatment on the mechanical characteristics of MWCNT-reinforced cementitious composites, Fig. 6.14 displays the stress–strain curves and the elastic modulus of plain cement paste and MWCNT-reinforced composites without heat treatment and after going through elevated temperatures of 300°C and 600°C [39]. Apart from the decreased ultimate compressive strength,

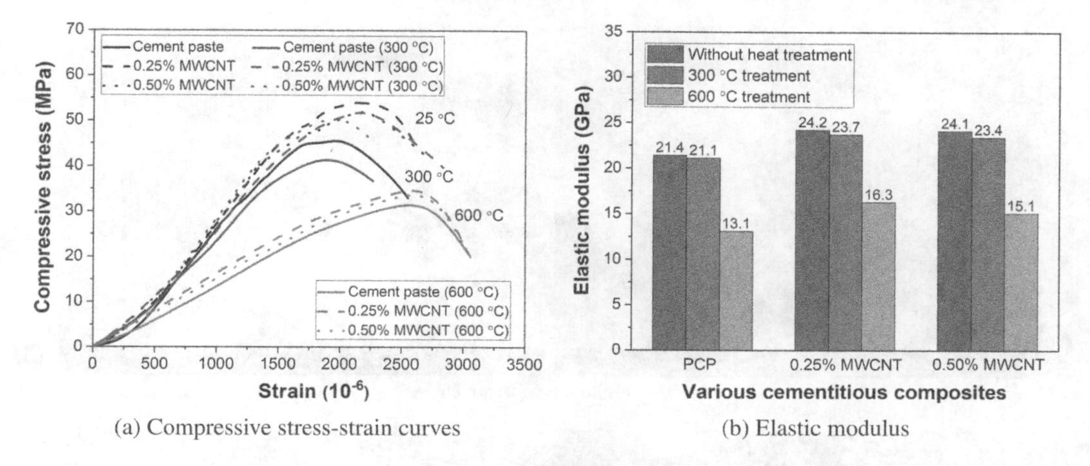

(a) Compressive stress-strain curves (b) Elastic modulus

Figure 6.14 Characteristics of cement paste and MWCNT/cementitious composites exposed to elevated temperature [39].

the peak strain increased, and elastic modulus decreased after elevated temperature treatment. Before the high-temperature treatment, the improved elastic modulus was mainly due to the dense microstructures and better ductility of composites reinforced by CNTs [43]. Besides, the reinforcement effects also played a positive impact even after the 300°C treatment, which demonstrated a relatively good cohesion between MWCNT and cement matrix at the temperature of 300°C [44]. However, when the temperature increased to 600°C, the elastic modulus was considerably decreased because of both damaged MWCNT and decomposition of cement hydration products, which contributed to the debonding between fibers and cement matrix, and lower the stiff of composites. Moreover, another potential reason for the decreased elastic modulus might be due to the thermal expansion of specimens, which will greatly decrease the ductility of specimens.

6.2.4.2 *Microstructural properties*

In this section, the microstructures of 0.25% MWCNT-reinforced cementitious composites without any heat treatment and after elevated temperature treatments of 300°C and 600°C could be investigated by the SEM analysis, to further demonstrate the mechanical properties and the later electrical properties in a microcosmic perspective. All the chosen regions were displayed in small and large magnifications for easier observation and comparison.

Fig. 6.15a illustrates the microstructures of composites at room temperature without any heat treatment [39]. Generally, intact microstructures could be detected both on the hydration products of C–S–H and the MWCNT. Just as many literatures mentioned reinforcements by MWCNT, the strengthen effect was mainly due to the MWCNT' bridging and filling effect [25, 43, 44]. In SEM images with small and large magnifications, it was observed that C–S–H and MWCNT maintained excellent connections, which were responsible for the improved compressive strength and elastic modulus for the MWCNT-reinforced composites. In addition, the nucleation effect of MWCNT could be deduced from the SEM images, due to different thicknesses of MWCNT with attached hydration products observed. For the composites after 300°C heat treatment, similar

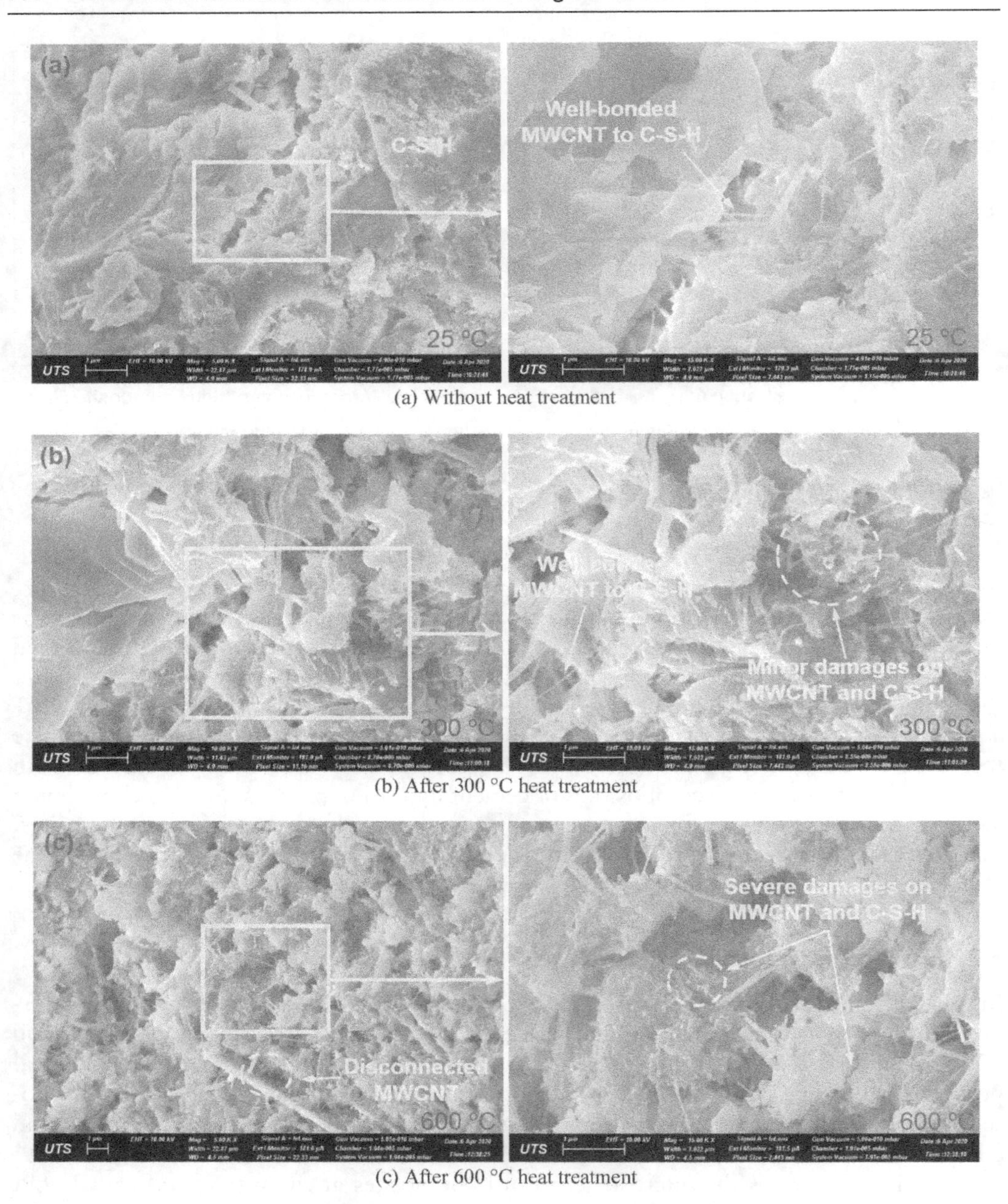

(a) Without heat treatment

(b) After 300 °C heat treatment

(c) After 600 °C heat treatment

Figure 6.15 Microstructure morphology of cementitious composites incorporating 0.25% MWCNTs [39].

microstructures of C–S–H and MWCNT could be observed in Fig. 6.15b. However, apart from the well-bonded MWCNT to C–S–H, minor damages could be observed on both the MWCNT and hydration products. In comparison to the composites without heat treatment, a small-scale destruction on the MWCNT and C–S–H demonstrated the microstructures interference by the heat treatment and the induced slight mechanical properties reduction. Furthermore, the MWCNT with similar thickness and morphology indicated that several hydration products and impurities attached on the surface of MWCNT might be eliminated or detached after heat treatment, which lays the basis for the better conductivity between connected MWCNT and the ameliorated piezoresistive properties. In terms of the 600°C heat-treated composites, severe destructions on the MWCNT and the cement hydration products could be detected in Fig. 6.15c, where worse connections between MWCNT and hydration products were found because of the disconnected MWCNT and damaged C–S–H. The phenomenon is mainly due to the decomposition of hydration products under elevated temperatures [45]. It indicates that the decomposition of hydration products directly led to the decreased mechanical properties, while the residual well-bonded MWCNT to hydration products slightly alleviated the negativity. Moreover, it could be detected that a few MWCNT might be broken to several smaller MWCNTs under elevated temperatures because of their initial microstructural instability and defects [36], which was why the MWCNT-reinforced composites greatly decreased their conductivity and only possessed a little higher compressive strength and elastic modulus than the plain cement paste. In addition, even though the broken MWCNT could greatly increase the electrical resistivity of the composites, it was concluded that the heat treatment also contributes to the improved piezoresistivity, which will be mentioned particularly later.

6.2.4.3 Physicochemical phases

Fig. 6.16 shows the pyrolysis characteristics of MWCNT-reinforced cementitious composites with/without heat treatments in the atmosphere of nitrogen by TG analysis [39]. The results of 0.25% MWCNT cementitious composites with the heat treatment of 300°C were similar to the counterparts with 0.50% MWCNT which were between the results of composites without heat treatment and with 600°C. Therefore, only the composites without heat treatment and either 0.25% or 0.50% MWCNT cementitious

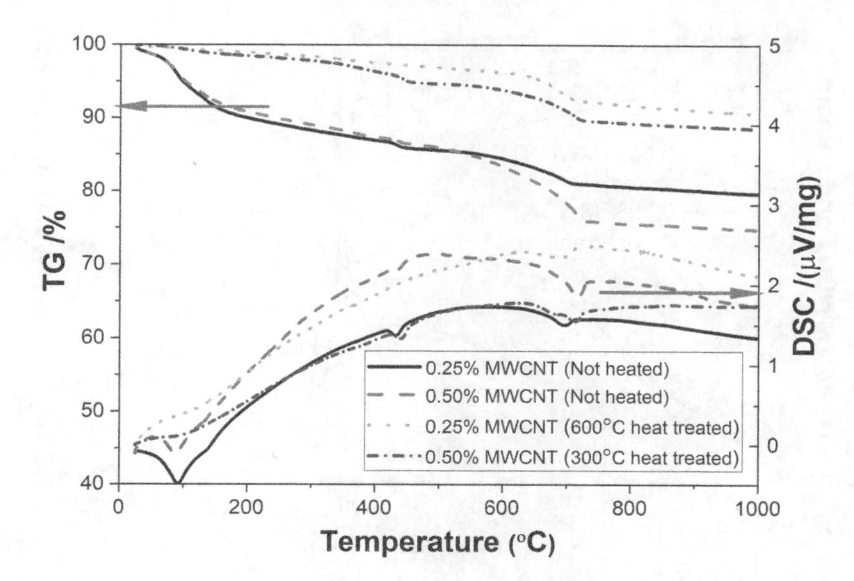

Figure 6.16 Thermogravimetric analysis of MWCNT/cementitious composites under heat treatments in nitrogen atmosphere [39].

composite subjected to 300°C and 600°C were tested in this study, to compare their differences more clearly. The X coordinate axis means the increase of temperature, and the left and right Y coordinate axes present the rate of loss weight and values of differential scanning calorimetry (DSC), respectively. In comparison to the composites without heat treatments, the composites after heat treatment had much lower weight losses, and especially for composites after exposure to 600°C heat treatment. This is easy to understand because of the decomposition of ettringite [45], which was observed in the first peak from approximately 60°C to 120°C for the composites without treatment (red and black lines). The second peak around the temperature of 450°C was the decomposition of calcium hydroxide [46]. However, it could not be observed for the composites after 600°C heat treatment, because of the calcium hydroxide destruction. As for the third peak in the range of 650–700°C, all the composites with/without heat treatment could be observed, it was considered that the main pyrolyzation was the decomposition of calcium carbonate, which produced calcium oxide and carbon dioxide, and was responsible for the dramatic strength reduction. Since the composites were in the atmosphere of nitrogen, the MWCNT could not become oxidized and decompose in the heating process.

On the other hand, the thermal gravity performances in the air atmosphere for the 0.50% MWCNT-reinforced cementitious composites are represented in Fig. 6.17, accompanied by the results of as received MWCNT powders without any cement materials [39]. The coordinate axes present identical meanings to Fig. 6.17, while the outer axes mean the results by as received MWCNT powders (red lines) and the inner axes are for the cementitious composites with MWCNT. In the air atmosphere, it was observed that the majority of MWCNT was decomposed from approximately 550–680°C, with only small amount of impurities got oxidized or remained outside the ranges, which was consistent to the studies on the purification of MWCNT by gas-phase oxidation [47]. In addition, different from the decomposition of cementitious composites which always absorb extra

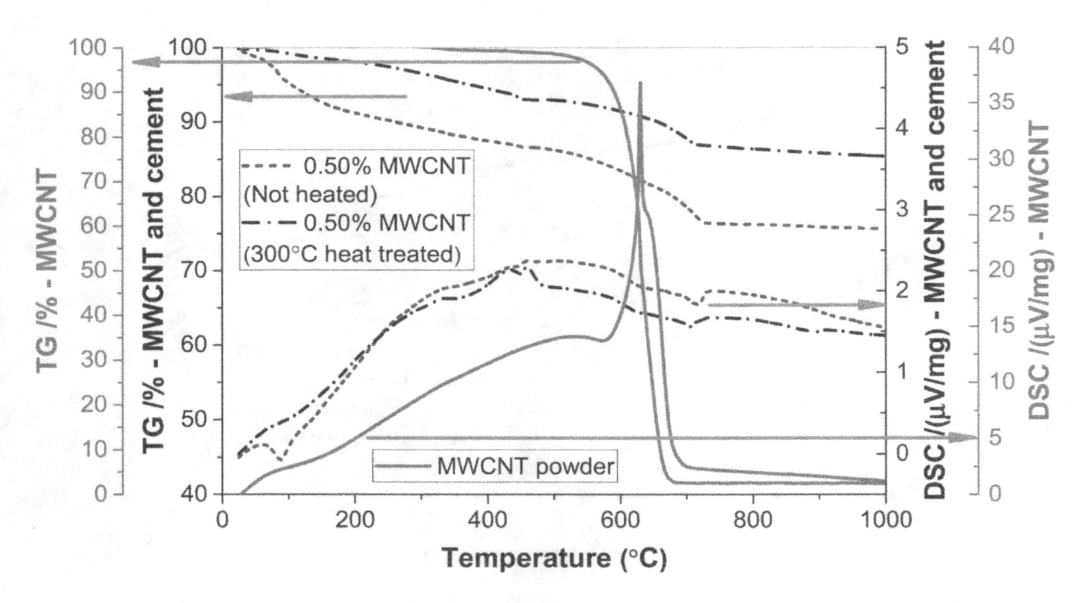

Figure 6.17 TG analysis of MWCNT powder and MWCNT/cementitious composites under heat treatments in air atmosphere [39].

energy, the oxidation of MWCNT greatly released heat energy which illustrated as a reverse peak for the DSC values (red curve in the range of 590–680°C). For the cementitious composites with MWCNT in the air atmosphere, since the content of MWCNT was much less to the weight of cementitious composites (0.50% to weight of binder and <0.50% to weight of composite), the release of heat energy was not clear in the heating process. Nevertheless, there are two peaks different from the curves of composites at nitrogen atmosphere. The first one is the energy absorption before temperature of 400°C where the mass began to lose for the MWCNT powders, which may be due to the oxidation of impurities and broken MWCNT. The second occurred before the decomposition of calcium hydroxide, where a small peak existed at the temperature slightly higher than 600°C for the MWCNT/cementitious composites under heat treatments. This could be due to the heat energy absorption for the preparation of MWCNT oxidation, and the later small opposite peak is due to the exothermic oxidation of MWCNT. The reason for the relatively higher temperature for the MWCNT oxidation in the cementitious composites is due to the lower heat transfer efficiency for cement hydration products.

Fig. 6.18 presents the XRD results of 0.50% MWCNT-reinforced cementitious composites without heat treatment and under different temperature heat treatments, in order to evaluate the deterioration of crystalline phases [39]. For easier observation on the crystalline phases in different conditions, the curves treatment of baseline correction was neglected. It was observed that the ettringite completely decomposed in the process of heat treatment at the temperatures of 300°C and 600°C, which were consistent to the TG results in this study and the literature that the decomposition temperature of ettringite was under 120°C [48]. In addition, based on the TG analysis, the results show that the decomposition temperature of calcium hydroxide was nearly 450°C, the bonds between OH^- and Ca^{2+} were gradually impaired with the increase of temperature, and that was why the crystal of calcium hydroxide decreased after heat treatment. Since the XRD tests could not quantitatively assess the number of crystals in the composites, OPC clinkers such as C2S or C3S could be observed for both composites with/without heat treatments.

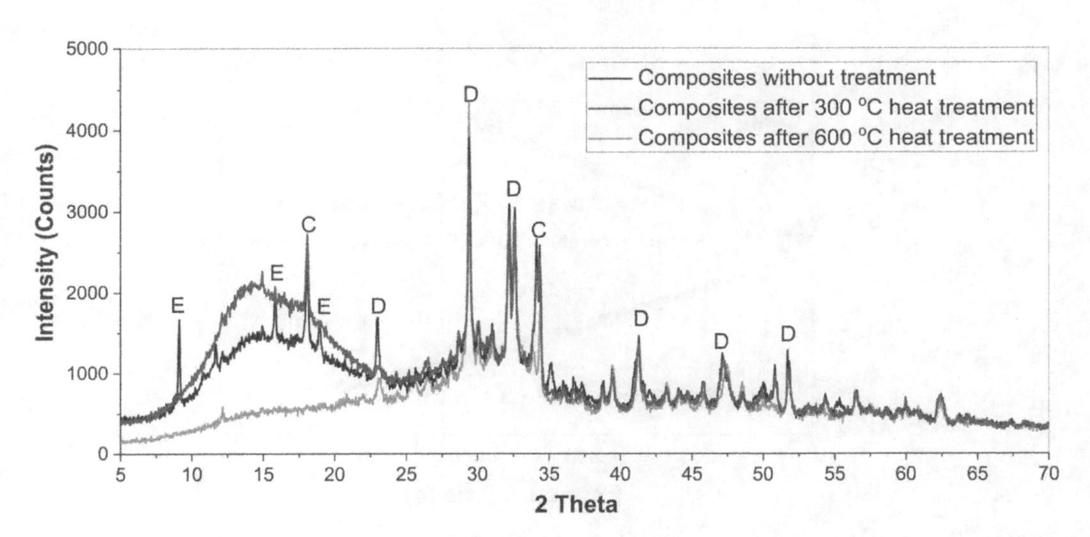

Figure 6.18 XRD results of MWCNT/cementitious composites under different heat treatments (E = ettringite; C = calcium hydroxide; D = OPC clinkers) [39].

6.3 IMPACT OF VARIOUS ENVIRONMENTS ON SENSORY PERFORMANCE

6.3.1 Temperature

6.3.1.1 Temperature to resistance

Fig. 6.19 illustrates the FCR variation of CB cementitious composites exposed in the ambient environment of 20°C and 60% relative humidity without any external forces [49]. It represented an opposite tendency among the heated and the frozen composites, with the increased resistivity by the heated specimens and the decreased resistivity by the frozen ones. Based on the theory of thermal exchange, this phenomenon is owing to their different exchange patterns. Since the composites under high temperature tended to emit heat into the ambient and decreased temperature until reaching equilibrium, the resistivity of composite decreased gradually with the exposure time. On the contrary, the frozen composites had a trend to absorb heat from the ambient environment and increased the temperature to decrease their electrical resistivity. Generally, curves of the relationship between FCR and exposure time from both thermal radiation and absorption of the CB/cementitious composite are in linear relationship, with the slope fluctuation as low as 10^{-5}. After 15 min of air exposure, the FCR changes for the thermal absorption reached higher than 30%, while the values were even higher for the composites under thermal radiation and achieved nearly 40%. The results are probably out of their different degree of temperature deviations to the room temperature, which reaches at 80°C for the heated composites (under 100°C) and only 40°C for the frozen ones (under –20°C). In other words, the fractional changes of resistivity for the CB/cementitious composites highly depend on the temperatures. Thus, unless the compression tests on these high/subzero temperature CB/cementitious composites were carried out in the incubator with identical environmental temperature, the FCR alterations induced by the thermal exchanges was not excluded from the next section of piezoresistive test on the CB/cementitious composite under various temperatures.

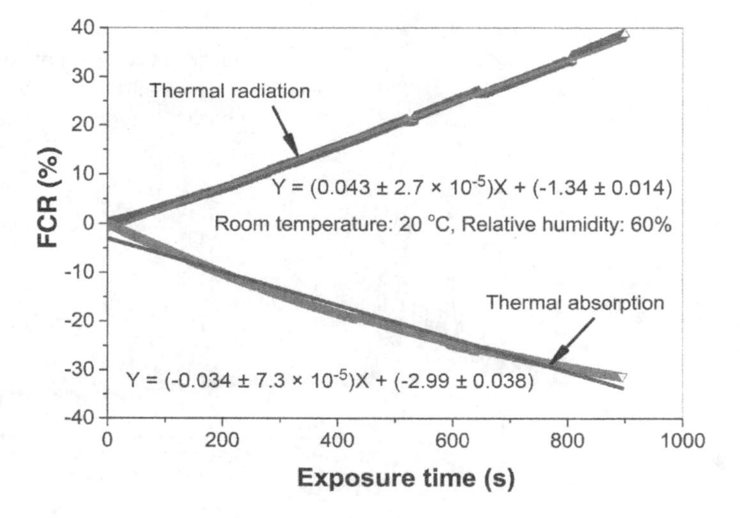

Figure 6.19 Relationship between thermal radiation/absorption and FCR of dry CB cementitious composites under temperature of 20°C and relative humidity of 60% [49].

6.3.1.2 Piezoresistivity during various temperature

Under the loading regime of three cycles and the stress magnitude of 2 MPa as loading limit, Fig. 6.20 shows the FCR and strain development of composites at different temperatures of –20°C, 20°C, 60°C, and 100°C [49]. The calibrated FCR is measured by removal of resistivity changes out of temperature reduction during compression. It was observed that only the specimens at the room temperature of 20°C exhibited excellent repeatability on electrical resistance, while all the other cementitious composites at the higher or lower temperatures performed gradual increase or decrease in the loading process and led to the irreversible FCR changes. Based on the aforementioned results, the original FCR changes (dotted red lines) were generated from the combined action of both cyclic compressive stress and the thermal exchange. Therefore, to eliminate the effect of thermal radiation and absorption on the irreversible resistivity of cementitious composites, the calibrated FCR (red lines) on the basis of the decreasing or increasing FCR of untreated specimens in the ambient environment were carried out. Overall, after the temperature calibration of FCR to remove the effect of thermal exchange, all the heated or frozen CB-filled cementitious composites showed good synchronicity and repeatability to compressive stress. The results illustrated that the temperatures themselves from –20°C to 100°C cannot affect the repeatability of CB/cementitious composites. However, under the extreme circumstances with fire attacks, concrete infrastructures embedded with CB

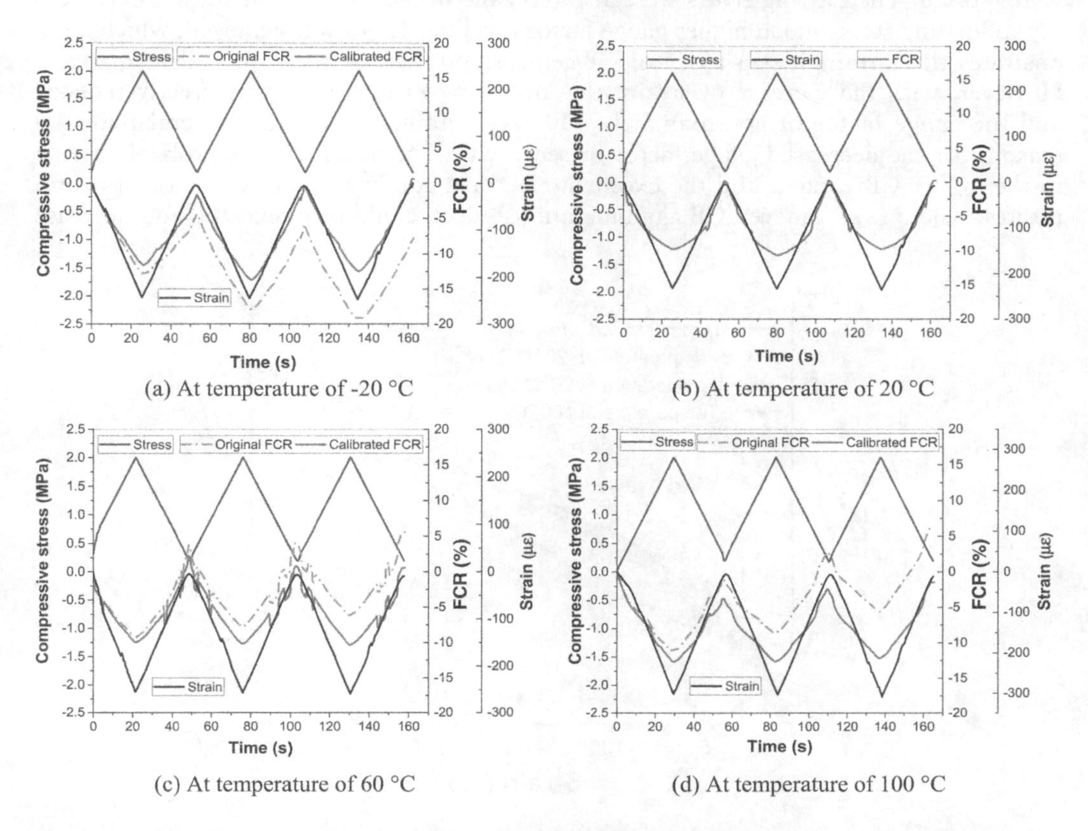

Figure 6.20 The FCR of dry CB cementitious composite at different temperatures under cyclic compression [49].

cementitious sensors might undergo temperature higher than 450°C. The repeatability of CB/cementitious composites is still unknown for both the reduced conductor content because of the oxidized carbon black particles and the damaged C–S–H gels [50].

In terms of the sensitivity of CB cementitious composites under different temperatures, assessments were conducted through the FCR alterations by per unit strain which is also known as the gauge factor. It indicates that during a loading process, both the changes of FCR and compressive strain are of importance to the sensitivity of cementitious composites. Fig. 6.21 compares the relationship between their strain and FCR changes among CB/cementitious composites at different temperatures [49]. The composites under 100°C were provided with largest FCR of 12.6% as well as producing the largest strain values. However, both smallest FCR changes and compressive strain were generated for the composites at the room temperature. Even though there were small differences between composites at different temperatures, two sections could be divided according to the curves of FCR to strain. It seems that higher fractional changes of resistivity occurred at low strain, while the resistivity changes decreased when the compressive strain reached the limit of 30×10^{-6}. This is consistent to the investigations by Pang et al. [51] who proposed that the reduced interfacial distance between conductive fillers and cement matrix was responsible for the swift resistivity alterations during initial loading. To describe their differences on the gauge factor and sensitivity, the fitting formulas of these curves in the two sections were plotted. Their fitting errors were also followed in the formulas. It shows that in the initial loading stage, much higher gauge factor reaching 1,240 was achieved, which demonstrates the ultrahigh sensitivity of CB/cementitious sensor with small deformations. However, with the increase of strain, the increasing rate of FCR was greatly reduced and the gauge factor of approximately 330 was obtained. There are several potential reasons for the decreased gauge factor or sensitivity. The first reason is probably owing to the higher CB content and the excellent electrical conductivity of cementitious composites, where exists uneven CB agglomerations which could be connected and decrease

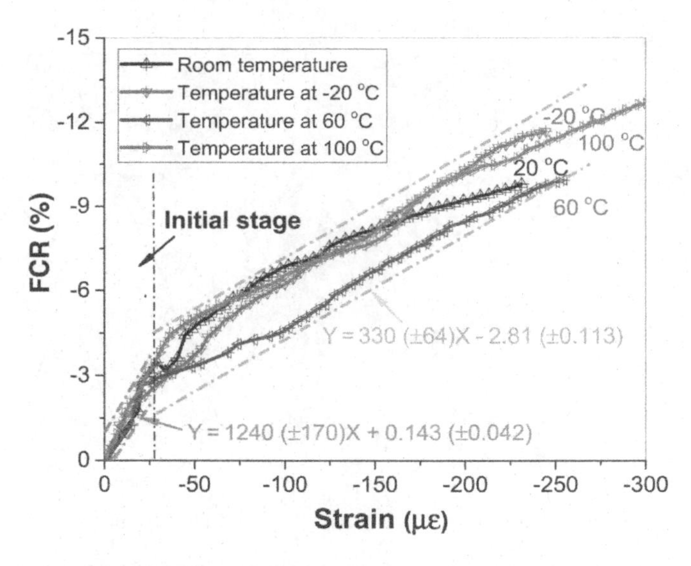

Figure 6.21 Relationship between FCR and strain for CB cementitious composites under various temperatures [49].

the resistivity with very small deformations. The second is because of the minor pores in the CB/cementitious composites, where physical and electrical properties were non-linear and vulnerable to the deformations and led to the damaged microstructures and decreased gauge factor. Furthermore, as mentioned above, the sudden resistivity reduction was likely due to the easily reduced interfacial distance between CB particles and cement matrix [51].

Overall, the experimental results illustrate that both the repeatability and sensitivity of 3% CB-filled cementitious composites make no differences on account of different application temperatures from –20°C to 100°C. However, the temperature changes or thermal exchange during the service of CBSs affected its electrical characteristics and must be calibrated. In addition, the sensitivity was influenced by different loading stages, with the much higher sensitivity at the initial loading stage and the relatively lower but more stable sensitivity in the later loading stage.

6.3.1.3 Piezoresistivity after elevated temperature treatment

Fig. 6.22a–c illustrates the cyclic compressive stress, strain, and the fractional changes of resistivity of 0.25% MWCNT-reinforced composites without treatment, after 300°C and 600°C heat treatments, respectively [39]. Generally, a decreased electrical resistivity in the loading process and an increased resistivity in the unloading process could be observed for all composites. For the composites without heat treatment, the fractional changes of

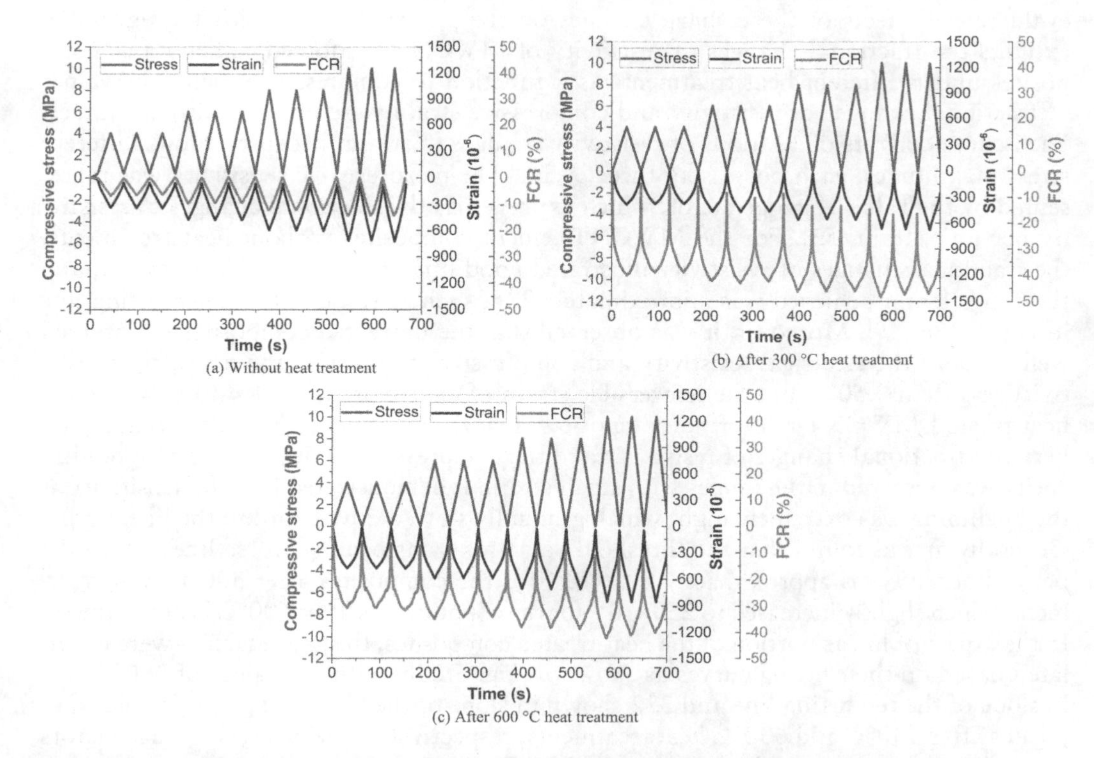

Figure 6.22 Fractional changes of resistivity of 0.25% MWCNT/cementitious composites related to compressive stress and strain [39].

resistivity reached 6.2%, 9.4%, 12.1%, and 14.5%, respectively, at the stress magnitudes of 4 MPa, 6 MPa, 8 MPa, and 10 MPa, exhibiting acceptable linearity and excellent reversibility in Fig. 6.22a. It demonstrated that the MWCNT-reinforced cementitious composites are capable to serve as CBSs for SHM [52]. However, for the elevated temperatures treated composites, considerable increases of fractional changes of resistivity were observed in Fig. 6.22b and c, with resistivity plunges and leaps at the beginning and end of loading. Afterward, the fractional changes of resistivity were similar to that of untreated MWCNT/cement composites and showed almost linear alterations with compressive stress. In the loading peak, it was seen that the sharp reduction of resistivity reached from 25% to 33% for the composites in different stress magnitudes after 300°C heat treatment, which greatly overwhelmed the later fractional changes of resistivity with increase of loadings, since the total fractional changes of resistivity were in the range of 40–50%. Furthermore, the irreversibility was gradually increased with loading cycles and stress magnitudes, with the average permanent resistivity reduction by 5.2%, 8.9%, 13.5%, and 16.2% in the stress magnitudes of 4 MPa, 6 MPa, 8MPa, and 10 MPa. In terms of the composites after 600°C heat treatment, similar results were observed on the sudden resistivity changes when small forces were applied and the later more stable and linear resistivity changes. Nevertheless, both the total fractional changes of resistivity and the sharp alterations were less in comparison to the counterparts after 300°C heat treatment, with the former in the scope of 30–40% at the stress peak and the latter within 20–30% based on different stress magnitudes.

To further understand the alteration patterns of fractional changes of resistivity and evaluate the effects of these sharp changes on the piezoresistive sensitivity, Fig. 6.23a exhibits the fractional changes of resistivity of MWCNT-reinforced cementitious composites under different heat treatments as a function to compressive strain. The values of fractional changes of resistivity and compressive strain were both in absolute values. Each curve consists of 12 loops of resistivity changes to strain, because of four different stress magnitudes, each possessing three cycles. The piezoresistive sensitivity was represented by the value of gauge factor, which expresses as the fractional changes of resistivity per unit strain [53]. For the MWCNT/cement composites without heat treatments, the fractional changes of resistivity illustrated good linearity to compressive strain, and the gauge factor achieved was approximately 228, such as slope of the green fitting line in Fig. 6.23a [39]. Moreover, it was observed that the untreated composites maintained well on both the electrical resistivity and compressive strain, with the irreversible resistivity less than 0.50% and the irreversible strain of approximately 50×10^{-6}. As for the heat-treated MWCNT-reinforced composites, before the reluctantly linear relationship between fractional changes of resistivity and the compressive strain, a significant nonlinearity was observed in the ranges with small strain. It demonstrated that the sensitivity at the beginning was extremely high, which gradually slowed down until to the linear part. Generally, it was found that the threshold value for transforming worse linearity to the better linearity was approximately 200×10^{-6} for the composites after 300°C heat treatment, which slightly increased to 250×10^{-6} for the counterparts after 600°C heat treatment. In the aspect of linear portion of the heat-treated composites, the gauge factors were calculated based on their fitting curves as shown in Fig. 6.23b, with the values of 380 shown in slope of the red fitting line and 322 shown in slope of the blue fitting line for the composites after 300°C and 600°C heat treatments, respectively. Obviously, the gauge factors of the linear portion for the heat-treated MWCNT-reinforced cementitious composites were higher than that without heat treatment, demonstrating the better piezoresistive

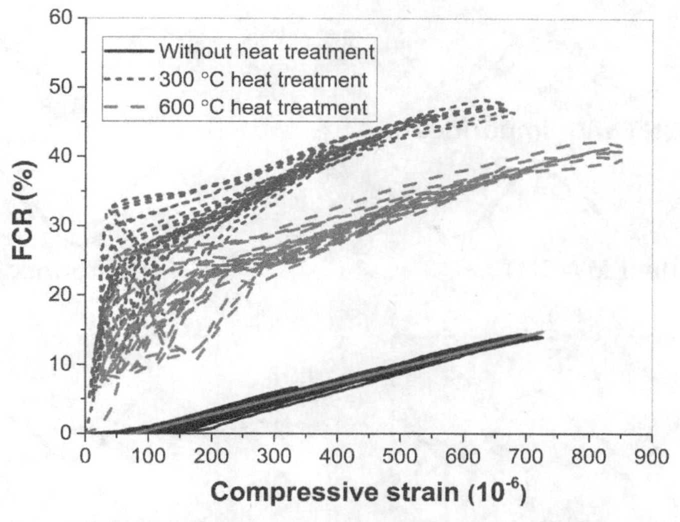

(a) Complete curves of FCR to compressive strain

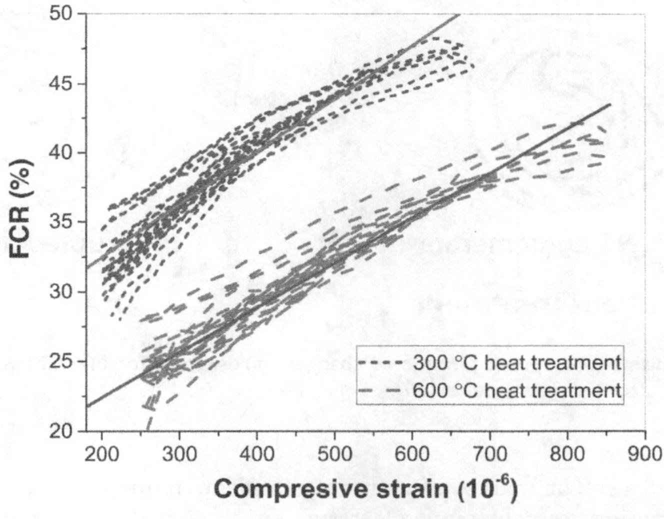

(b) Linear part of FCR to compressive strain

Figure 6.23 Fractional changes of resistivity as a function to compressive strain of cementitious composites incorporating 0.25% MWCNT after heat treatments [39].

sensitivity in the larger strain regions for the composites after the heat treatment. The results indicate that the heated MWCNT–composites at the temperature of 300°C were provided with best sensitivity.

6.3.1.4 Mechanism discussion

To elucidate the sharp alterations and increased fractional changes of resistivity of the heat-treated MWCNT-reinforced composites, Fig. 6.24 displays four possible transformations

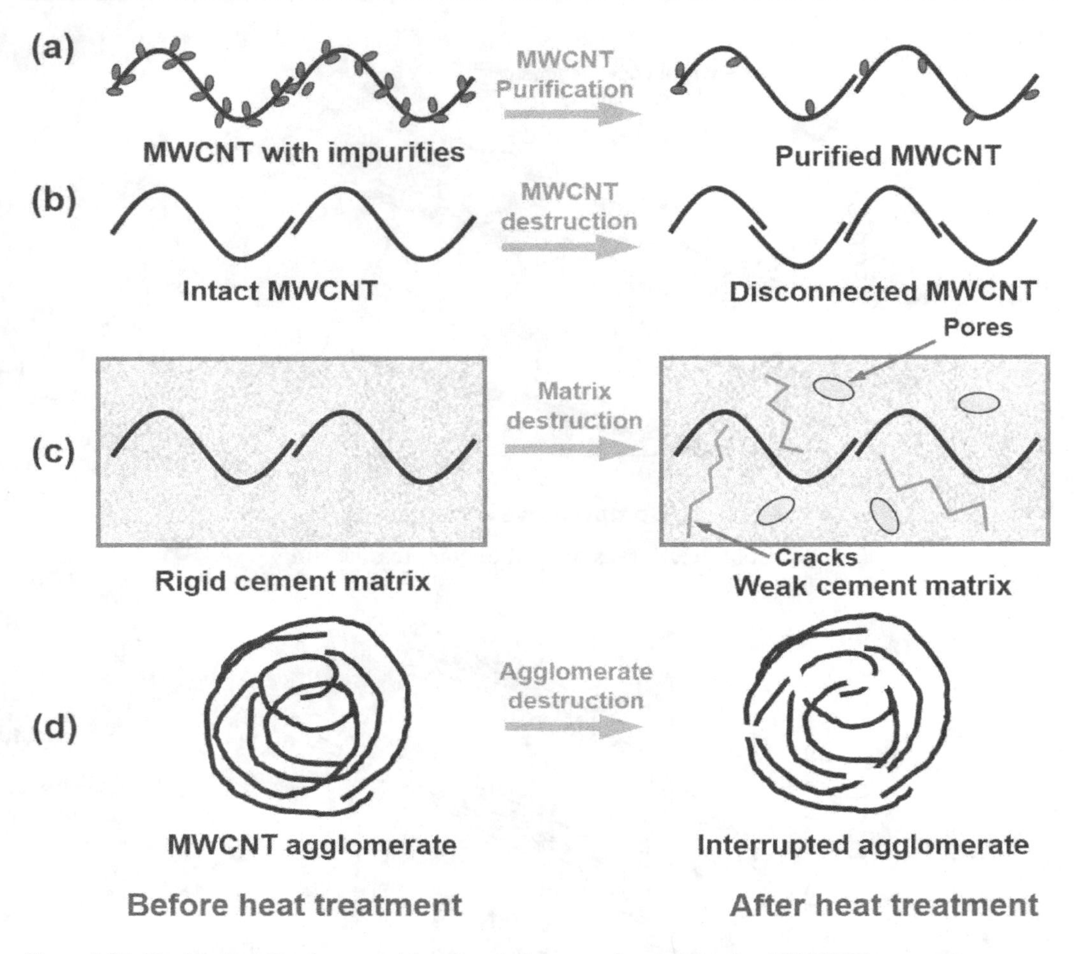

Figure 6.24 Mechanisms for improved fractional changes of resistivity for MWCNT/cementitious composites after heat treatments [39].

of MWCNT and cement matrix after heat treatment, named MWCNT purification, MWCNT destruction, cement matrix destruction, and agglomerates destruction [39]. Since the cementitious composites had been dried before heat treatment, the effects of pore solutions and conductive ions that significantly affect the electrical resistivity and piezoresistivity will not be mentioned here. The manufacturing of MWCNT is always accompanied by introducing metal catalysts, and the final MWCNT products often possess more or less impurities, such as residual metal particles, carbon nanofibers, graphite, and amorphous carbon, which somewhat influence their electrical properties. The weight loss of as received CNTs at high temperature includes the oxidation of carbonaceous impurities [47]. It has been proposed that the temperature ranges of gas-phase oxidative purification of CNTs are from 225°C to 760°C under an oxidizing atmosphere [54]. The temperatures of 300°C and 600°C in this study did have the capacity to eliminate a proportion of carbonaceous impurities and affect the electrical conductivity of MWCNT. In addition, by means of TG analysis and electron dispersion X-ray (EDX), Porro et al. [55]

observed the presence of iron and oxygen in CNTs due to the generation of Fe_2O_3. Since these conductive metal particles were attached into the surface of nanotube, the oxidation in nanotube bodies must affect the electrical conductivity. This could be further demonstrated by the different electrical conductivity for the annealed MWCNT with different weight losses [56]. Since the intact MWCNT have higher oxidation temperature, less impurities were found on the mixtures after the severe oxidation process by heat treatment over 500°C. As shown in Fig. 6.24a, the impurities attached in the external surface of MWCNT might block the electrical contact between nearby MWCNT, while the heat treatment could eliminate some of impurities and make the MWCNT easier to contact with each other, which to some extent contribute to the higher fractional changes of resistivity under same loadings.

As for the MWCNT destruction in Fig. 6.24b, based on the SEM morphology of MWCNT without and after the heat treatments, significant MWCNT destruction was observed, especially for the composites subjected to 600°C heat treatment. The disconnected MWCNT may be due to its initial structural defects or the weakened properties induced during ultrasonication [57]. Even though the disconnected MWCNT could cause higher electrical resistivity for the cementitious composites, more contact points were created to generate the conductive passages between new formed MWCNT. It means that small external loading might reconnect these disconnected MWCNT and decrease electrical resistivity, and that was one of the reasons for the generation of sharp resistivity changes in the beginning and end of loading after heat treatments. In other words, the nonlinearity of the fractional changes of resistivity to compressive strain in the beginning and end of loading was partially due to the reconnection of disconnected MWCNT.

As shown in Fig. 6.24c, the third potential factor that influenced the piezoresistivity is the cement matrix destruction during heat treatment. It has been reported that the high pressure steam could be generated inside the composites during heat treatment, which caused inner stress to larger the cracks and pores [42]. In addition, the decomposition of C–S–H gel could also soften the MWCNT–cement composites under high-temperature treatment, which could be supplemented by the SEM images and the increased compressive strain under the same loading. Both of these factors caused the destruction of cement matrix and lower the strength of composites, and the "softer" composites might contribute to the easily altered fractional changes of resistivity during the compression.

Another potential factor that may affect the piezoresistive performance is the features of MWCNT agglomerates after the heat treatment. As briefly illustrated in Fig. 6.24d, the morphology of MWCNT agglomerates before the heat treatment looked like a firm and completely conductive cluster. However, owing to the disconnected MWCNT or other oxidized conductive impurities-induced cracks or gaps, the agglomerates with defects have possibility to become nonconductive phase or the clusters with much higher resistivity. It was proposed that crystal particles could be vaporized from CNTs agglomerates in high heating rates for CNTs purification [58]. Hence, it was deduced that the electrical resistivity of MWCNT agglomerates could be affected after high-temperature treatment. In addition, this could be further demonstrated by the disconnection of MWCNT after heat treatment in Fig. 6.25, which could be occurred in the agglomerates to increase the electrical resistivity. Since the agglomerates possessed worse resistance to external loadings [59], a small compressive stress/strain might cause the reconnection between agglomerated MWCNTs and greatly reduce the fractional changes of resistivity. Furthermore, the MWCNT used in this study was relatively in lower qualities (purity higher than 95 wt.% in Table 6.1), for the sake of lower costs especially when extensively applied in

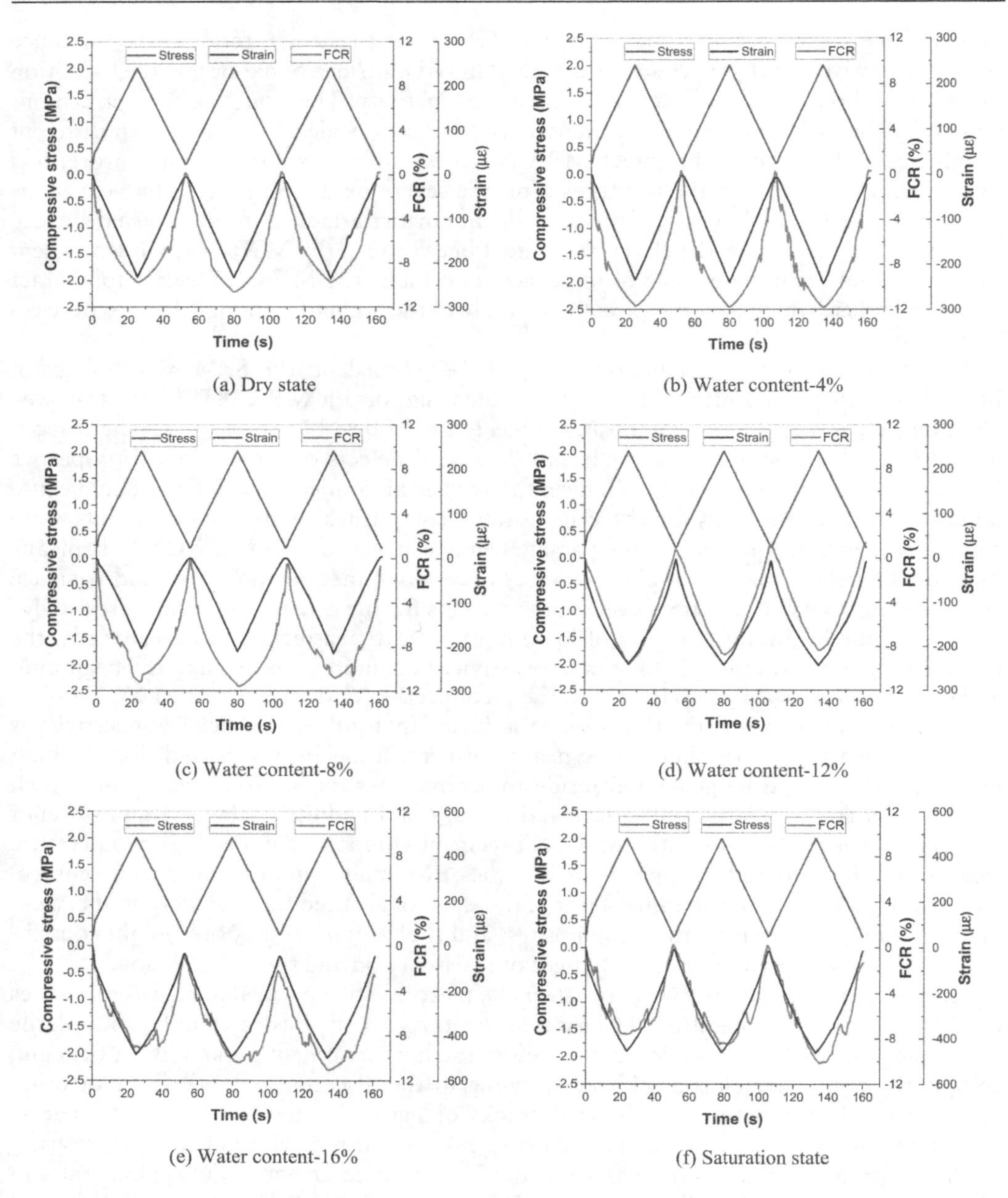

(a) Dry state

(b) Water content-4%

(c) Water content-8%

(d) Water content-12%

(e) Water content-16%

(f) Saturation state

Figure 6.25 The FCR of CB/cementitious composites with different water contents under cyclic compression [49].

engineering project, there inevitably existed impurities in the MWCNT and agglomerates. It can be deduced that the gaps/cracks due to the oxidization and disconnection of impurities or MWCNT were large enough to ensure the separation between some of MWCNTs in the unloading process, because of the sharp increases on the resistivity on this stage.

6.3.2 Water content

6.3.2.1 Normal CBS with various water contents

Fig. 6.25 depicts the FCR and compressive strain changes of CB/cementitious composites at different water contents under the same cyclic compression [49]. It could be observed that the FCR changes showed excellent repeatability when the CB cementitious composites contained water contents lower than 12%, which was only accompanied with a small fluctuation on resistivity changes. Also, it represents that the CB/cementitious composites containing water content of 4% and 8% expressed larger FCR changes than that with 12% water content and the dry specimens. However, for the saturated specimens and the composites containing water content of 16%, it depicts that the composites not only showed lower FCR changes but also more volatile and fluctuated FCR alterations. One reason for these fluctuations is the polarization effect in pore solutions, where exists in the movements of ions in electric field until reaching electrical potential balance [45]. Another reason is that the water content affects the concentration and the number of solution-filled pores, the touch and detach of nearby pore solutions also increased the curve volatility [31].

Fig. 6.26 plots the dependence of FCR to the compressive strain for cementitious composites with different water contents [49]. It was observed that the sensitivity of CB/cementitious composites first increased and then decreased with rise of water content. For the composites at the saturation state, the piezoresistivity experienced the lowest sensitivity with the gauge factor of approximately 150. Then the gauge factor slightly increased to nearly 175 for the composites cooperating 16% water content. The composites at dry state and that with 12% water content showed nearly identical gauge factor of 400, with small discrepancy on the rapid growth in the early stage for the former and in the late stage for the latter. The rough evaluation of optimal water content for CB/cementitious composites was approximately at 8%, with the gauge factor reaching the

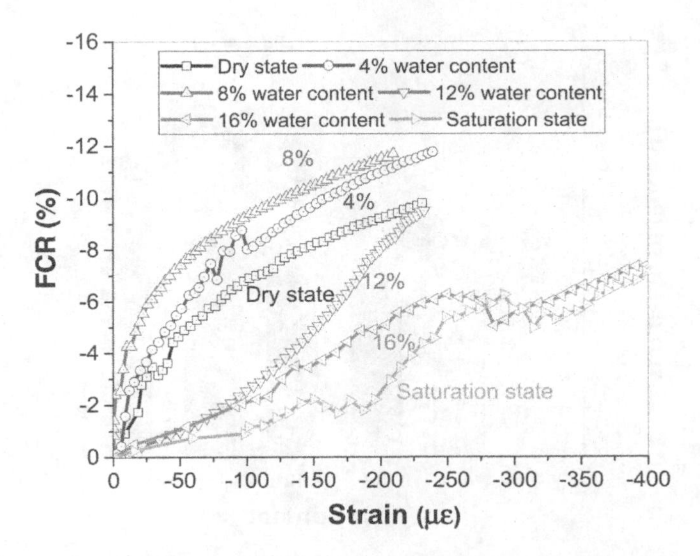

Figure 6.26 The FCR of CB/cementitious composites at different water contents as a function of compressive strain [49].

value of 488. To acquire more precise optimum value of water content in the CB/cementitious composites, more tests are needed to narrow down the interval of water content in the composites. In terms of the tendency of individual curve, it was seen that for the dry specimens and the composites with water content of 4% and 8%, their growth rates of FCR (gauge factor) gradually decreased with the loading process. However, for the composites at saturated state and that with 12% and 16% water content, their FCR growth rates were well-maintained (saturated specimens and that with 16% water content) and even gradually increased (composites with 12% water content). This is probably out of the easier contact between CB particles without the interferences from water content, and the detailed explanation will be presented in the next section. In summary, the water content influenced both repeatability and sensitivity of CB/cementitious composites. Based on the current experimental results and previous studies [31], the potential mechanisms will be discussed in the next section.

Fig. 6.27 shows the electrical resistance changes of composites with different water contents, and relative positions between CB particles in different stages on account of water content, striving to explain the relationship between the CB particles, water content, electrical resistance and the aforementioned piezoresistivity [49]. Different stages from A to D represent the dry stage, less-water stage, medium-water stage, and the large-water stage, respectively. The CB particles enclosed by water layer demonstrated transformation among different conductive mechanisms, due to the increase of water content in the CB/cementitious composites. It was observed that the electrical resistance of cementitious composites increased with the water content, from approximately 102 Ω for the dry composites to nearly 620 Ω for the saturated ones. One possibility for the increased electrical resistance was due to the decreased conductivity of the water-enclosed CB particles. Also, the higher water content resulted in severe polarization effect during the resistance measurement and caused larger measuring errors, especially for the cementitious composites under high temperatures [60].

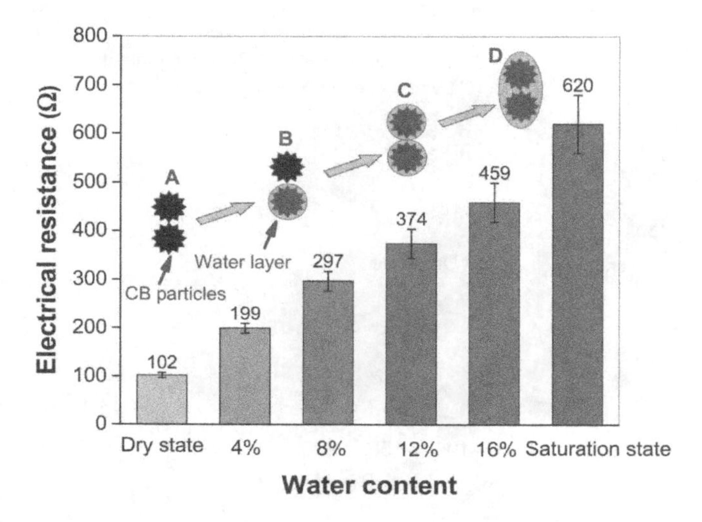

Figure 6.27 Resistance of CB cementitious composites to water content and the relative positions between CB particles on account of water layers [49].

It is widely known that the electrical conductivity includes electronic conductivity which is generated from the movement of free electrons, and the ionic conductivity that produced out of ions' movement [61]. In the dry stage (Fig. 6.27 [A]), the electrical conductivity mainly came from the contacts between CB particles, which possessed the electrical resistivity of lower than 0.43 Ω·cm. It seems that the direct contact between CB particles could provide a very low resistance to the electrons' movement and improve the conductivity of cementitious composites. As for the piezoresistivity, the FCR changes were only due to the contact between CB particles to reduce electrical resistance, while the cement hydration products with worse conductivity rarely altered the resistance and the FCR.

With the increase of water content, the composites came to the less-water stage (Fig. 6.27 [B]), where some CB particles were enclosed by water layers. It blocked the direct connections among CB particles by thin water film to reduce the electrical conductivity. Interestingly, the piezoresistivity in this stage was improved, just like the composites with 4% and 8% water content, respectively. This is because the FCR changes were due to not only the contact of CB particles but also the resistance reduction because of the ions movements such as Ca^{2+}, OH^{-}, and SO_4^{2-} in the water layers [62]. Owing to the limited water content, most of the FCR changes still came from the CB particles, and brought the largest gauge factors.

The following medium-water stage (Fig. 6.27 [C]) had higher electrical resistance just as the composites with 12% and 16% water content. In this stage, increased CB particles were covered by the water layers, which considerably increased the contact resistance between CB particles. Although the FCR changes from the ions conductivity were stimulated, the electrical responses from the contact of CB particles were weakened due to the water layer–decreased conductivity. According to the reduced gauge factor, it was deduced that the negative effect on the electronic conductivity of CB particles and the induced piezoresistivity overwhelmed the positive effect brought by water content–introduced ionic conductivity.

In the case of saturated cementitious composites, it was found from Fig. 6.27 [D] that the CB particles or small CB agglomerations were entirely surrounded by water layers. In this circumstance, the ions conductivity had predominance on the electrical conductivity than the electrons conductivity and led to highest electrical resistivity [63]. Since the CB particles are totally inside the water layers, the applied load first caused the deformation of composites and then delivered to the deformation of water layers, and finally transmitted to the location alterations of CB particles. As plotted, the saturated composites and the composites with water content of 12% and 16% had lower gauge factor at the initial loading stage and relatively higher gauge factor in the later stage, just because the FCR changes were first induced by ion conductivity and then produced by the electron's conductivity. Even if the FCR changed by water layers increased, it greatly decreased the FCR changes by the contact of CB particles, and then reduced the whole sensitivity. Moreover, the intensity of ionic conductivity mainly depended on the ion concentration of pore solutions and water layers, which was hardly altered during the compression process and failed to significantly improve the FCR changes and piezoresistivity.

6.3.2.2 Hydrophobic CBS

Fig. 6.28 shows the FCR of the cementitious composite without SHP subjected to cyclic compression, before and after 1 day water immersion treatment [26]. For the specimen

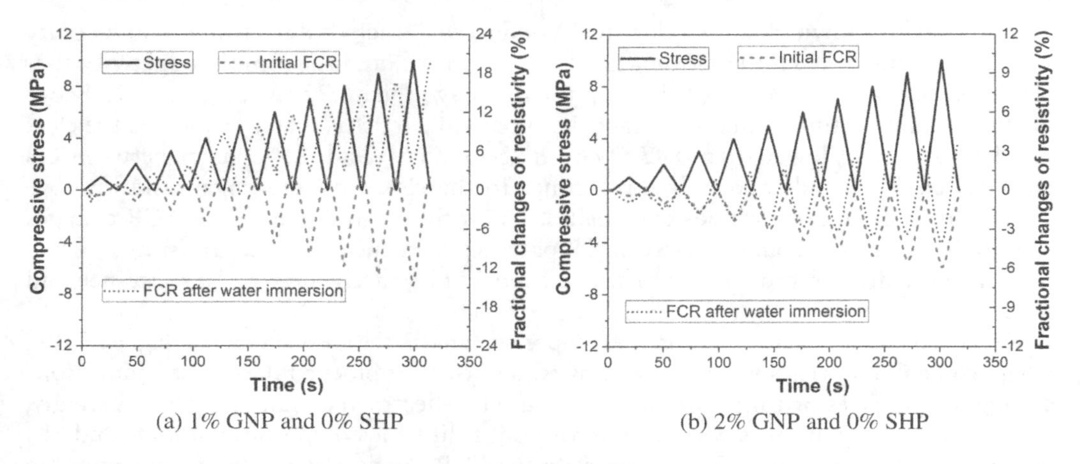

(a) 1% GNP and 0% SHP

(b) 2% GNP and 0% SHP

Figure 6.28 FCR of cementitious composite without SHP before/after water immersion treatment [26].

filled with 1% GNP in Fig. 6.28a, the initial FCR exhibited an excellent consistency to compressive stress, and the largest FCR reached approximately 14.4% at the stress magnitude of 10 MPa. As for the piezoresistive repeatability, the irreversible FCR was as low as 2.0% after ten cycles of compression. The irreversibility was partially due to microdefects in the cementitious composite, which retarded the electrons' transport and increased the electrical resistivity of specimen. Moreover, because of the movement of free ions in residual solutions under the electric field, which was also known as polarization effect, the electrical resistivity of specimen could gradually increase with the duration of measurement [62]. In this study, the specimens were dried previously before the piezoresistive test; hence, the polarization effect was considered not significant. However, after the immersion treatment, the fractional changes of resistivity showed different patterns by a considerable increase in resistivity during the piezoresistive test. The resistivity reduction still occurred in the process of loading, but the decreasing rate was overwhelmed by the increasing rate. It led to the positive values of FCR in compression test, except the values in the beginning. Moreover, it was observed that the irreversibility of FCR reached 19.5% which was almost ten times higher than that of the dried specimen. This phenomenon was mainly due to the effect of water content, which penetrated into the specimen and brought the polarization effect. Fig. 6.28b illustrates the FCR alterations of the specimen filled with 2% GNP. Previous results demonstrated that the specimen possessed similar water absorption to the counterparts with 1% GNP, and it was found to influence the FCR values because of the introduction of water content. The irreversible FCR accounted for 0.5% and 3.5%, respectively, before and after water immersion, which is magnified by seven times. In addition, not only the piezoresistive repeatability, but also the FCR intensity altered as well under the influence of water content. The original FCR value was approximately 5.9% at the stress magnitude of 10 MPa, while the new value increased to approximately 7.8%. Dong et al. [49] explored the effect of water content on the piezoresistive sensitivity, and found that there was optimal water content to achieve the best piezoresistivity. In this case, the absorption of water content improved the piezoresistive sensitivity for the dried cementitious composite filled with 2% GNP.

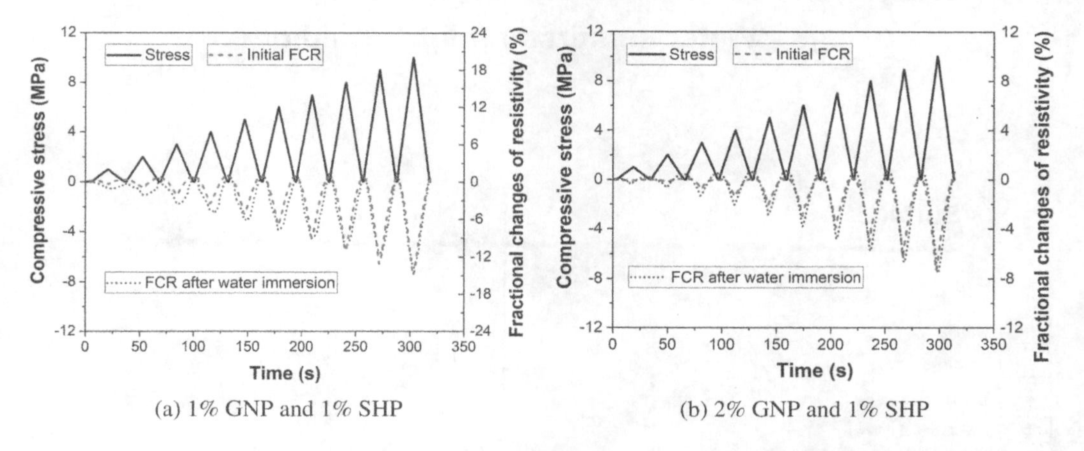

(a) 1% GNP and 1% SHP

(b) 2% GNP and 1% SHP

Figure 6.29 FCR of cementitious composite with 1% SHP before/after water immersion treatment [26].

The piezoresistive performance of specimen filled with 1% SHP before and after water immersion is shown in Fig. 6.29 [26]. To estimate the hydrophobicity and water repellency, the specimens were water immersed for seven days. For the specimen with 1% GNP, it could be observed that the FCR after water immersion was approximately identical to the original values, except for the alterations at the beginning of compression. Similarly, in comparison to the specimen without SHP, a much low discrepancy on the FCR occurred for the specimen filled with 2% GNP. The results should be attributed to the failed entrance of water since the water absorption was greatly decreased from 4 mm to 0.5 mm after the application of SHP in the cementitious composite. However, a proportion of water had the potential to penetrate the near-surface area or remain in the surface of specimen and affect the generation of conductive passages. That is why the piezoresistive performances of the cementitious composite slightly fluctuated after the water immersion. Overall, in comparison to the piezoresistive performance of specimen without SHP after one day of water immersion, more stable and consistent piezoresistivity could be found for the specimen with SHP even after seven days of water immersion.

The affected piezoresistive behaviors by water immersion were consistent well with the performance of water absorption of the cementitious composite. In the case of water immersion of cementitious composite, the aforementioned results indicated that the amount of penetrated water played a critical role in the development of piezoresistivity. Fig. 6.30 schematically depicts the functionality of GNP and SHP in cementitious composite, and the potential channels of water penetration, attempting to elucidate the mechanism behind the cementitious composite reinforced with incorporated GNP and SHP [26]. For the plain cement paste, because of the microcracks and pores, water can easily penetrate the cement paste and lead to high water absorption. With the addition of GNP, the pores and cracks are reduced to resist the water absorption, and the plate-like structures of GNP also benefit the water impermeability. It was found that the GNP has the capacity to block the water channel and retard the further penetration of water content. Furthermore, according to the previous studies, the length of water penetration passages can be increased with the intervention of GNP which also weakens the ability of water absorption [15]. However, combining the results of water absorption of the cementitious composites only filled with GNP, the conclusion can be drawn that a large

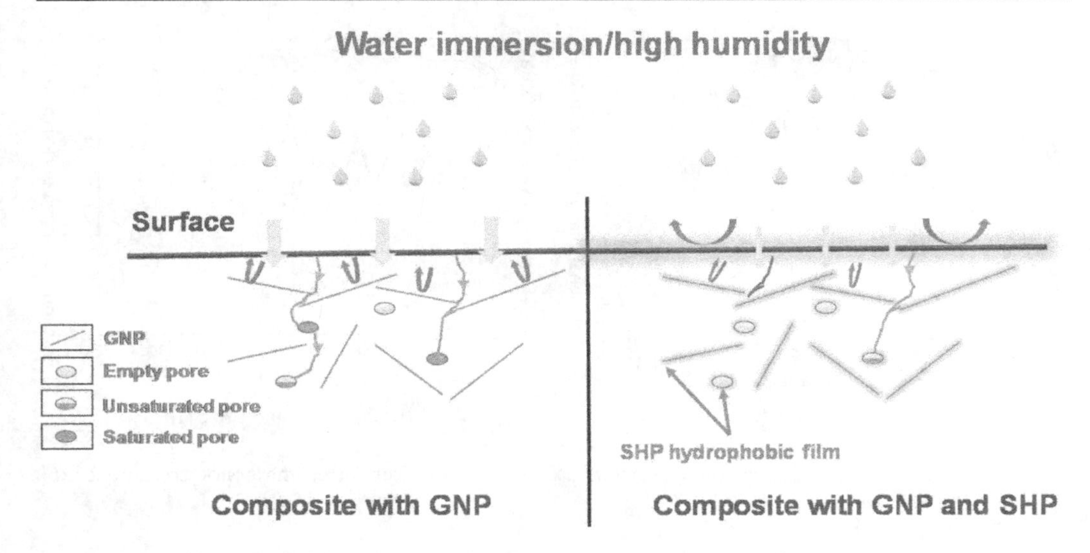

Figure 6.30 Schematic diagram of water penetration into GNP/cementitious composite with/without SHP [26].

proportion of water penetrates the specimen even the water absorption is decreased by more than 30%. In other words, some pores might not be penetrated from the external water due to the GNP, but there exist water-saturated pores close to the surface and unsaturated pores because of the super long channels. The electrical resistivity of specimen is very sensitive to the water content changes, and that is why the piezoresistive behaviors of specimen filled with GNP altered considerably after the water immersion. For the cementitious composite filled with combined GNP and SHP, in addition to the isolation effect of GNP, the SHP helps to establish a hydrophobic film in the surface of specimen, GNP, pores, and cracks. The hydrophobic film on the surface of pores and cracks represents that the cement hydration products possess hydrophobicity. Moreover, the water channels that might work in the specimen with GNP are further blocked by the enclosed hydrophobic film. Even if the water is successful in penetrating through the channels, the attached hydrophobic film can greatly limit its entrance into the pores. This contributes to the small water absorptions and stable piezoresistive performances of the cementitious composite.

6.3.3 Chloride penetration

The piezoresistivity of varied contents of CWA- and SHP-incorporated cementitious composites after 0, 1, 7, 14, and 90 days of immersion in sodium chloride solution are displayed in Fig. 6.31 [23]. For the CB/cementitious composites without additives, the FCR values were very close and reached nearly 36% when the immersion duration was within 14 days. After 90 days of immersion, the FCR decreased to 25%. Compared to the counterparts immersed in freshwater, the cementitious composites immersed in sodium chloride solution kept a larger FCR value. For the dried CB/cementitious composites, the electronic conduction predominated over the ionic conduction and led to piezoresistivity. After immersion in sodium chloride solution, the penetrated water and sodium

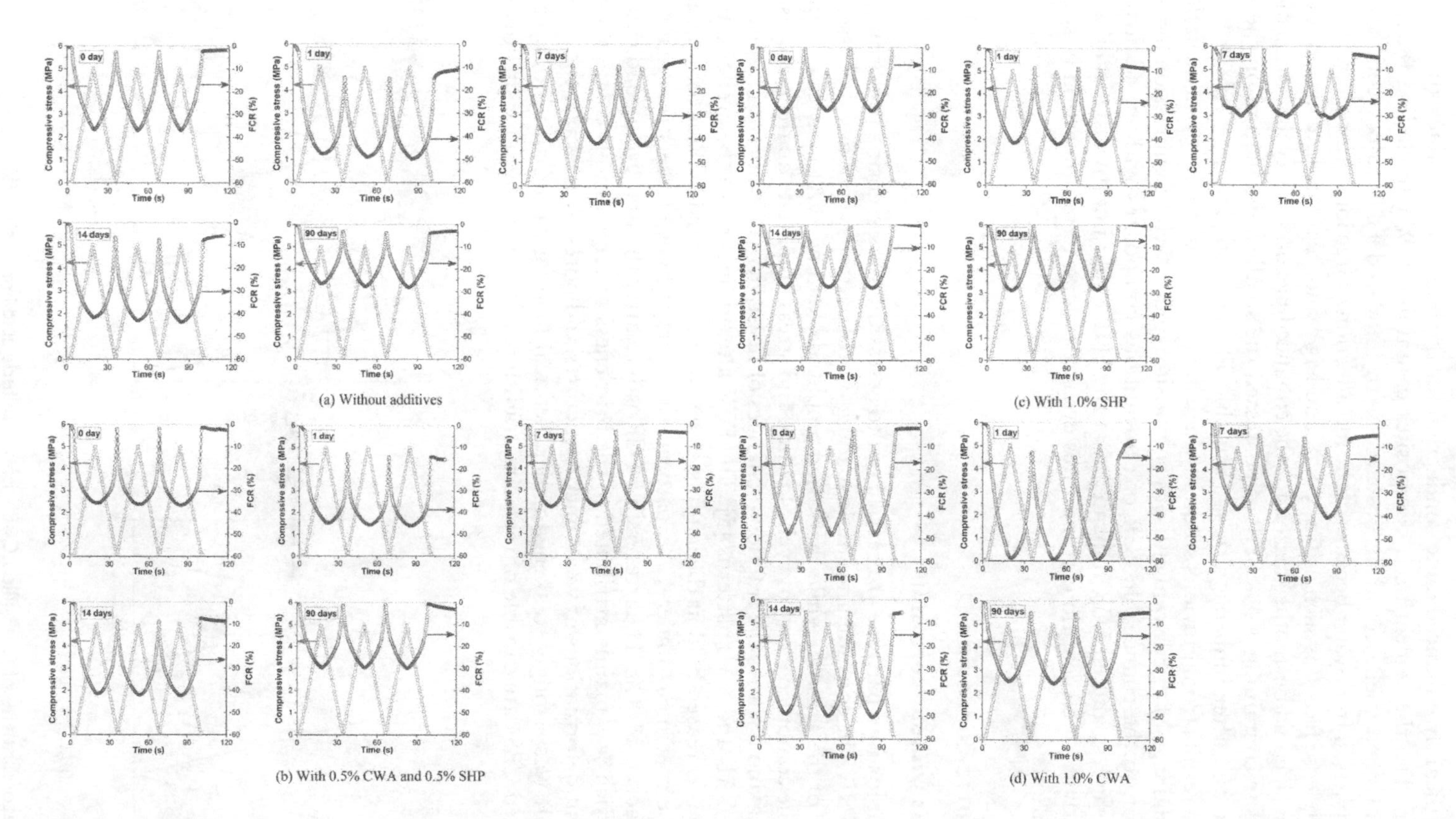

Figure 6.31 Piezoresistivity of CB/cementitious composites containing multiple additives after immersion in solution of sodium chloride for 0, 1, 7, 14, and 90 days [23].

chloride significantly enhanced ionic conduction for the higher FCR values after 90 days of immersion. For the cementitious composites containing CWA (including 0.5% and 1.0%), similar phenomena could be observed with improved FCR values compared to their counterparts in freshwater because of the introduction of free ions. Moreover, given the CWA might release free ions such as Ca^{2+}, Na^{+}, and Mg^{2+} and increase the concentration of pore solutions, it could be another reason for the enhanced piezoresistivity. For the cementitious composite containing 1.0% SHP, it possessed the smallest fluctuation on the FCR values before and after immersion in sodium chloride solution. The irreversible FCR is similar to the results from previous studies, which are mainly due to the permanent deformation and damage of brittle cementitious composites under cyclic compression [64–66]. In comparison to the changeable FCR of cementitious composites without additives or with CWA, the less altered FCR indicated that the SHP could improve the stabilization of piezoresistivity of cementitious composites even after long-term immersion in sodium chloride solution.

6.3.4 Sulfuric corrosion

6.3.4.1 CBSs without corrosion

Two different loading patterns are carried out to assess the stress-sensing performance of CBSs with GNP. The constant amplitude-loading (CAL) pattern is conducted to assess the repeatability of the sensors, while the varied amplitude-loading (VAL) pattern is applied to estimate the combined effect of cyclic loading and stress amplitude on the piezoresistivity. Fig. 6.32a and b shows the fractional changes of resistivity (FCR) of initial specimens subjected to CAL and VAL patterns [5]. The FCR altered linearly to the cyclic compression, with the decreased FCR at the loading stage and the increased FCR at the unloading stage. For the specimens under CAL pattern, the average FCR at the stress magnitude of 2 MPa reached 11.8%. The irreversible FCR values after the first cyclic load could be observed, which was mainly attributed to the brittleness of cementitious composites and the permanent deformations of weak microstructures and pores. Afterward, the irreversibility of FCR was strongly weakened under identical loading patterns. It indicates that the GNP-filled CBSs are capable of replacing traditional sensors to monitor the forces on

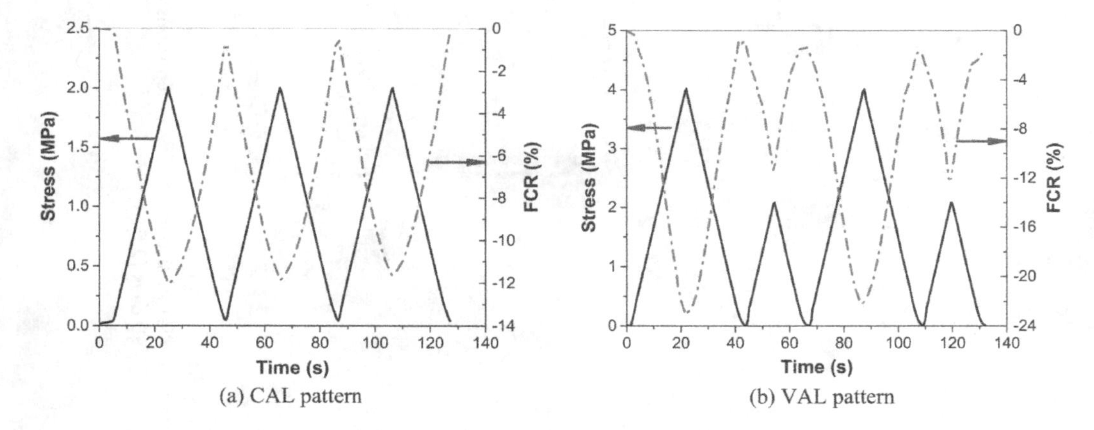

Figure 6.32 Piezoresistivity of GNP-filled CBS without sulfuric acid immersion [5].

structures, as long as the irreversible performance of piezoresistivity can be eliminated by applying preloading. Moreover, the FCR values at the zero stress slightly increased with the cyclic load because of the gentle polarization during the test. It demonstrated that some conductive solutions with movable ions were still preserved in the closed pores after three days of drying. This implied that the CBSs should be dried entirely before practical application. Otherwise, the AC rather than DC should be applied to capture the electrical signal alterations during the piezoresistivity test. Similar FCR to compressive stress changes could be found for the initial specimens subjected to VAL pattern. The largest FCR values at the stress peaks of 4 MPa and 2 MPa reached 22.8% and 11.9%, respectively. The linearity of specimens was very similar, with the FCR values achieved being 11.8% and 11.9%, respectively, under CAL and VAL patterns. However, compared to the specimens under the CAL pattern, the larger applied stress amplitude led to higher irreversible FCR values. The stable irreversibility of FCR occurred after several cycles of loadings, implying that CBS with GNP monitoring high stress should be previously loaded with large cyclic loadings to achieve both the stable linearity and reversibility.

Fig. 6.33 displays the relationship between FCR and the compressive stress for the initial specimens under CAL and VAL patterns. The linear fitting curves were also plotted, based on the correlation of FCR values to compressive stress [5]. The slope of fitting curves indicates the FCR changes under unit compressive stress, which also means stress-sensing efficiency. It was found that the two fitting curves were almost parallel to each other, with slopes of –6.08 and –5.98. The small discrepancy of the slopes is mainly due to the unstable or weak cementitious structures that possess different bearing capacities under multiple loading amplitudes. It leads to changes in the microstructures and conductive passages, and slightly affects the piezoresistivity. Anyways, it showed that the stress sensing efficiency of GNP-filled CBSs was nearly constant under the CAL and VAL patterns. Further, the deviations of these fitting curves R_C^2 and R_V^2 show the larger value of 0.973 for the CAL pattern and a lower value of 0.966 for the VAL pattern, demonstrating

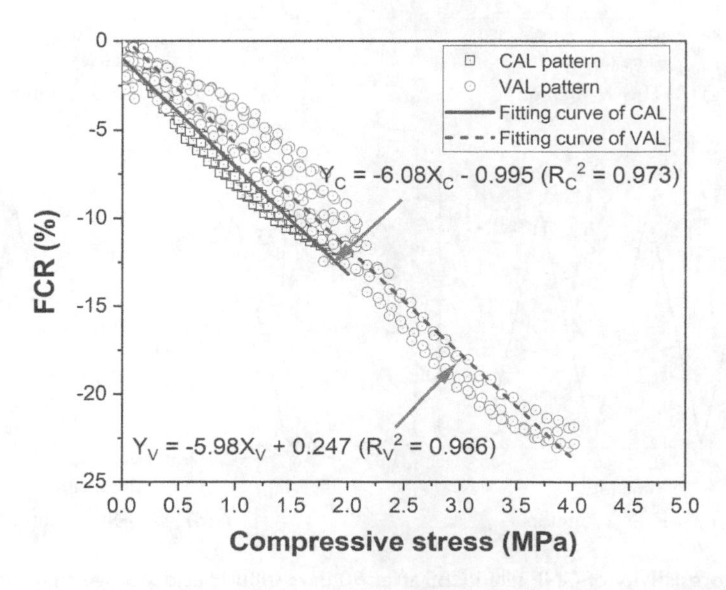

Figure 6.33 FCR as a function to compressive stress for the initial CBS under CAL and VAL patterns [5].

the relatively poor linearity for the specimens subjected to VAL pattern. The VAL pattern consists of higher stress amplitudes, higher loading rates, and more complicated stress conditions than those of CAL pattern. The poorer linearity of specimens under the VAL pattern indicated that the linearity of piezoresistivity might worsen with the complexity of external loading forces. Although the linearity of specimens after the VAL pattern was still acceptable, much more in-depth analysis of the minor changes of piezoresistivity under complicated loadings should be further explored.

6.3.4.2 CBSs after corrosion

Fig. 6.34 illustrates the piezoresistive performance of cementitious composites after 90 days of storage in sulfuric acid solutions, subjected to CAL patterns for the piezoresistivity test [5]. The FCR values showed satisfactory linearity to the compressive stress for the specimens in 0% H_2SO_4. In comparison to the initial counterpart without treatment, the water-stored specimens showed smaller FCR values of 9.4% at the stress peak of 2 MPa. As mentioned previously, the further hydration of specimens could occur during long-term water immersion. The slightly lower FCR values might

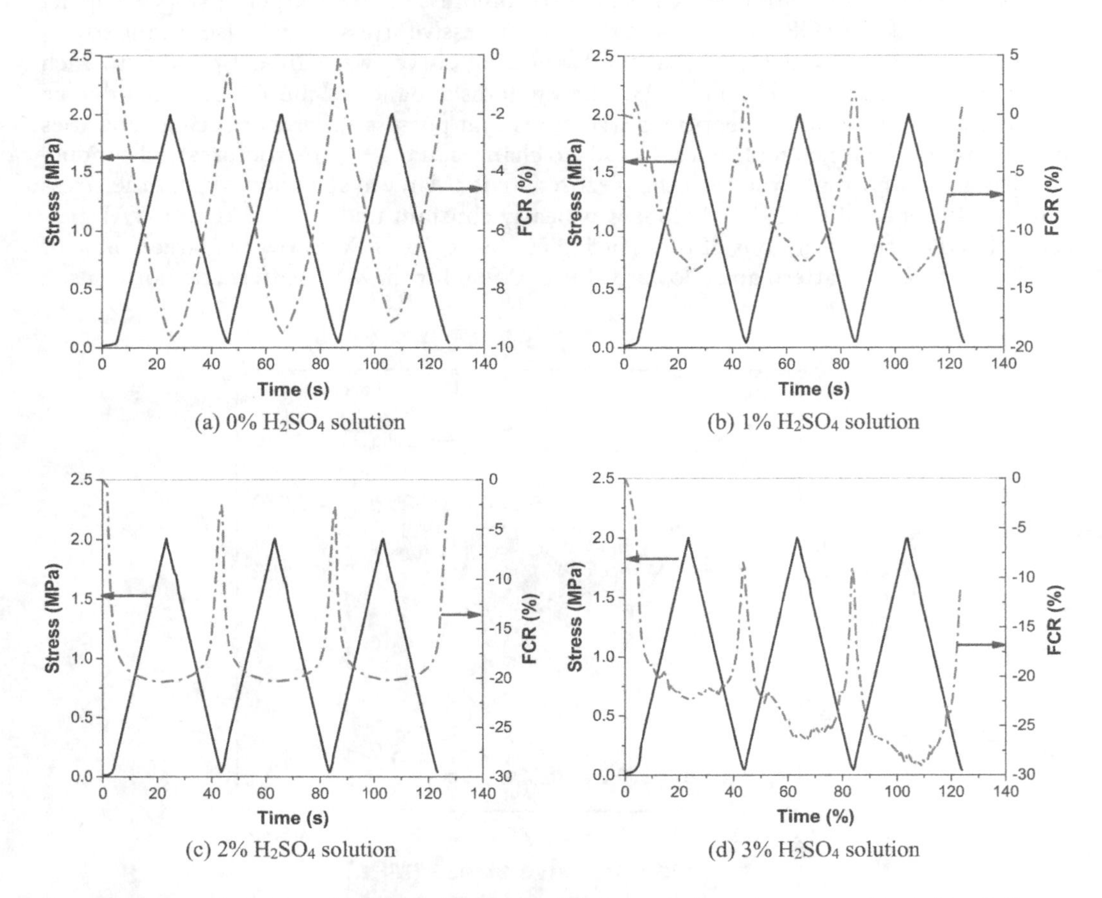

Figure 6.34 Piezoresistivity of GNP filled CBS after 90 days sulfuric acid storage subjected to CAL pattern [5].

be due to the denser structures of specimens after water storage. For the specimens in 1% H_2SO_4 solutions, the piezoresistive stability seemed to be slightly affected, with the small fluctuation on FCR values. This is because the erosion of the cement matrix led to some weak areas and increased the inhomogeneity of CBSs. Moreover, it could be found that the largest FCR increased to approximately 13.5%. The erosion of cementitious composites by H_2SO_4 is similar to a specimen softening process, making the conductive filler of GNP easier to connect with each other to create a larger electrical resistivity change under the same specific force. Even so, the linearity of FCR to compressive stress could still be observed for the specimens in 1% H_2SO_4. In terms of the specimens immersed in 2% and 3% H_2SO_4, the linearity among FCR and compressive stress was wholly destroyed, as plotted in Fig. 6.34c and d. Take the specimens in 2% H_2SO_4 for example, the FCR at the stress peak of 2 MPa finally attained to 20.5%, while the values at the first 1 MPa nearly reached 19.7%, indicating that the later applied forces almost made no differences on the electrical resistivity of the CBSs. Further, it was found that the FCR values could mostly recover to the initial values, implying the moderate repeatability of the CBSs after erosion treatment. With this, the conclusion can be proposed that the piezoresistive linearity possesses no relationship to the repeatability. The excellent repeatability of CBSs cannot ensure acceptable piezoresistivity, linearity, and serviceability. In addition to the worse linearity, Fig. 6.34d shows the gradually decreased FCR values with cyclic compressive stress. It demonstrated that the corroded specimens under external compression were subjected to permanent damages on the microstructures. For instance, CBS might be forced to close their pores and cracks to decrease the electrical resistivity. Also, the conductive GNP could get closer and connected in this process.

The piezoresistivity of cementitious composites after 180 days of storage in sulfuric acid solutions subjected to VAL patterns is shown in Fig. 6.35 [5]. Being different from the specimens after 90 days of water storage, both the linearity and largest FCR values were altered after 180 days of immersion in 0% H_2SO_4. It indicates that long-term storage of CBSs in water can greatly affect the piezoresistivity. It seems that the water can also change the micro-conductive passages in the CBSs through slow but continuous hydration of cement. The denser microstructures caused the FCR values to strongly decrease even subjected to the larger compressive stress of 4 MPa. It implied that the future practical application of CBSs should ensure their excellent hydrophobicity and water repellency, to get rid of the effects of humidity or water vapor from working environments. For the specimen exposure to 1% and 2% H_2SO_4, worse piezoresistive repeatability and linearity were observed compared to the specimens after 90 days of storage. This is because the more severe damaged microstructures and GNP formed conductive networks, leading to the more suddenly changed FCR values. The phenomenon is significant for the specimens exposed to 3% H_2SO_4 for 180 days. The FCR arrived at nearly 67.0% when the stress reached 1 MPa, which then slowly increased to 75.5% at the stress magnitude of 4 MPa. Previous studies have mentioned that the formation of sulfate compounds, e.g., gypsum and ettringite, possesses less resistance to external forces than the cement matrix. At the beginning of loading, the soft compositions are first compressed to alter the electrical resistivity; this leads to the swift changes on the FCR values. Afterward, the rigid cement matrix starts to hold the increased external forces to decrease the alteration of FCR values. In summary, the experimental results implied that CBS with GNP were damaged both on the piezoresistive sensitivity and repeatability when they were subjected to a highly corrosive environment.

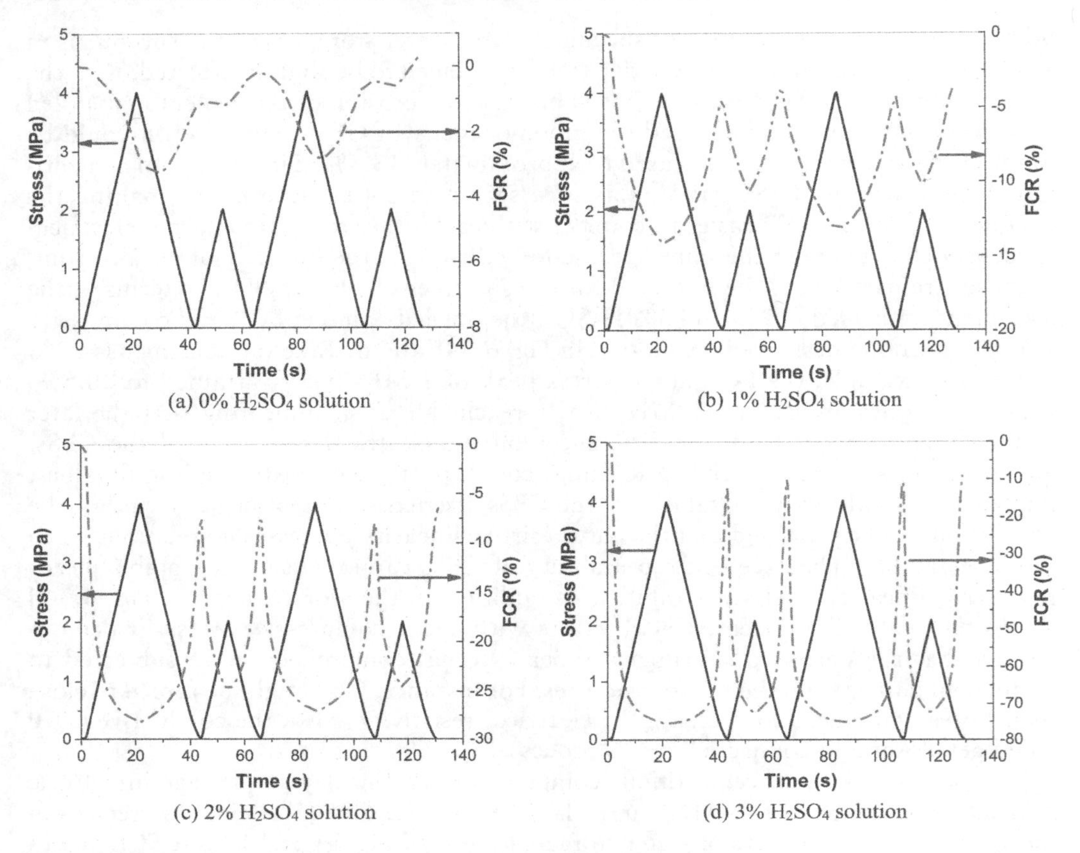

Figure 6.35 Piezoresistivity of GNP-filled CBS after 180 days of sulfuric acid immersion subjected to VAL pattern [5].

6.3.4.3 Failure stress detection

As for the specimens after 180 days, Fig. 6.36a illustrates the electrical resistivity changes of specimens under compression to failure [5]. It could be found that all composites went through a decrease in electrical resistivity during compression. When the stress reached the ultimate compressive strength, there existed a sudden increase in the electrical resistivity. Similar results were obtained by previous studies that the rapid electrical resistivity changes occurred at the flexural failure point [67]. This is because the complete damage of specimens destroys the conductive passages to increase the measured electrical resistance. In other words, the performance of the electrical resistivity of CBSs at the failure point could be the sign to reveal the structural health conditions. As for the piezoresistive linearity, only the specimens immersed in 0% H_2SO_4 solutions displayed moderate linearity to compressive stress. Similar to the performance under cyclic compression, the corroded specimens showed fast changes in the FCR values at the beginning of loading. Then the electrical resistivity kept constant almost with the increase of compressive stress. To clearly show the FCR changing rate, Fig. 6.36b shows the first derivation results of FCR alterations in the first 50 s loading process. In the later loading stage, the slope of curves is almost unchanged, and that is why only the first 50 s loading is chosen to analyze the FCR changing rate. The results showed the largest FCR changing rate of specimens

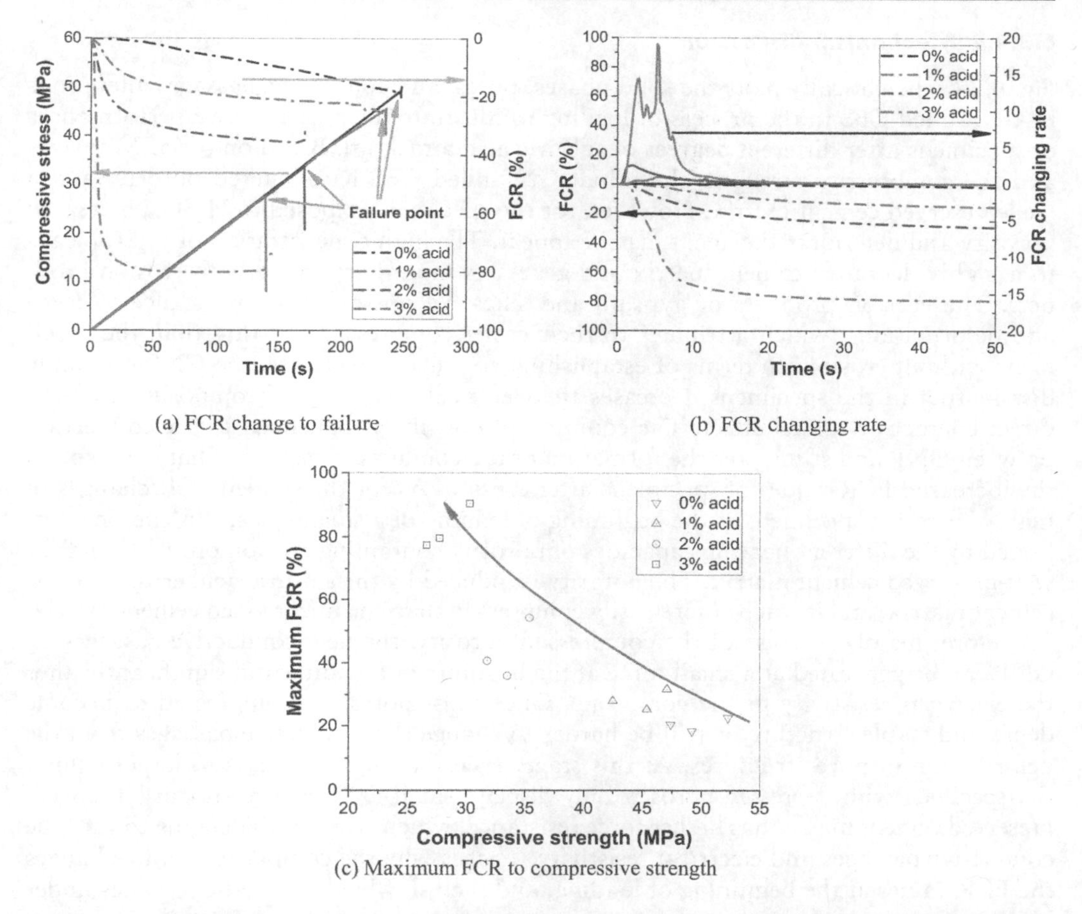

(a) FCR change to failure

(b) FCR changing rate

(c) Maximum FCR to compressive strength

Figure 6.36 FCR changes of GNP-filled cementitious composite subjected to compression after 180 days of sulfuric acid immersion [5].

immersed in 3% H_2SO_4 solutions, followed by their counterparts in 2% and 1% H_2SO_4 solutions. The specimens without acid erosion showed an almost constant FCR changing rate. Compared to the smooth alteration changes of FCR changing rate, randomly varied FCR changing rate indicated severe damages on the microstructures of specimens exposed to 3% H_2SO_4 solutions. Because the erosion products and cement matrix had different stiffness, the highly corroded specimens with a high content of erosion products led to the varied FCR changing rate with increasing loading. In addition, the destruction of erosion induced micropores and microcracks that had the capacity to change the electrical resistivity of specimens suddenly [68, 69]. Fig. 6.36c plots the relationship among the FCR values at failure point and the compressive strength of the specimens after different contents of erosion. For the intact specimens without sulfuric acid erosion, the largest compressive strength and smallest FCR values were observed. Contrarily, the smallest strength and largest FCR values were obtained for the specimens after exposure to 3% H_2SO_4 solutions. It demonstrated the softening effect of H_2SO_4 solutions on the GNP-filled cementitious composites, thus leading to reduced mechanical strength and piezoresistivity. Overall, for the identical cementitious composites, the FCR values could be another index to assess the erosion degree of specimens, in addition to the compressive strength.

6.3.4.4 Mechanism discussion

Fig. 6.37 schematically plots the solid phases, pores, and conductive passages alterations in GNP-filled CBS in the process of loading to illustrate the piezoresistive performances of specimens after different degrees of sulfuric acid attack [5]. Based on the SEM images and the compressive strength, the initial GNP-filled CBS have limited porosity and a well-preserved cement matrix. However, for the specimens exposed to H_2SO_4, increased porosity and deteriorated cement matrix appear. The high concentration of H_2SO_4 leads to a highly degraded cement matrix and generates more pores filled with erosion products. The erosion products of gypsum and silica gel possess weak resistance to force and deformations, which attributes to their easier compressed porosity than the specimens without erosion. In terms of establishing conductive passages, the GNP randomly distributing in the specimens decreases the electrical resistivity of composites by their direct contact or tunnel effect. The compressed porosity is beneficial to the connection between GNP and stimulates the formation of the conductive passage. That can explain the increased FCR values of specimens after erosion. As for the sudden FCR changes of highly corroded specimens at the beginning of loading of a small force, this can be interpreted by the different bearing capacities of porosity containing erosion products and the well-preserved cement matrix. The porosity is induced by the sulfuric acid erosion of the cement matrix, and it can be more easily compressed than the uncorroded cement matrix. Therefore, mainly because of the compressed porosity, the new conductive passages by GNP can be generated at a small force at the beginning of loading and significantly alter the electrical resistivity of the composite. Once most pores are compressed to become dense and stable structure, it will be harder to change the conductive passages from the deformation of pore structures. At this stage, even the force increases to larger values, the specimen with depressed porosity only displays small FCR changes because the well-preserved cement matrix has higher force resistance, which is more challenging to vary the conductive passages and electrical resistivity. That is why the composite rapidly changes the FCR values at the beginning of loading, and then slowly alters the FCR values under high-stress magnitude.

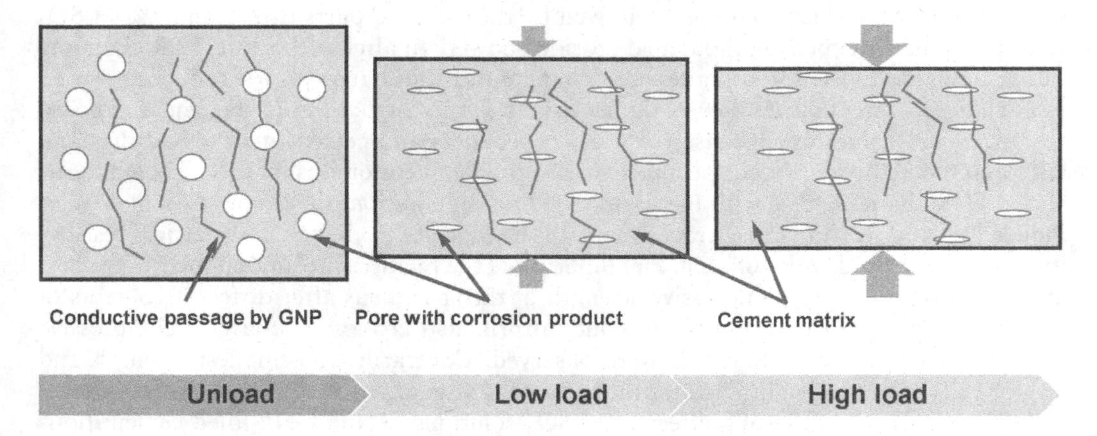

Figure 6.37 Schematic diagram of conductive passage alteration of highly porous GNP-filled cementitious composite under loading [5].

6.3.5 Freeze-thaw cycles

To further understand the effects of freeze-thaw cycles and ice generation on the resistance and piezoresistivity of CB cementitious composites, two groups of heating and cooling cycles were represented. The first one was higher than the freezing point of water at 5°C, 20°C, 40°C, 60°C, 80°C, and 100°C, and the second one was –20°C, 5°C, 20°C, 40°C, 60°C, 80°C, and 100°C. Each group experienced ten cycles of heating and cooling treatments [49]. Because the water content played a vital role in the process of freeze-thaw treatment, the freezing water expanded its volume which might cause inner stress inside the CB/cementitious composites to damage the microstructures and influence the electrical resistivity. Therefore, in this section, the dried and saturated specimens were prepared to investigate their electrical resistivity and piezoresistivity under the effect of freeze-thaw cycles. Different from other studies to apply the temperatures from –20°C to 20°C [70] or –20°C to 52°C [71], the chosen temperatures for freeze-thaw cycles are from –20°C to 100°C, to more closely imitate the environmental conditions where the cementitious composites might receive higher temperature with long-term exposure to intensive sunshine weather.

6.3.5.1 Resistivity development under different cycles

Fig. 6.38 illustrates the resistance changes of dry CB cementitious composites in two patterns of heating and cooling cycles [49]. The small diagram at top right shows the heating and cooling processes without water freezing, and the larger diagram involves the subzero temperatures called the freeze-thaw cycles. Both patterns included ten cycles of heating and cooling processes. It was found that only a small discrepancy was observed among electrical resistances in different cycles at the same temperature and showed good repeatability even in the process of freeze-thaw cycles. For the dry specimens without free water, the resistance changes of CB/cementitious composites under various temperatures mainly

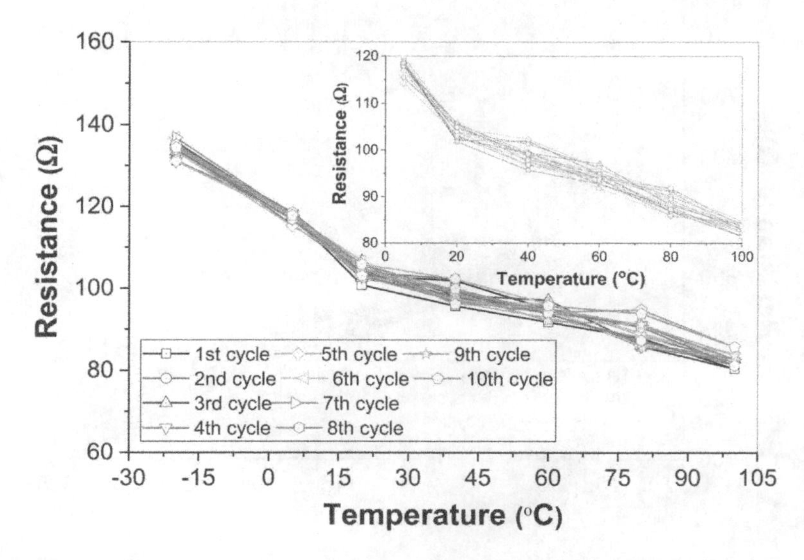

Figure 6.38 Electrical resistance development of dry CB cementitious composites with or without subzero temperature exposures [49].

come from the bound water [72]. It could be altered in viscosity and ionic activity of solutions with temperatures to change the electrical resistance of cementitious composites. However, the bound water cannot be eliminated under the drying condition of 100°C [73], hence it can be deduced that the resistance repeatability of dry CB/cementitious composites are immune from the freeze-thaw cycles, regardless of the heating and cooling processes with or without the freezing stage. This phenomenon is easy to understand for the lack of free water which fails to get frozen and expand its volume to damage the microstructures of cementitious composites.

Fig. 6.39 shows the resistance development of CB cementitious composites under the same treatment of heating and cooling patterns in the saturation state [49]. Compared to that of dry specimens, the electrical resistance of saturated cementitious composites was more sensitive to the temperature changes, because of the great amount of free water in the composites. The resistance drops with an increase of temperature were partially due to the worse viscosity of solution resistance, while the main reason is because of the activated ions in the pore solution to greatly reduce the electrical resistance [74]. For the cycles without the subzero temperature, small divergence on the resistance was measured at the same temperature and showed good repeatability. In comparison to the similar results from the dry specimens, the deduction could be proposed that the temperature cycles above zero had no impacts on the electrical resistance of CB cementitious composites, no matter how much water content they contained. In terms of the freeze-thaw cycles on the saturated specimens, an obviously larger resistance discrepancy was generated with the increase of freeze-thaw cycles, especially the resistance of composites at the temperature of –20°C. This can be explained by the effect of water freezing and expansion on the micro-conductive network in the CB/cementitious composites, which might be permanently destroyed and result in higher resistance with freeze-thaw cycles.

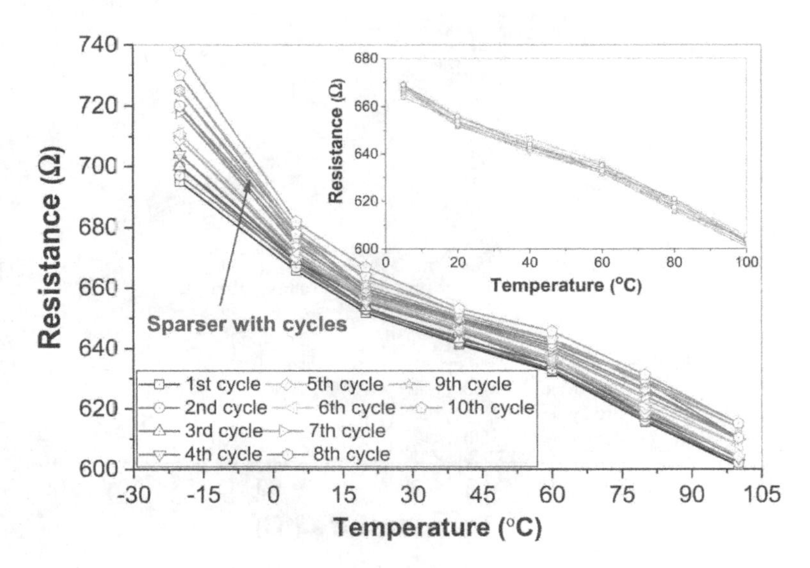

Figure 6.39 Electrical resistance development of saturated CB/cementitious composites with or without subzero temperature exposures [49].

6.3.5.2 Piezoresistive properties after freeze-thaw cycles

Fig. 6.40 illustrates the changes of compressive strain and FCR before and after ten cycles of freeze-thaw treatment for the dry CB/cementitious composites under the same cyclic compression [49]. The dotted lines are the results before the freeze-thaw cycles and the solid lines represent the same composites after the freeze-thaw cycles. It was observed that the composites before the freeze-thaw cycles exhibited excellent repeatability and sensitivity with compressive strain, while for the composites after the freeze-thaw cycles, slightly higher FCR were observed in the second cycle and a bit lower FCR occurred in the first and third cycles. The average of the largest FCR was similar before and after freeze-thaw cycles, and the small discrepancy could be attributed to the minor differences in microstructures between composites. Hence, it could be considered that the freeze-thaw cycles made no differences on the electrical resistivity of the dry CB cementitious composite, which was consistent well with the repeatable resistance. On the other hand, as for the compressive strain of CB cementitious composites, higher deformation by compression was observed after the freeze-thaw cycles, with the ultimate strain increasing by approximately 11.9%. As a result, the sensitivity parameter of gauge factor was decreased owing to the maintained FCR values but the increased strain. Generally, the piezoresistive repeatability of the dry CB cementitious composites was well-preserved with gentle fluctuation after the freeze-thaw cycles. As the piezoresistive sensitivity depended on both the electrical and mechanical properties, and freeze-thaw cycles might weaken the mechanical properties of composites by enlarging the compressive strain [75], the dry CB/cementitious composites still expressed a reduced gauge factor by a rough approximation of 11.9% (equal to the increased strain) after the freeze-thaw cycles.

In the aspect of saturated CB cementitious composites, their FCR and compressive strain changes before and after the freeze-thaw cycles were more complicated than that

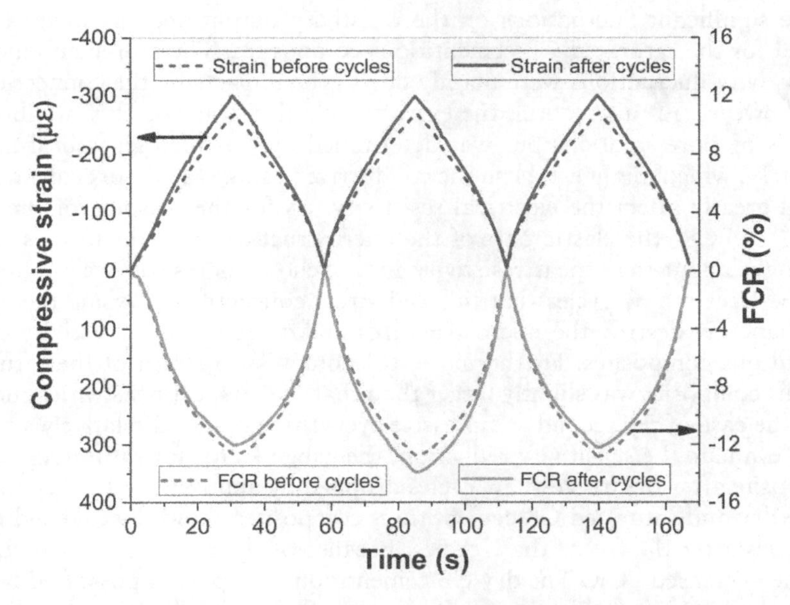

Figure 6.40 The FCR changes of dry CB cementitious composites with compressive strain before and after the freeze-thaw cycles [49].

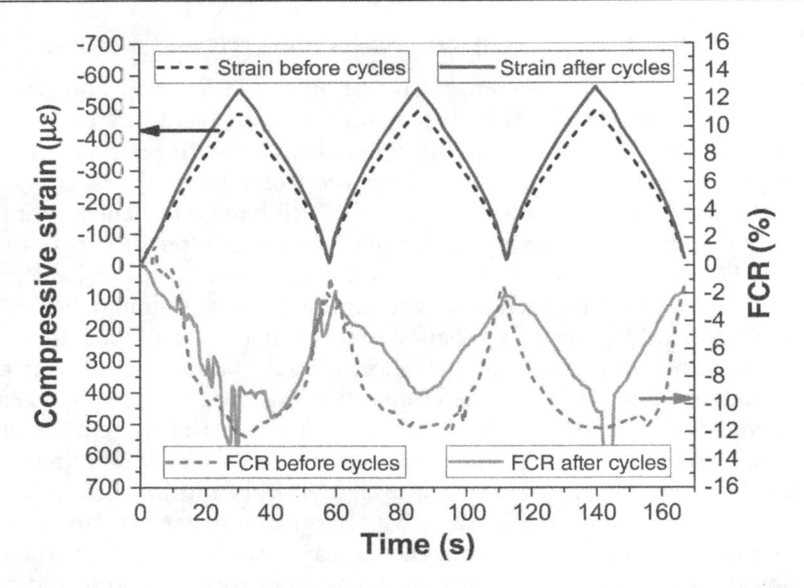

Figure 6.41 The FCR changes of saturated CB cementitious composites with compressive strain before and after the freeze-thaw cycles [49].

of dry CB/cementitious composites, since the volume expansion of pore solutions generated stress concentration and damaged the microstructures of composites. As mentioned before, the saturated CB/cementitious composites without freeze-thaw cycles could provide a relatively high FCR but with worse linearity and repeatability, as shown by the pink dotted line in Fig. 6.41 [49]. However, after the treatment of freeze-thaw cycles, much more significant fluctuations on the resistivity output and the lower FCR could be observed for the saturated CB cementitious composites. The aforementioned reasons for the resistivity fluctuations were mainly due to the impacts by the connection of pore solutions; however, in this section, the extreme volatility was not only attributed to the interferences by pore solutions but was determined by the microstructural damages of cement matrix, which might establish the conductive passages again in cementitious composites and greatly affect the electrical resistivity. As for the reasons for the decreased peak values of FCR, the elastic part of the microstructures in cementitious composites which brought about the repeat resistivity in a cyclic compression was vulnerable and limited. The freeze-thaw cycles that induced stress concentration inside the composites could permanently destroy the microstructures and caused lower resistivity changes of the cementitious composites. Furthermore, the compressive strain of the saturated CB/cementitious composite was slightly larger than that of dry specimens which increased by 14.6%. In the case of the second compressive cycle that possessed relatively smooth FCR changes to evaluate the sensitivity reduction, the gauge factor for saturated CB/cementitious composite after the freeze-thaw cycles dropped by approximately 30.7%.

Both the dry and saturated CB cementitious composites showed weakened piezoresistive properties after the freeze-thaw cycles, because of the either increased compressive strain or the decreased FCR. The dry CB/cementitious composites possessed better resistance to the freeze-thaw cycles by smaller variations on the electrical resistivity, while the saturated CB/cementitious composites were sensitive to the freeze-thaw cycles due to the damages caused by volume shrinkage and expansion of pore solution. Overall, it demonstrated that the CB/cementitious sensors have limitations on the real applications in the environment with subzero temperatures, because of the relatively weak mechanical properties and the poor frost resistance of CB particles.

6.4 SUMMARY

This chapter summarizes the durability of CBSs (nanomodified cementitious materials with excellent conductivity) subjected to chemical attack, transport property, shrinkage deformation, and high-temperature environments, and then investigates their self-sensing capacity to environmental factors of temperature and humidity, followed by the alteration of electrical and piezoresistive performance of deteriorated CBSs in various environments. The key conclusions are as follows:

1. The CBSs after nanomodification exhibit superior resistance to chemical corrosion, dense pore structure, and high-temperature durability compared to regular concrete. They also hold their own durability compared to conventional metal or polymer-based sensors. This validates that CBSs are well-equipped to be utilized in various harsh environmental conditions, thereby serving the purpose of replacing conventional sensors.
2. The impact of temperature changes on the electrical resistance of CBSs is not constant. Usually, an increase (or decrease) in temperature results in a decrease (or increase) in resistance. In the environment with temperature fluctuations, changes in resistance are influenced not only by external load variations but also by temperature fluctuations. Therefore, CBSs also require temperature compensation circuits to eliminate the effects of temperature. CBSs subjected to high-temperature treatment (e.g., after a fire) undergo material softening and a significant increase in resistance due to the damage to the nanofillers and the cement matrix.
3. The electrical conductivity and piezoresistive performance of CBSs vary significantly with different moisture levels. CBSs exhibit an optimal moisture content for achieving the best piezoresistive sensitivity, which is not a fixed value, in contrast to dry and saturated states. As for the impact of external moisture on the sensing performance of CBSs, it can be mitigated by developing hydrophobic or superhydrophobic CBSs. The results demonstrate that waterproof particles on the pore surfaces can effectively isolate moisture ingress and reduce its impact on electrical performance.
4. The influence of chloride ions on CBSs primarily stems from the pore solution and the resulting ionic conductivity changes. Therefore, chloride ion penetration can also be assessed by monitoring changes in electrical conductivity. In addition, the CBSs incorporating waterproofing additives exhibit better piezoresistive stability compared to the counterparts with only conductive fillers.
5. The intact GNP-filled cementitious composite before erosion exhibited excellent piezoresistivity for SHM as CBSs. However, after 90 and 180 days of sulfuric acid erosion, the piezoresistivity was strongly affected with less linearity and repeatability. In particular, after subjected to 3% H_2SO_4 solutions, the FCR changes under low and high loads presented extremely high and low piezoresistivity, respectively. The mechanism of the altered piezoresistivity is due to the mechanical and microstructural properties of corroded GNP-filled cementitious composite. The

erosion-induced porous structure filled with erosion products was easily compressed under low load, which led to the connection of GNPs and decreased the electrical resistivity. With the increase of load, the compressed porosity maintained unchanged and the intact cement matrix took responsibility for the further deformation, and then the FCR changes decreased under high load.

6. The freeze-thaw cycles exhibited limited impacts on the electrical resistivity of dry CB/cementitious composites, but the induced higher compression strain still caused reduced gauge factor and slightly lower piezoresistive sensitivity with reduction of 11.9%. However, both the electrical and mechanical properties of saturated CB/cementitious composites were greatly affected by the freeze-thaw cycles, which resulted in the decreased piezoresistive sensitivity by 30.7%.

REFERENCES

1. W. Dong, W. Li, Z. Tao, K. Wang, Piezoresistive properties of cement-based sensors: Review and perspective, Construction and Building Materials 203 (2019) 146–163.
2. M. Adresi, F. Pakhirehzan, Evaluating the performance of self-sensing concrete sensors under temperature and moisture variations: A review, Construction and Building Materials 404 (2023) 132923.
3. Z. Deng, W. Li, W. Dong, Z. Sun, J. Kodikara, D. Sheng, Multifunctional asphalt concrete pavement toward smart transport infrastructure: Design, performance and perspective, Composites Part B: Engineering 265 (2023) 110937.
4. D. Jang, J. Bang, H. Jeon, Impact of silica aerogel addition on the electrical and piezo-resistive sensing stability of CNT-embedded cement-based sensors exposed to varied environments, Journal of Building Engineering 78 (2023) 107700.
5. W. Dong, W. Li, K. Vessalas, X. He, Z. Sun, D. Sheng, Piezoresistivity deterioration of smart graphene nanoplate/cement-based sensors subjected to sulphuric acid attack, Composites Communications 23 (2021) 100563.
6. J.-K. Chen, M.-Q. Jiang, Long-term evolution of delayed ettringite and gypsum in Portland cement mortars under sulfate erosion, Construction and Building Materials 23(2) (2009) 812–816.
7. C. Evju, S. Hansen, Expansive properties of ettringite in a mixture of calcium aluminate cement, Portland cement and β-calcium sulfate hemihydrate, Cement and Concrete Research 31(2) (2001) 257–261.
8. J. Tang, H. Cheng, Q. Zhang, W. Chen, Q. Li, Development of properties and microstructure of concrete with coral reef sand under sulphate attack and drying-wetting cycles, Construction and Building Materials 165 (2018) 647–654.
9. M. Muthu, N. Ukrainczyk, E. Koenders, Effect of graphene oxide dosage on the deterioration properties of cement pastes exposed to an intense nitric acid environment, Construction and Building Materials 269 (2021) 121272.
10. M. Muthu, E.-H. Yang, C. Unluer, Resistance of graphene oxide-modified cement pastes to hydrochloric acid attack, Construction and Building Materials 273 (2021) 121990.
11. L. Sabapathy, B.S. Mohammed, A. Al-Fakih, M.M.A. Wahab, M. Liew, Y.M. Amran, Acid and sulphate attacks on a rubberized engineered cementitious composite containing graphene oxide, Materials 13(14) (2020) 3125.
12. T. Tong, Z. Fan, Q. Liu, S. Wang, S. Tan, Q. Yu, Investigation of the effects of graphene and graphene oxide nanoplatelets on the micro-and macro-properties of cementitious materials, Construction and Building Materials 106 (2016) 102–114.
13. S. Sharma, S. Arora, Economical graphene reinforced fly ash cement composite made with recycled aggregates for improved sulphate resistance and mechanical performance, Construction and Building Materials 162 (2018) 608–612.
14. H. Du, S. Dai Pang, Enhancement of barrier properties of cement mortar with graphene nanoplatelet, Cement and Concrete Research 76 (2015) 10–19.

15. H. Du, H.J. Gao, S. Dai Pang, Improvement in concrete resistance against water and chloride ingress by adding graphene nanoplatelet, Cement and Concrete Research 83 (2016) 114–123.
16. B. Wang, R. Zhao, Effect of graphene nano-sheets on the chloride penetration and microstructure of the cement based composite, Construction and Building Materials 161 (2018) 715–722.
17. A. Mohammed, J.G. Sanjayan, W. Duan, A. Nazari, Incorporating graphene oxide in cement composites: A study of transport properties, Construction and Building Materials 84 (2015) 341–347.
18. W. Li, F. Qu, W. Dong, G. Mishra, S.P. Shah, A comprehensive review on self-sensing graphene/cementitious composites: A pathway toward next-generation smart concrete, Construction and Building Materials 331 (2022) 127284.
19. Y. Alharbi, J. An, B.H. Cho, M. Khawaji, W. Chung, B.H. Nam, Mechanical and sorptivity characteristics of edge-oxidized graphene oxide (EOGO)-cement composites: Dry-and wet-mix design methods, Nanomaterials 8(9) (2018) 718.
20. X. Li, Z. Lu, S. Chuah, W. Li, Y. Liu, W.H. Duan, Z. Li, Effects of graphene oxide aggregates on hydration degree, sorptivity, and tensile splitting strength of cement paste, Composites Part A: Applied Science and Manufacturing 100 (2017) 1–8.
21. F. Matalkah, P. Soroushian, Graphene nanoplatelet for enhancement the mechanical properties and durability characteristics of alkali activated binder, Construction and Building Materials 249 (2020) 118773.
22. M. Krystek, D. Pakulski, M. Górski, L. Szojda, A. Ciesielski, P. Samorì, Electrochemically exfoliated graphene for high-durability cement composites, ACS Applied Materials & Interfaces 13(19) (2021) 23000–23010.
23. W. Dong, W. Li, Y. Guo, F. Qu, K. Wang, D. Sheng, Piezoresistive performance of hydrophobic cement-based sensors under moisture and chloride-rich environments, Cement and Concrete Composites 126 (2022) 104379.
24. Y. Zha, J. Yu, R. Wang, P. He, Z. Cao, Effect of ion chelating agent on self-healing performance of cement-based materials, Construction and Building Materials 190 (2018) 308–316.
25. M.J. Al-Kheetan, M.M. Rahman, D.A. Chamberlain, Moisture evaluation of concrete pavement treated with hydrophobic surface impregnants, International Journal of Pavement Engineering 21(14) (2020) 1746–1754.
26. W. Dong, W. Li, X. Zhu, D. Sheng, S.P. Shah, Multifunctional cementitious composites with integrated self-sensing and hydrophobic capacities toward smart structural health monitoring, Cement and Concrete Composites 118 (2021) 103962.
27. J. Hodul, N. Žižková, R.P. Borg, The influence of crystalline admixtures on the properties and microstructure of mortar containing by-products, Buildings 10(9) (2020) 146.
28. Y. Zhao, Y. Liu, T. Shi, Y. Gu, B. Zheng, K. Zhang, J. Xu, Y. Fu, S. Shi, Study of mechanical properties and early-stage deformation properties of graphene-modified cement-based materials, Construction and Building Materials 257 (2020) 119498.
29. Z. Lu, X. Li, A. Hanif, B. Chen, P. Parthasarathy, J. Yu, Z. Li, Early-age interaction mechanism between the graphene oxide and cement hydrates, Construction and Building Materials 152 (2017) 232–239.
30. Y. Xu, D. Chung, Improving silica fume cement by using silane, Cement and Concrete Research 30(8) (2000) 1305–1311.
31. Z. Chen, Y. Xu, J. Hua, X. Wang, L. Huang, X. Zhou, Mechanical properties and shrinkage behavior of concrete-containing graphene-oxide nanosheets, Materials 13(3) (2020) 590.
32. W. Yao, X. Yidong, Z. Juqing, P. Zhihong, L. Mingming, Influence of graphene oxide on autogenous shrinkage of cement-based composites, Journal of Functional Materials/Gongneng Cailiao 51(3) (2020).
33. X. Zhu, X. Kang, J. Deng, K. Yang, L. Yu, C. Yang, A comparative study on shrinkage characteristics of graphene oxide (GO) and graphene nanoplatelets (GNPs) modified alkali-activated slag cement composites, Materials and Structures 54(3) (2021) 106.
34. B. Wang, D. Shuang, Effect of graphene nanoplatelets on the properties, pore structure and microstructure of cement composites, Materials Express 8(5) (2018) 407–416.
35. Y. Gao, D.J. Corr, M.S. Konsta-Gdoutos, S.P. Shah, Effect of carbon nanofibers on autogenous shrinkage and shrinkage cracking of cementitious nanocomposites, ACI Materials Journal 115(4) (2018) 615–622.

36. A. Mohammed, J. Sanjayan, A. Nazari, N. Al-Saadi, Effects of graphene oxide in enhancing the performance of concrete exposed to high-temperature, Australian Journal of Civil Engineering 15(1) (2017) 61–71.
37. H.-Y. Chu, J.-Y. Jiang, W. Sun, M. Zhang, Effects of graphene sulfonate nanosheets on mechanical and thermal properties of sacrificial concrete during high temperature exposure, Cement and Concrete Composites 82 (2017) 252–264.
38. A. Mohammed, N.T.K. Al-Saadi, R. Al-Mahaidi, Bond behaviour between NSM CFRP strips and concrete at high temperature using innovative high-strength self-compacting cementitious adhesive (IHSSC-CA) made with graphene oxide, Construction and Building Materials 127 (2016) 872–883.
39. W. Dong, W. Li, K. Wang, B. Han, D. Sheng, S.P. Shah, Investigation on physicochemical and piezoresistive properties of smart MWCNT/cementitious composite exposed to elevated temperatures, Cement and Concrete Composites 112 (2020) 103675.
40. L. Zhang, M. Kai, K. Liew, Evaluation of microstructure and mechanical performance of CNT-reinforced cementitious composites at elevated temperatures, Composites Part A: Applied Science and Manufacturing 95 (2017) 286–293.
41. X. Li, Y. Bao, L. Wu, Q. Yan, H. Ma, G. Chen, H. Zhang, Thermal and mechanical properties of high-performance fiber-reinforced cementitious composites after exposure to high temperatures, Construction and Building Materials 157 (2017) 829–838.
42. M. Şahmaran, E. Özbay, H.E. Yücel, M. Lachemi, V.C. Li, Effect of fly ash and PVA fiber on microstructural damage and residual properties of engineered cementitious composites exposed to high temperatures, Journal of Materials in Civil Engineering 23(12) (2011) 1735–1745.
43. W. Khaliq, V. Kodur, Thermal and mechanical properties of fiber reinforced high performance self-consolidating concrete at elevated temperatures, Cement and Concrete Research 41(11) (2011) 1112–1122.
44. B. Persson, Fire resistance of self-compacting concrete, SCC, Materials and Structures 37 (2004) 575–584.
45. G.-F. Peng, W.-W. Yang, J. Zhao, Y.-F. Liu, S.-H. Bian, L.-H. Zhao, Explosive spalling and residual mechanical properties of fiber-toughened high-performance concrete subjected to high temperatures, Cement and Concrete Research 36(4) (2006) 723–727.
46. W. Sha, G. Pereira, Differential scanning calorimetry study of ordinary Portland cement paste containing metakaolin and theoretical approach of metakaolin activity, Cement and Concrete Composites 23(6) (2001) 455–461.
47. P. Ajayan, T. Ebbesen, T. Ichihashi, S. Iijima, K. Tanigaki, H. Hiura, Opening carbon nanotubes with oxygen and implications for filling, Nature 362(6420) (1993) 522–525.
48. Y. Shimada, J.F. Young, Structural changes during thermal dehydration of ettringite, Advances in Cement Research 13(2) (2001) 77–81.
49. W. Dong, W. Li, N. Lu, F. Qu, K. Vessalas, D. Sheng, Piezoresistive behaviours of cement-based sensor with carbon black subjected to various temperature and water content, Composites Part B: Engineering 178 (2019) 107488.
50. M. Cao, X. Ming, H. Yin, L. Li, Influence of high temperature on strength, ultrasonic velocity and mass loss of calcium carbonate whisker reinforced cement paste, Composites Part B: Engineering 163 (2019) 438–446.
51. S. Dai Pang, H.J. Gao, C. Xu, S.T. Quek, H. Du, Strain and damage self-sensing cement composites with conductive graphene nanoplatelet, Sensors and Smart Structures Technologies for Civil, Mechanical, and Aerospace Systems 2014, SPIE, 2014, pp. 546–556.
52. A.L. Materazzi, F. Ubertini, A. D'Alessandro, Carbon nanotube cement-based transducers for dynamic sensing of strain, Cement and Concrete Composites 37 (2013) 2–11.
53. B. Han, B. Han, J. Ou, Experimental study on use of nickel powder-filled Portland cement-based composite for fabrication of piezoresistive sensors with high sensitivity, Sensors and Actuators A: Physical 149(1) (2009) 51–55.
54. P.-X. Hou, C. Liu, H.-M. Cheng, Purification of carbon nanotubes, Carbon 46(15) (2008) 2003–2025.
55. S. Porro, S. Musso, M. Vinante, L. Vanzetti, M. Anderle, F. Trotta, A. Tagliaferro, Purification of carbon nanotubes grown by thermal CVD, Physica E: Low-Dimensional Systems and Nanostructures 37(1–2) (2007) 58–61.

56. R.R. Kohlmeyer, M. Lor, J. Deng, H. Liu, J. Chen, Preparation of stable carbon nanotube aerogels with high electrical conductivity and porosity, Carbon 49(7) (2011) 2352–2361.
57. B. Zou, S.J. Chen, A.H. Korayem, F. Collins, C.M. Wang, W.H. Duan, Effect of ultrasonication energy on engineering properties of carbon nanotube reinforced cement pastes, Carbon 85 (2015) 212–220.
58. A. Meyer-Plath, G. Orts-Gil, S. Petrov, F. Oleszak, H.-E. Maneck, I. Dörfel, O. Haase, S. Richter, R. Mach, Plasma-thermal purification and annealing of carbon nanotubes, Carbon 50(10) (2012) 3934–3942.
59. H. Kim, I. Park, H.-K. Lee, Improved piezoresistive sensitivity and stability of CNT/cement mortar composites with low water–binder ratio, Composite Structures 116 (2014) 713–719.
60. J. Cao, D.D.L. Chung, Electric polarization and depolarization in cement-based materials, studied by apparent electrical resistance measurement, Cement and Concrete Research 34(3) (2004) 481–485.
61. S. Wen, D. Chung, The role of electronic and ionic conduction in the electrical conductivity of carbon fiber reinforced cement, Carbon 44(11) (2006) 2130–2138.
62. H. Li, H. Xiao, J. Ou, Electrical property of cement-based composites filled with carbon black under long-term wet and loading condition, Composites Science and Technology 68(9) (2008) 2114–2119.
63. L. Zhang, S. Ding, B. Han, X. Yu, Y.-Q. Ni, Effect of water content on the piezoresistive property of smart cement-based materials with carbon nanotube/nanocarbon black composite filler, Composites Part A: Applied Science and Manufacturing 119 (2019) 8–20.
64. W. Dong, W. Li, L. Shen, Z. Sun, D. Sheng, Piezoresistivity of smart carbon nanotubes (CNTs) reinforced cementitious composite under integrated cyclic compression and impact, Composite Structures 241 (2020) 112106.
65. W. Dong, W. Li, K. Wang, K. Vessalas, S. Zhang, Mechanical strength and self-sensing capacity of smart cementitious composite containing conductive rubber crumbs, Journal of Intelligent Material Systems and Structures 31(10) (2020) 1325–1340.
66. W. Dong, W. Li, Z. Luo, Y. Guo, K. Wang, Effect of layer-distributed carbon nanotube (CNT) on mechanical and piezoresistive performance of intelligent cement-based sensor, Nanotechnology 31(50) (2020) 505503.
67. W. Dong, W. Li, Z. Luo, G. Long, K. Vessalas, D. Sheng, Structural response monitoring of concrete beam under flexural loading using smart carbon black/cement-based sensors, Smart Materials and Structures 29(6) (2020) 065001.
68. D.G. Meehan, S. Wang, D. Chung, Electrical-resistance-based sensing of impact damage in carbon fiber reinforced cement-based materials, Journal of Intelligent Material Systems and Structures 21(1) (2010) 83–105.
69. S. Wang, D. Chung, J.H. Chung, Impact damage of carbon fiber polymer–matrix composites, studied by electrical resistance measurement, Composites Part A: Applied Science and Manufacturing 36(12) (2005) 1707–1715.
70. H. Wang, X. Gao, J. Liu, Effects of salt freeze-thaw cycles and cyclic loading on the piezoresistive properties of carbon nanofibers mortar, Construction and Building Materials 177 (2018) 192–201.
71. J. Cao, D. Chung, Damage evolution during freeze–thaw cycling of cement mortar, studied by electrical resistivity measurement, Cement and Concrete Research 32(10) (2002) 1657–1661.
72. J. Cao, D.D.L. Chung, Role of moisture in the Seebeck effect in cement-based materials, Cement and Concrete Research 35(4) (2005) 810–812.
73. M. Mouret, A. Bascoul, G. Escadeillas, Study of the degree of hydration of concrete by means of image analysis and chemically bound water, Advanced Cement Based Materials 6(3–4) (1997) 109–115.
74. W.J. McCarter, Effects of temperature on conduction and polarization in Portland cement mortar, Journal of the American Ceramic Society 78(2) (1995) 411–415.
75. V. Penttala, F. Al-Neshawy, Stress and strain state of concrete during freezing and thawing cycles, Cement and Concrete Research 32(9) (2002) 1407–1420.

Chapter 7

CBSs with integrated capacities

7.1 INTRODUCTION

Cement-based sensors (CBSs), classified as a type of semiconductor material with commendable conductivity, can be utilized to further develop multifunctional CBSs [1]. These CBSs possess additional features such as self-heating based on the Joule effect and the electromagnetic shielding properties [2–4]. The self-heating capacity provides various practical functionalities such as temperature adjustment and deicing for buildings and infrastructures, particularly in cold climates and environments with specific temperature control requirements [5, 6]. The electromagnetic shielding capacity furnishes buildings and facilities with electromagnetic protection and information security functionalities by diminishing or blocking electromagnetic radiation [7]. It can be applied in laboratories, medical facilities, and electronic equipment manufacturing plants where these features are essential. In addition, due to the brittle nature and susceptibility to cracking of cement-based materials, self-healing capability has also been integrated into CBSs. It can autonomously repair minor cracks and damages in concrete, preventing the propagation of cracks and thereby reducing the need for maintenance and repairs. Simultaneously, to ensure independent operation of the sensor unaffected by external interfering factors, such as temperature and humidity interferences, researchers have developed a superhydrophobic CBS with excellent waterproofing properties. Even when applied in humid or marine environments, the integrated capacity can guarantee the normal operation of CBSs. Additionally, research on self-powering sensors is beginning to emerge, resulting in the feasible exploration of CBSs with self-powering capacity. The self-powering CBS, through internally integrated self-powering technology, can autonomously generate electrical energy without the need for an external power source [8, 9]. This capability renders it particularly useful in locations where conventional power sources are unavailable or in applications requiring prolonged independent operation.

This chapter summarizes CBSs with integrated autogenous self-healing capacity from the perspective of the experimental process, methodology, self-sensing, and self-healing performances. Similarly, the preparation and performance of the superhydrophobic CBS are detailed. Subsequently, due to the relatively limited number of articles studying the performance of self-powering CBSs, a brief overview is provided here, along with proposed directions for future research. Research on CBSs that combine self-heating performance is even scarcer, with most studies tending to focus on one of these functions. After all, for CBSs, temperature is an indispensable influencing factor, and sensors combining self-heating capacity will become more complex.

DOI: 10.1201/9781032663685-7

7.2 INTEGRATED SELF-SENSING AND SELF-HEALING PROPERTIES

7.2.1 Experimental program

Multifunctional CBSs with integrated self-healing and self-sensing properties are developed using microencapsulation of nano-carbon black (CB) to enclose healing agent slaked lime (SL) [10]. Based on the principle, the conductive CB nanoparticles were divided equally into two groups, with the first group being well-dispersed for excellent conductivity and piezoresistivity of the composites, and the second group coating on the surface of SL particles as microencapsulation. The specimen manufacturing procedures include three separate steps: (a) preparation of well-dispersed CB suspension; (b) preparation of SL particles encapsulated by CB; and (c) the composites mixing, casting, and curing. Previous studies have demonstrated that excellent dispersion of CB nanoparticles could be achieved by using ultrasonication and superplasticizer [11]. The preparation of SL particles enclosed with CB was carried out in a rigidly sealed container to avoid dust leakage. The mixture of SL and CB was mechanically stirred for 5 min with a speed of 200 rpm, to ensure the CB nanoparticles attached to the surface of SL (see Fig. 7.1). To make a cementitious composite, cement, silica fume, and encapsulated SL with CB were first dry-mixed in the mixer. Then, the well-dispersed CB suspension was added as mixing water, following the standard of ASTM C305-14 (Standard Practice for Mechanical Mixing of Hydraulic Cement Pastes and Mortars of Plastic Consistency). Given the added suspension consisted of a moderate dosage of CB, limited CB on the surfaces of SL particles would enter into the mixture so that encapsulation of SL could be maintained during this mixing process. Subsequently, the SL particles were partially protected from the carbonation process. The mix proportions of the specimens are listed in Table 7.1, to display

Figure 7.1 Microstructural morphology of carbon black, slaked lime, and their dry mixture [10].

Table 7.1 Mix proportions of CB/cementitious composites with varied contents of SL [10]

Cement	*Silica fume*	*Carbon black (CB) (%)*	*Slaked lime (SL) (%)*	*Water*	*Superplasticizer (%)*
0.9	0.1	1.0	0	0.45*	0.8
0.9	0.1	1.0	2.0	0.45	0.8
0.9	0.1	1.0	3.0	0.45	0.8
0.9	0.1	1.0	4.0	0.45	0.8

Note: Figures under the cement, silica fume, carbon black, slaked lime, water, and superplasticizer represent their ratios to the weight of binder (sum of cement and silica fume), e.g., 0.45* under water means the water-to-binder ratio of 0.45.

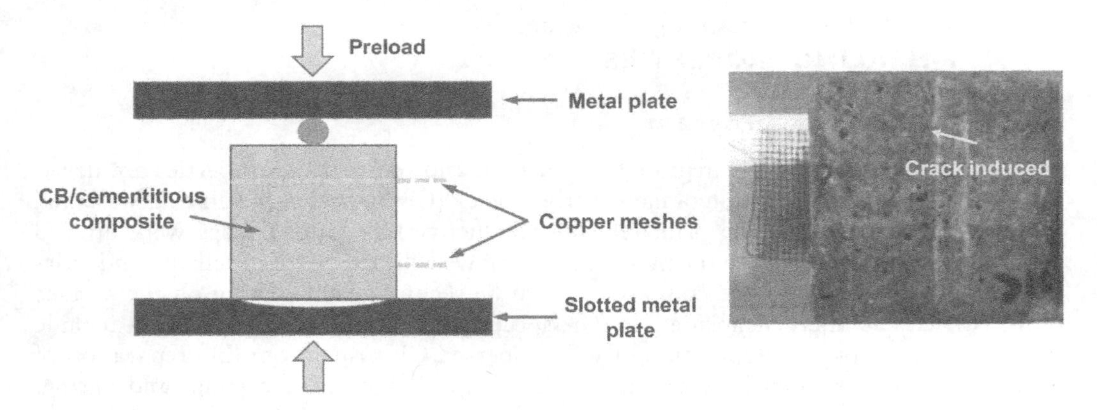

Figure 7.2 Schematic diagram of the experimental setup for crack generation of CB/cementitious composites [10].

the multiple self-sensing and self-healing properties of CB/cementitious composites under various contents of SL.

To evaluate the self-healing and cracks closure behaviors of CB/cementitious composites, small cracks were generated in the cementitious composite before the healing procedure. Fig. 7.2 plots the experimental setup regarding the formation of cracks on the CB/cementitious composites [10]. The slotted metal pate incorporating a cylindrical steel bar provided a concentrated load on specimens to generate cracks. The load provided by the compression machine gently increased until the formation of small cracks. To reduce the interference by autogenetic healing of remained unhydrated cement, the self-healing test was conducted on the CB/cementitious composites after a three-month curing process. The self-healing performances of the CB/cementitious composites were carried out by the cracks self-healing test. First, the specimens were cracked by the compression equipment AGX50 to induce minor cracks as aforementioned. Afterward, an additional ultrasonication by water bath on the pre-cracked specimens was conducted. Since the SL was premixed with CB nanoparticles which possessed high surface energy, the ultrasonic treatment was to remove the attached CB on the surface of the SL. Subsequently, the pre-cracked specimens were used for the self-healing test. In particular, the self-healing procedure included the water spray and standard curing in the chamber with a temperature of 23 ± 2°C and relative humidity of 90%. The water spray which was conducted twice a day during the self-healing process was to wet CB/cementitious composites and provide extra water for SL dissolution, transport, and carbonation.

7.2.2 CB dispersion with SL

Fig. 7.3 shows the sedimentation of CB with time in the superplasticizer–water solutions with different contents of SL [10]. The CB showed excellent uniform distribution within 160 min in the superplasticizer–water solution without SL, indicating that the incorporated dispersion methods of superplasticizer and ultrasonication can successfully disperse CB nanoparticles. However, it could be observed that the addition of SL considerably improved the settlement of CB and weakened its uniform dispersion. Previous studies have proposed that $Ca(OH)_2$ would affect the dispersion of carbon nanotube in superplasticizer–water suspension and lead to reagglomeration [12]. They attributed the

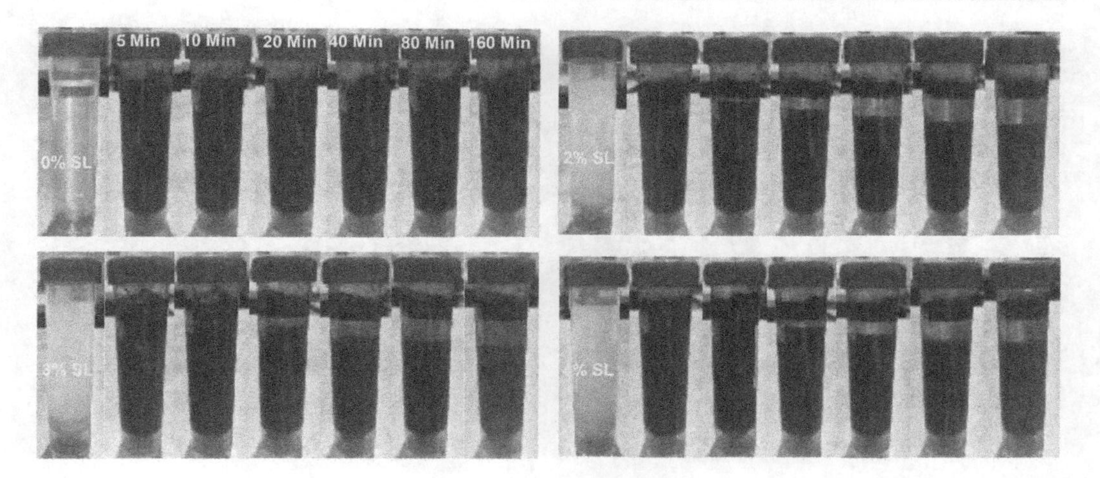

Figure 7.3 Sedimentation of CB with duration in superplasticizer–water solutions with varied SL contents [10].

phenomenon to the alkaline environments that affected the electrostatic repulsion between nanoparticles and superplasticizer molecules. In this study, the pH of polycarboxylate-based superplasticizer is 5 ± 1. Based on the description that below pH of 8, the polycarboxylate-based superplasticizer molecules could absorb on the surface of nanoparticles to separate the individual nanoparticle [13]. This can explain the well-dispersed CB in the water solution with the only superplasticizer. With the increasing SL, the pH of superplasticizer/SL/water solutions increased to larger than 12 in the ambient temperature and affected the attachment efficiency of superplasticizer molecules on the surface of CB. In other words, the $Ca(OH)_2$ from slaked lime can retard the electrostatic repulsion between CB nanoparticles and superplasticizer molecules, to affect the dispersion efficiency of CB suspension. In addition to the increased pH, the precipitation of supersaturated SL or $Ca(OH)_2$ might be another factor to affect the CB dispersion. Given the CB nanoparticles tend to absorb on the surface of SL, the precipitation of oversaturated SL could stimulate the settlement of CB nanoparticles because of the lower solubility of SL in superplasticizer–water solution. Moreover, the decreased CB dispersion with SL demonstrated that an increasing amount of CB was enclosed on the surface of SL, which indicated the preserved SL for the self-healing operation.

7.2.3 Self-healing performance

7.2.3.1 Compressive strength recovery

The compressive strength of CB/cementitious composites incorporating various concentrations of SL is displayed in Fig. 7.4a [10]. The initial compressive strength of approximately 33.5 MPa was found for the plain CB/cementitious composite. With the addition of 2%, 3%, and 4% SL, the strengths showed an increasing tendency and the values reached 38.3 MPa for the CB/cementitious composites with 4% SL. The enhancement might be due to the addition of SL that increased the alkali content in the CB/cementitious composites, which improved the cement hydration because the calcium hydroxide could react with active silica from added silica fume [14]. The strengths of pre-cracked specimens and

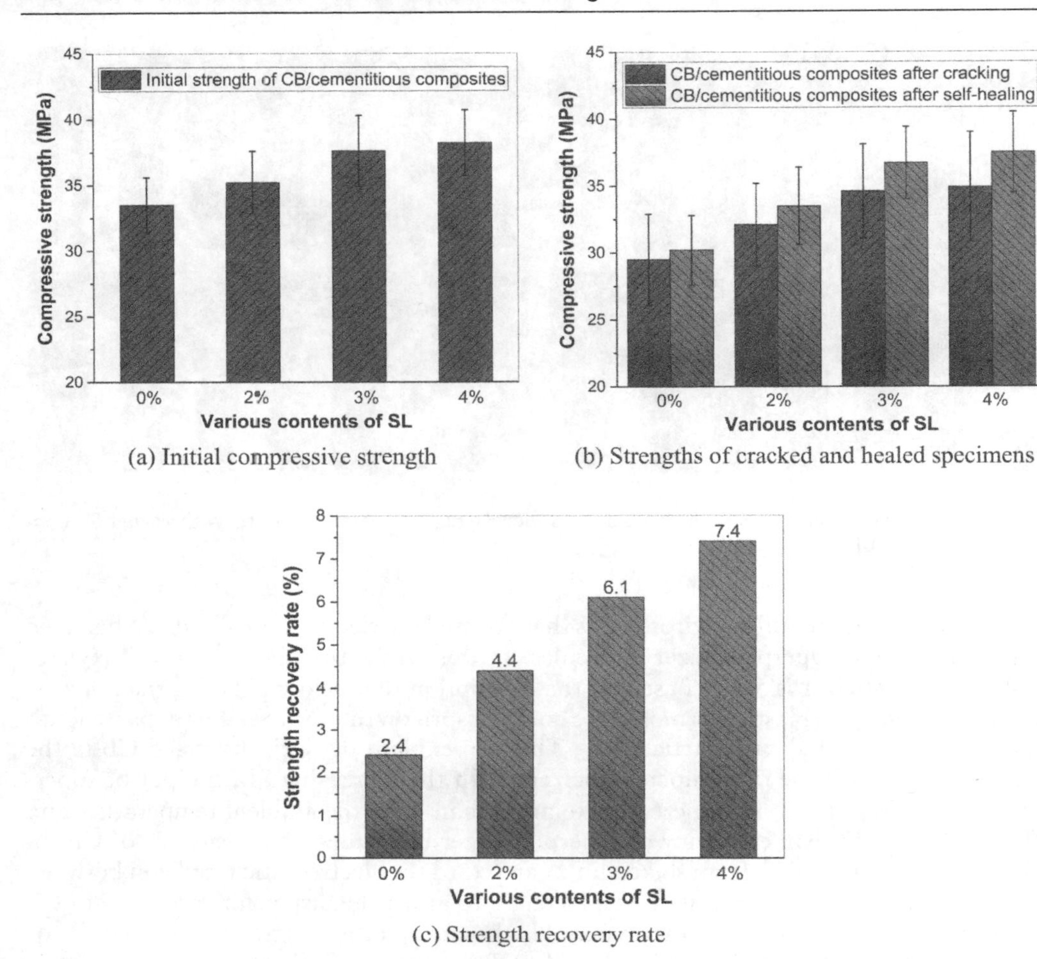

(a) Initial compressive strength

(b) Strengths of cracked and healed specimens

(c) Strength recovery rate

Figure 7.4 Compressive strength of CB/cementitious composites with various SL contents [10].

the specimens after 28 days of healing procedures were illustrated in Fig. 7.4b [10]. It could be found that the strengths of pre-cracked specimens possessed smaller values than those of corresponding intact specimens. They followed the same trend by the increased compressive strength with the increasing SL. The compressive strength of cracked plain CB/cementitious composites was approximately 29.5 MPa, which increased to 34.9 MPa for the composites filled with 4% SL. After 28 days of the healing process, the strengths of pre-cracked specimens increased because of the further cement hydration or the filling effect of calcium carbonate. As displayed in Fig. 7.4c, it can be found that the strength recovery rate increased from 2.4% to 7.4% with the addition of SL [10]. For the plain CB/cementitious composites, the strength enhancement was mainly due to the further cement hydration, because the unreacted binder was exposed to the moist ambiance when the specimens cracked. It showed limited strength recovery efficiency for the autogenous healing of cementitious materials, although the carbonation of calcium hydroxide sourced from cement hydration also enhanced the autogenous healing [15]. However, for the CB/cementitious composites with SL, in addition to the low-efficient autogenous healing, the preserved SL stimulated the generation of calcium carbonate and improved the autonomous healing efficiency of cementitious materials. Previous studies have proved that the main self-healing product was calcium carbonate which had the capacity to seal the cracks [16]. In the study, the SL enclosed with CB could be partially preserved in the process of cement hydration; the more the amount of SL applied, the more the remaining SL in the composites. These SL could be released when the cracks were generated, and the carbonation process and formation of calcium carbonate were beneficial for the sealing of cracks.

7.2.3.2 Cracks closure behaviors

Fig. 7.5a–d illustrates the crack closure characterizations of CB/cementitious composites with different contents of SL at the healing ages of 0, 7, 14, and 28 days [10]. For the pure CB/cementitious composites without SL, it could be observed that only a small proportion of cracks got partially healed, mainly the cracks with a small width. The average width of healed cracks was approximately 0.121 mm, which is mainly because of the formation of cement hydration products or calcium carbonate given the continual cement hydration or the carbonation of generated calcium hydroxide (CH). However, for the cracks with a large average width of 0.198 mm, they were unhealed. It implied that the pure CB/cementitious composite possessed limited self-healing capacity with small cracks being partially healed. In addition, the morphology of healed cracks was almost kept unchanged from 7 days to 28 days, indicating the termination of generated hydration or carbonation products after 7 days of the healing process.

Fig. 7.5b shows clear closure of cracks after the healing process for the CB/cementitious composites filled with 2% SL. In addition to the hydration of unhydrated cement particles, hydration of SL played a significant role in cracks healing. This has been testified by the specific SEM and EDX results presented in the next section. It could be observed that the healing efficiency strongly depended on the healing age and the width of cracks. With the increase of healing age, the cracks gradually filled and closed by the increasing amount of calcium carbonate. For example, the cracks were still clearly seen at 7 days. After 28 days of healing, most cracks were completely filled with calcium carbonate, especially fine cracks. There existed unhealed cracks in the CB/cementitious composites containing 2% SL, probably due to the uneven distribution of SL.

For the composites with 3% SL, Fig. 7.5c shows that after seven days of healing, some self-healing products (calcium carbonate) and cement hydration products already started to generate in some specific cracked areas. These areas either possess fine cracks or abundant SL and unhydrated cement particles. After 14 days of healing, the continuous self-healing products along with the cracks could be observed in the composites. The crack closure continued in the composites at 28 days of healing, it was found that the pre-induced cracks were almost filled with calcium carbonate, and only very few areas with open cracks were observed. The results suggest that 3% SL provided the CB/cementitious composites with slightly higher healing efficiency than those with 2% SL.

For the CB/cementitious composites with 4% SL, Fig. 7.5d shows some well-healed cracks with carbonation products, as well as some cracks with poor healing efficiency. Similar to the previously mentioned composites, the reason for the failed self-healing of secondary induced cracks might be due to the agglomeration of SL. Probably due to the excessive SL, it could be found that calcium carbonate also formed in some surface areas of composites without cracks. Moreover, with the increase of SL, it was possible that the SL was not well-coated with a designed weight of CB, which could be another reason

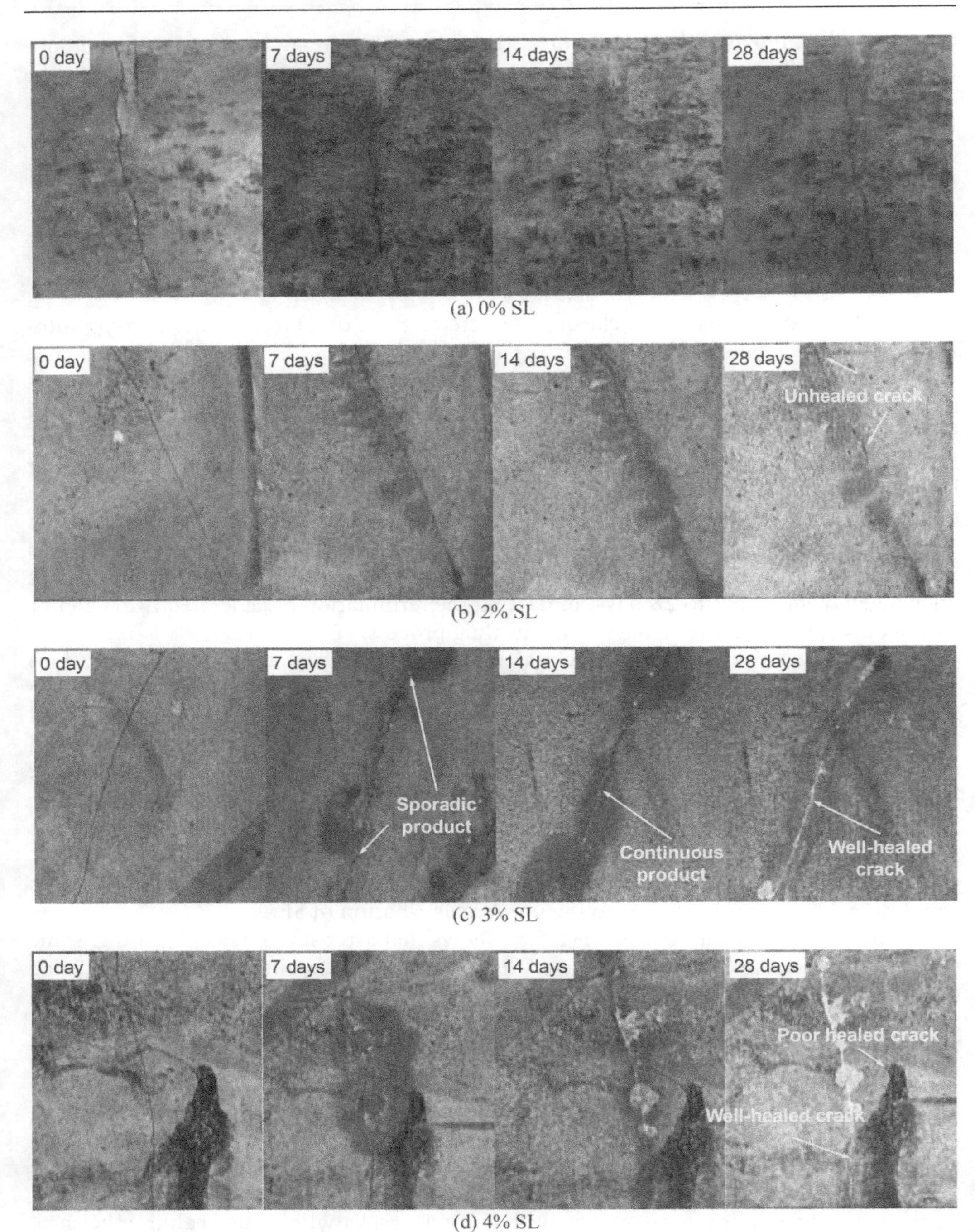

Figure 7.5 Crack closure characterizations of CB/cementitious composites with SL at various curing age [10].

for the poor healing performance. In summary, it can be concluded that the addition of SL does have the capacity to close the cracks in the CB/cementitious composites. For the composites filled with 2% and 3% SL, the majority of the cracks were completely healed automatically with limited cracks being unhealed. For the 4% SL filled composites, some cracks were poorly healed even though the main cracks were well-healed.

7.2.3.3 Multiple self-healing efficiency

Above results indicated the potential of SL in the CB/cementitious composites for self-healing and crack closure performance. However, it should be noted that there also existed the challenges of unhealed cracks in some specific areas. As shown in Fig. 7.6, the initial cracks possessed similar widths of 0.151 mm, 0.125 mm, and 0.139 mm, respectively, in the regions of 1, 2, and 3 [10]. After seven days of healing, the cracks of region 1 were almost healed, while the region 2 was partially healed and the region 3 was slightly attached with a small number of self-healing products in the crack edges. The experimental results are radically different from the plain CB/cementitious composites which showed the healed cracks in the regions with small widths. It implied that the SL was not uniformly spreading in some regions, and region 1 had a higher concentration of SL than regions 2 and 3. Moreover, it possessed the possibility of still coated SL by CB nanoparticles, even after the pretreatment of ultrasonication. This might be another reason for the uneven distribution of self-healing products.

In addition to the cracks of a similar size, Fig. 7.7 illustrates the self-healing performances of CB/cementitious composites containing various widths of cracks [10]. It could be observed clearly that the self-healing efficiency of cracks with small width was better than that of wide cracks. One reason might be due to the distribution of SL that affects the formation of calcium carbonate. Moreover, the SL and unhydrated cement particles can be easily released in the water when the wide cracks are ultrasonicated to uncover

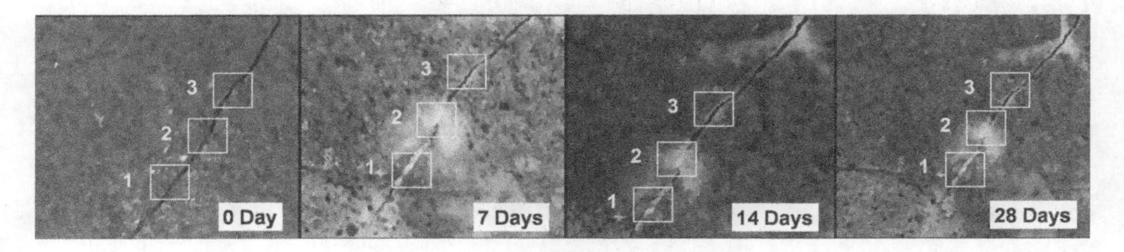

Figure 7.6 Crack closure efficiency of CB/cementitious composites because of uneven dispersion of SL [10].

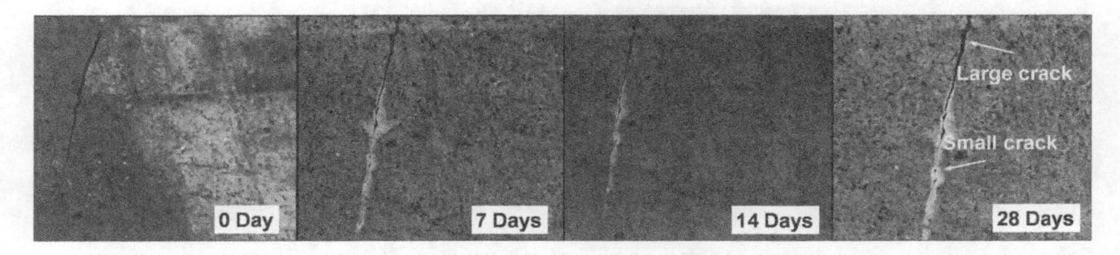

Figure 7.7 Crack closure efficiency of CB/cementitious composites because of different widths of cracks [10].

the SL particles, and that might be another reason for the poor healing efficiency of wide cracks.

7.2.3.4 Mechanism discussion

Fig. 7.8 schematically plots the effect of CB-coated SL on the self-healing capacity and the healing process in the cementitious composites [10]. The healing agent SL provides calcium ions, and the CB nanoparticle is to prevent SL from carbonation. In the case of pre-cracked CB/cementitious composites, the ultrasonication conducted is to eliminate the CB nanoparticles attached to the surface of SL and expose the SL to water spray and air atmosphere. Since calcium compounds usually possess a low solubility, the applied water spray is to transfer the calcium ions from SL to the cracks with a low concentration of calcium ions. The continual supplementary from SL increases the dosage of CH and moveable calcium ions until the solution reaches a saturation state. Without regard to the effect of impurities from SL, three potential self-healing products should be generated in the process of curing. First, the healing product (e.g., C—S—H and CH) is sourced from the remaining unhydrated cement particles. Second, the moveable calcium ions and hydroxide ions from SL gradually increase the concentration of calcium hydroxide; hence, the CH crystal can be formed in the cracks. It seems like the CH crystal is moved from SL to the cracks. Third, the formation of calcium carbonate due to CH carbonation is the main self-healing product, which can be demonstrated in the following section of microstructure and phase observation. Apart from the cement hydration, these reactions

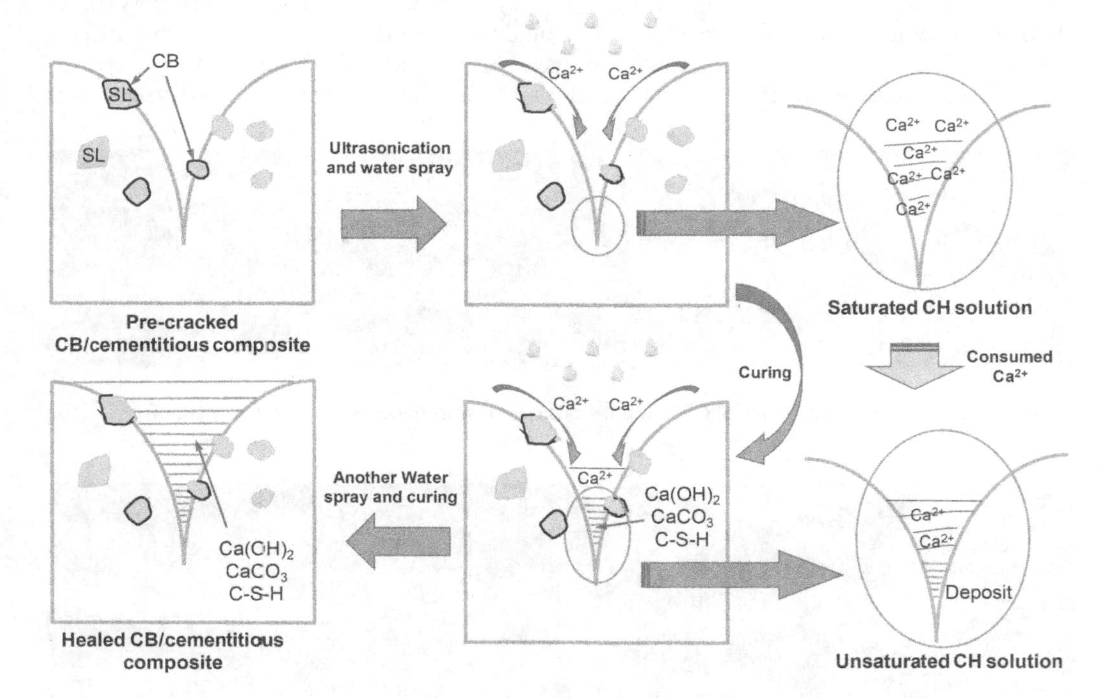

Figure 7.8 Dissolution, transport, and carbonation of calcium from SL in the CB/cementitious composites during self-healing progress [10].

consume calcium ions and result in unsaturated CH solutions in the cracks, given that self-healing is conducted after three months when the cement hydration is significantly weakened. Therefore, the conducted another water spray is to transfer the calcium ions from SL and continually supply the CH concentration.

7.2.4 Self-sensing performance

7.2.4.1 Effect of SL concentration

Fig. 7.9 shows the stress-sensing performance of CB-filled cementitious composites with various contents of SL [10]. For easy comparison, the experimental results of CB/cementitious composites with various contents of SL are illustrated together in one figure. For all composites, the fractional changes of resistivity (FCR) decreased with the increase of compressive stress to 5.0 MPa and returned to the initial values when the stress decreased to zero. The absolute values of FCR under the peak stress gradually increased with the increase of SL, reaching 0.98%, 1.25%, 1.41%, and 2.18%, respectively, for the composites without SL and with 2%, 3%, and 4% SL. It demonstrated that the stress-sensing sensitivity of CB cementitious composites was improved with the addition of SL. There are several possible reasons for the higher FCR values. First, in comparison to CB, the oversaturated SL particles are nonconductive phases in the composites. They can block the conductive passages established by CB and create more potential contact points between conductive passages. Under compression, the SL-induced discontinued conductive passages reconnect and lead to larger FCR values. Second, although CB dispersion becomes worse because of the SL precipitation in superplasticizer–water solution without binder, the next microstructures show that the addition of SL decreases the size of CB agglomerations and stimulates the dispersion of CB in cementitious composites. These two potential reasons result in increased stress-sensing efficiency.

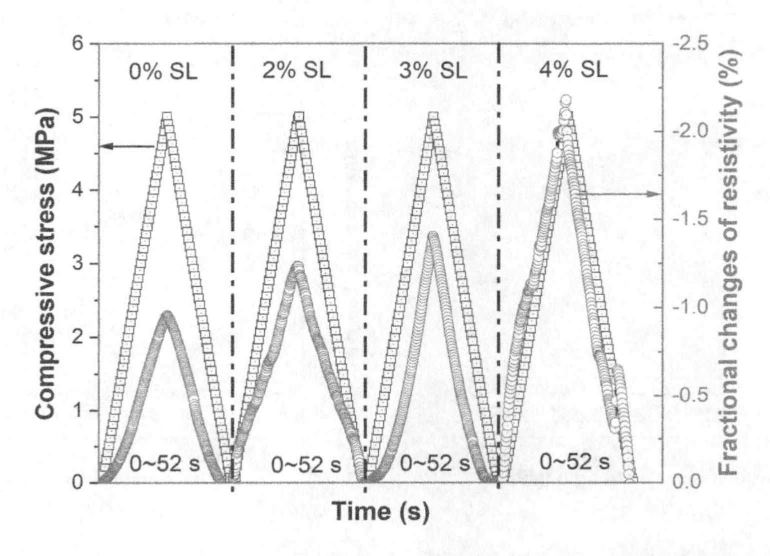

Figure 7.9 FCR under cyclic compression for CB/cementitious composites with different SL contents [10].

7.2.4.2 *Effect of self-healing process*

Different from the above specimens, the composites tested in this section were accompanied by pre-damaged cracks, in order to elucidate the effect of the healing process on the stress-sensing performances. Fig. 7.10 depicts the FCR changes under cyclic compression of CB/cementitious composites filled with different contents of SL without and after the healing process [10]. It was found that the FCR of pre-cracked composites was higher than that of intact composites. In accordance with the previous studies regarding the effect of impact-induced cracks on the piezoresistivity, a higher FCR value is possibly due to the compression of cracks that altered the conductive passages [17]. In addition, similar to the intact specimens, the pre-cracked CB/cementitious composites showed increased FCR values with increasing SL content. The resistivity changes reached 1.36%, 2.05%, 2.28%, and 3.89% for the pre-cracked CB/cementitious composites with 0%, 2%, 3%, and 4% SL, respectively. However, after 28 days of self-healing process, much higher FCR values occurred for the CB/cementitious composites, increasing to 3.24%, 3.76%, 6.23%, and 7.89%, respectively. There are three potential reasons for the FCR enhancement under

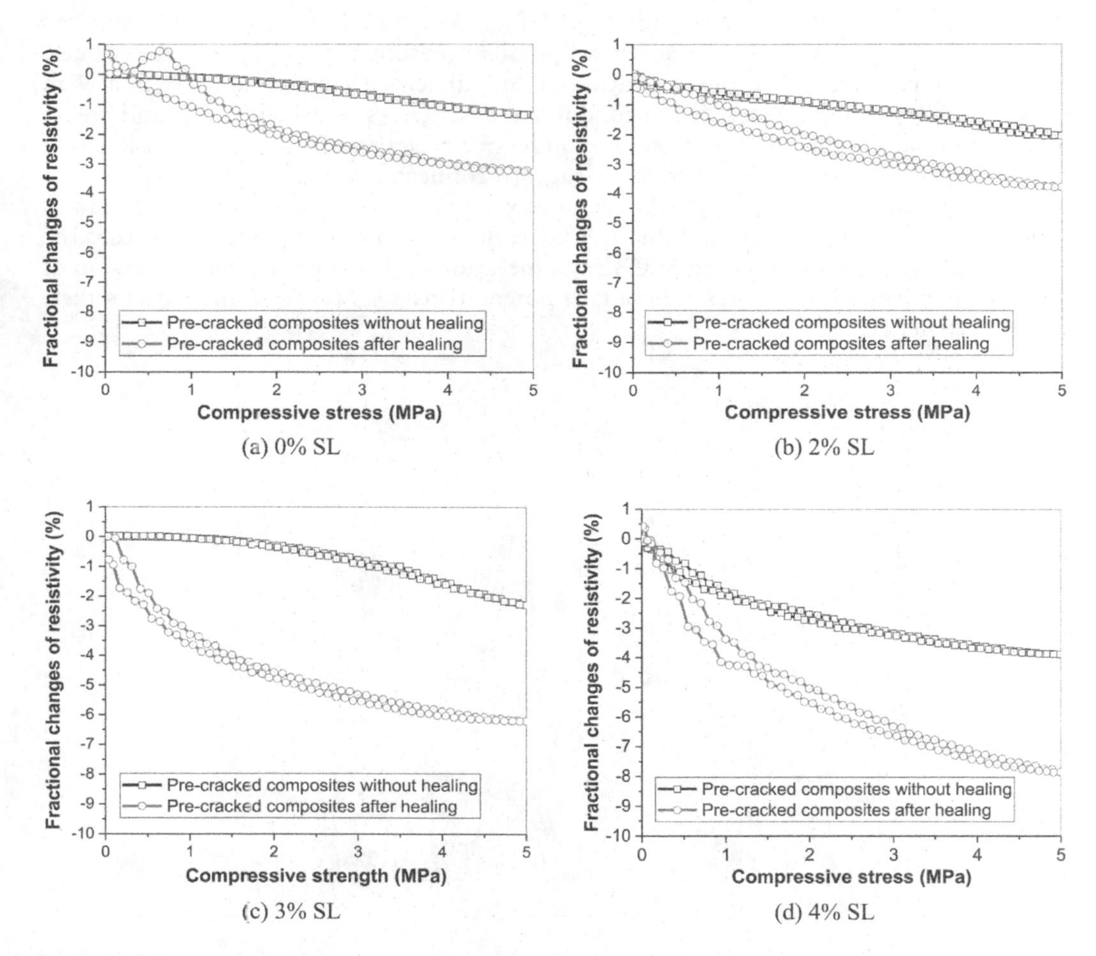

Figure 7.10 Stress-sensing behavior of CB cementitious composites with different SL contents [10].

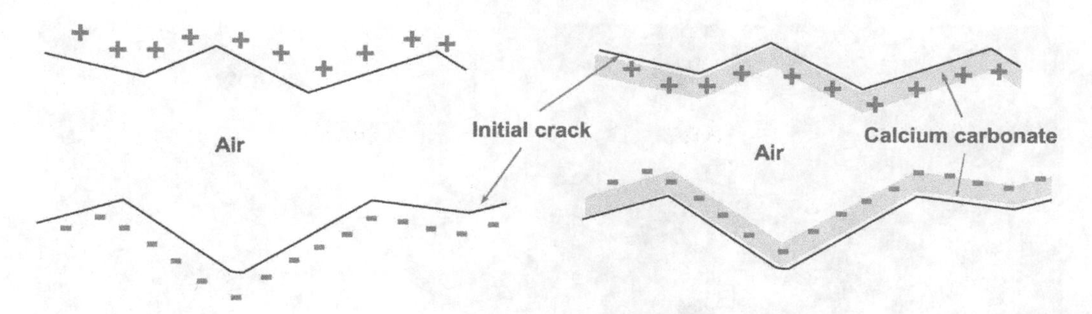

Figure 7.11 Schematic diagram of field emission effect on the pre-cracked CB/cementitious composites [10].

identical compressive stress. First, the self-healing process of the specimen involved water spray curing. Previous studies have demonstrated that the CBS with appropriate water content could more efficiently exhibit piezoresistivity [18], the improved FCR might be due to induced water content that enhanced the number of conductive contact points. Second, the formation of self-healing products stimulates the electrical resistivity changes under compression. As illustrated in Fig. 7.11, the cracks in the CB/cementitious composite could be considered as small capacitance, with a constant voltage among the "two plates" of capacitance provided by a multimeter [10]. For the CB/cementitious composites with SL, the distance between the two plates decreased because of the formation of calcium carbonate. It improved the intensity of electric field and enhanced the movement among conductive ions. Moreover, the field emission effect might also be enhanced because of the closed cracks and the induced water content [19]. Third, the contact resistance of conductive phases distributing on both sides of cracks could be altered because of the formation of nonconductive calcium carbonate. In terms of the piezoresistive stability and repeatability, the results showed that CB/cementitious composites without healing performed better than those after the healing process, and this phenomenon was much more significant for the composites without or with less content of SL. This might be due to the instability of self-healing products, thus affecting the electrical resistivity changes during compression.

7.2.5 Microstructural characterization

7.2.5.1 CB micromorphology

The self-sensing behaviors of CB/cementitious composites firmly relate to the electrical conductivity and the dispersion efficiency of CB nanoparticles. Fig. 7.12 shows the morphology of CB/cementitious composites without or with SL, from where the typical distribution of CB can be found [10]. The previous discussion on the CB dispersion efficiency in the superplasticizer–water with or without SL demonstrated a better uniform distribution of CB in the superplasticizer–water solution without SL than that in the superplasticizer–water solution with SL. The first reason would be due to the increased pH, and the second is because of the sedimentation of oversaturated SL that absorbed CB agglomerations. However, the microstructures showed totally opposite results with some large CB agglomerations in the CB/cementitious composites without SL, as illustrated in Fig. 7.12a. There are several potential reasons for the deteriorated CB dispersion in the cementitious materials, compared to its excellent dispersion in superplasticizer–water

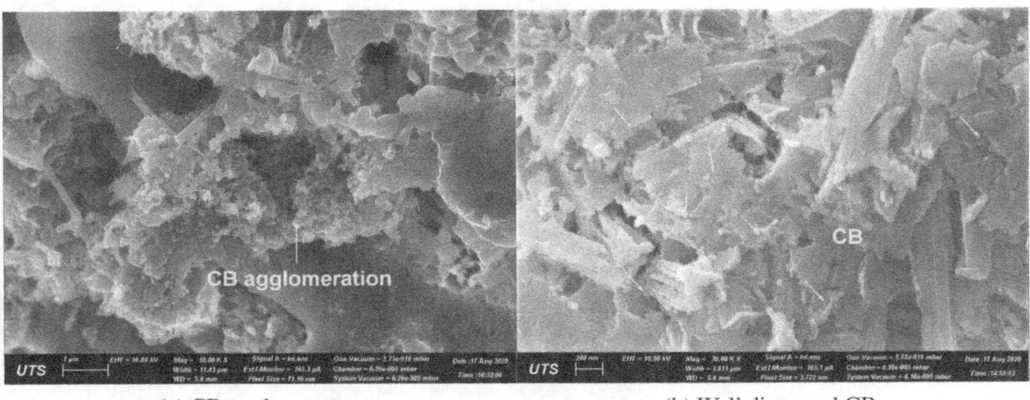

(a) CB agglomerates (b) Well dispersed CB

Figure 7.12 Microstructural morphology of CB in cementitious composites [10].

solution. First, the pH values of CB/superplasticizer–water suspensions increased with the addition of cement because of the fast cement hydration in the early stage. It weakened the effect of superplasticizer on the CB dispersion and stimulated the reagglomeration of CB nanoparticles. Second, the added binder of cement and silica fume provided a new point of attachment so that a proportion of CB nanoparticles were absorbed into the surface of the cement and silica fume. Given the size of cement and silica fume is much larger than that of CB, the CB agglomeration might be induced in the process of attachment to the surface of binder particles.

Different from the low CB dispersion efficiency in superplasticizer–water with oversaturated SL, Fig. 7.12b illustrates a relatively better CB distribution with a small volume of CB agglomerations in the cementitious composites with the process of premixing with SL. Three possibilities might be responsible for the improved CB dispersion. In the first 5 min, it could be found that the CB was still well-dispersed in the superplasticizer–water solutions with SL, which indicated the possibility of uniform distribution of CB with SL once the mixing of cementitious materials timely conducted. Moreover, the premixed CB with SL might avoid the reagglomeration of CB to the surface of the binders. In addition, the SL precipitation in superplasticizer–water solution would not occur in the cementitious composites that reduce the possibility of CB agglomeration. These phenomena imply that maybe the dispersion evaluation of nanomaterials in cementitious composites should not be simply replaced by the distribution performance in the solutions without any binders.

7.2.5.2 SL characterization

The critical factor that determines the self-healing performance of CB/cementitious composites is whether or not SL can be well-preserved in the composites from initial carbonation. Fig. 7.13 shows the typical morphology of SL in the CB/cementitious composites with increased magnification from 2,000× to 50,000×, given the vast difference between the sizes among SL and CB [10]. In particular, a large amount of SL could be found as a flaky crystal of calcium hydroxide in the cement matrix. Studies have demonstrated that the concentration of free CH sourced from binder and cement

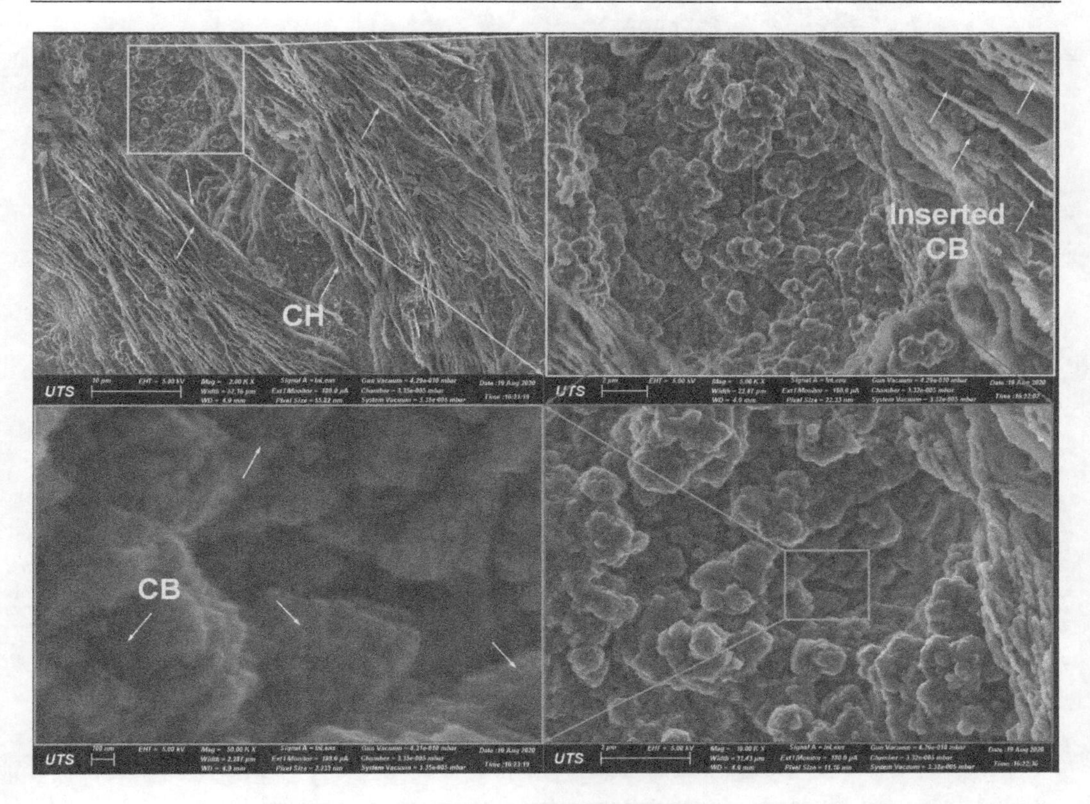

Figure 7.13 Enclosed SL or calcium hydroxide by CB in CB/cementitious composites at four months [10].

hydration was insufficient to establish self-healing capacity [20], while the added moderate amount of SL in this study improved the concentration of calcium and CH to enhance the self-healing capacity. Also, it could be observed that the CB with cement matrix inserted into the layers of CH, as well as CB nanoparticles attached to the surface of CH. The micromorphology supports the proposed assumption that partial SL can be preserved from air carbonation, given some of them are enclosed by CB nanoparticles during the SL and CB premixing process. Moreover, before the cementitious composites was cracked, the hardened cement matrix was another protective layer to maintain the SL.

7.2.5.3 Phase observation

Fig. 7.14 shows the microstructural morphology of the healed cracks with self-healing products [10]. Based on the shape of the crystal and the EDX results of oxygen and calcium, it could be demonstrated that the main products of the self-healing product were the calcium carbonate. With the addition of SL, rather than the enhanced formation of C—S—H, the exposed SL reacted with carbon dioxide and filled the cracks.

Fig. 7.15 shows thermogravimetric results of CB/cementitious composites at the age of four months filled with different contents of SL, to elucidate the phase generation characteristics [10]. It could be observed that the total weight loss of samples increased with

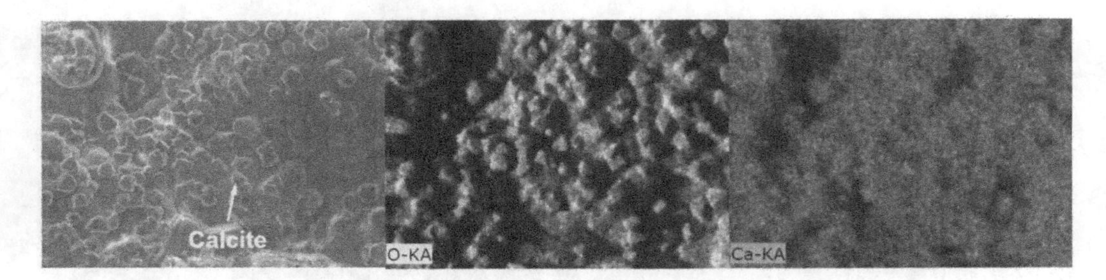

Figure 7.14 Formation of calcite in the cracks of cementitious composites [10].

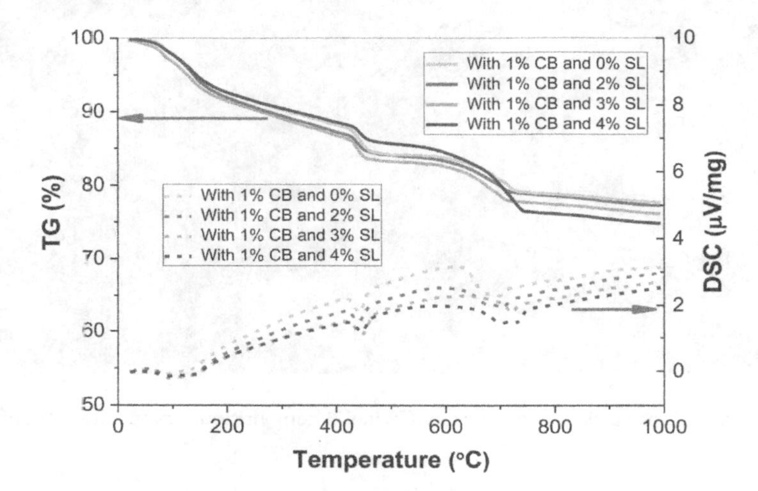

Figure 7.15 TG results of CB/cementitious composites filled with various contents of SL at four months [10].

the increase of SL, reaching 22.2%, 23.1%, 23.8%, and 25.3% for the composites with 0%, 2%, 3%, and 4% SL, respectively. Studies have demonstrated that the mass loss in the first 200°C mainly presented the decomposition of C—S—H and AFt [21, 22], it was found that the decomposition of C—S—H of CB/cementitious composites without SL or with 4% SL was weaker than the counterparts with 2% and 3% SL. It implied that the moderate concentration of SL improved the formation of C—S—H in the CB/cementitious composites because of the sufficient calcium and additional alkaline environment. It accelerated the pozzolanic reaction of silica fume in the composite. The decreased C—S—H for the composites with 4% SL might be due to the relatively high proportion of calcium carbonate in the samples. Moreover, the rapid weight changes around the temperature of 450–650°C are mainly due to the dehydration of calcium hydroxide and the decarbonation of calcium carbonate [23]. Incorporating the curve areas by DSC, the amount of calcium carbonate in the CB/cementitious composite with 4% SL was much higher than the other composites. The TG results are consistent with the previous conclusion of self-healing performance given the main product of self-healing of CB/cementitious composite is calcium carbonate.

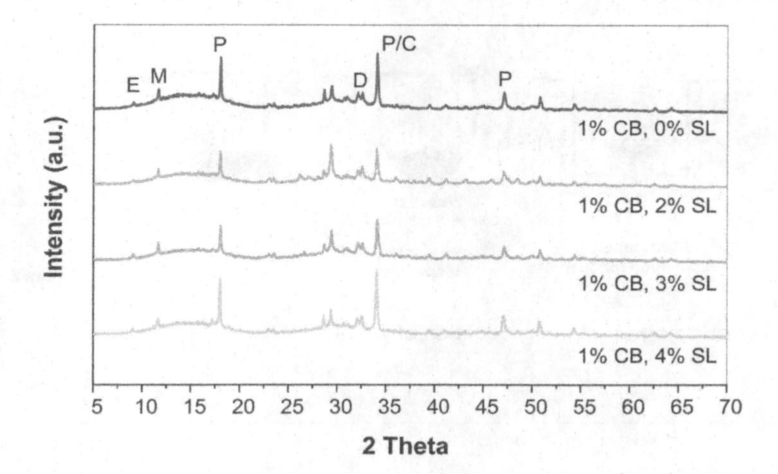

Figure 7.16 XRD results of CB/cementitious composites with varied contents of SL (E = ettringite; M = monocarboaluminate; P = portlandite; C = calcite; D = OPC clinkers) [10].

Fig. 7.16 shows the XRD results of CB/cementitious composites with varied contents of SL [10]. The addition of SL made limited conversion of ettringite to monocarboaluminate. However, it could be found that the portlandite first decreased and then increased with the increase of SL. This is consistent with the previous TG results, which might be due to the increased calcium ions and enhanced formation of C—S—H that consumed calcium hydroxide. For the increased portlandite of CB/cementitious composites with 4% SL, even with the consumed calcium hydroxide, a proportion of SL remained and showed slightly high content of portlandite. Besides, the TG results showed relatively higher C—S—H of CB/cementitious composites with 2% and 3% SL, indicating the less consumed calcium hydroxide for the CB/cementitious composites with 4%. Although the peak of calcite superposed the peak of portlandite, it could be found in the highest amount of calcite of the CB/cementitious composites with 4% SL.

7.3 INTEGRATED SELF-SENSING AND SUPERHYDROPHOBIC PROPERTIES

7.3.1 Experimental program

The preparation of CBSs follows ASTM C305-14 (Standard Practice for Mechanical Mixing of Hydraulic Cement Pastes and Mortars of Plastic Consistency). Fig. 7.17 schematically plots the manufacturing process of CBSs [24]. In particular, mold prepared for the CBS is 10 mm ×10 mm ×60 mm, to ensure sufficient surface modification in beakers. Four electrodes are embedded during casting to measure the electrical resistivity of the CBS by four-point method.

To remove the surficial impurities and smooth surfaces, CBSs are first polished and cleaned by sandpaper before coating. The surface modification follows the steps in Fig. 7.18 [24]. For the modification with isopropanol, the solvent of isopropanol is prepared in the measuring beaker, and 4% silane by volume of isopropanol is added, followed by 5 min mechanical

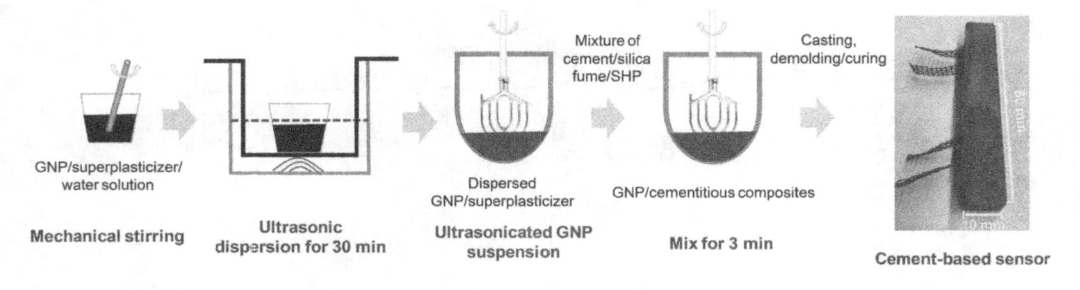

Figure 7.17 Schematic diagram of the production of CBS [24].

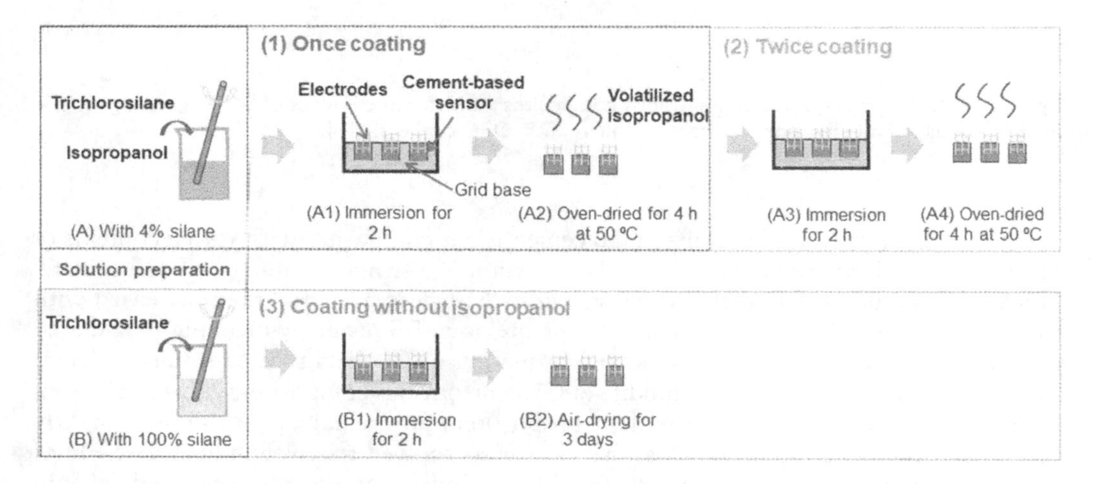

Figure 7.18 The procedures of multiple surface modifications of CBSs [24].

mixing to dissolve and disperse silane. For another method without isopropanol, the same volume of 100% silane is prepared in the beaker. The CBSs have been placed above a copper mesh in a plastic container with a distance of 5 mm, thus all of the surfaces have continual contact to the silane/isopropanol or silane solutions. Afterward, the mixed solution is gently poured into the container until the top surfaces of CBSs are just covered.

There are three different types of surface modifications, ranging from once coating, twice coating, and once coating without isopropanol. It should be noted that the electrodes of CBSs would not be immersed in the silane/isopropanol solution, to ensure the excellent conductivity of electrodes. For the modification with isopropanol, the container should be sealed by a plastic film to avoid the volatilization of isopropanol. The immersion of CBSs lasts for 2 h to ensure the thorough entrance of mixed solution into cracks and pores of cementitious materials. Then the CBSs are dried in an oven at the temperature of 50°C for 4 h to volatilize the isopropanol. To evaluate the effect of different times of coating, the studies on CBSs after one time and two times of the above-mentioned immersion and drying processes are conducted. For the modification without isopropanol, the immersion time is identical to that of the previous method, and the drying process is three days of air-drying due to the lack of isopropanol.

7.3.2 Hydrophobic behaviors

7.3.2.1 Water contact angle (CA) of surface

Fig. 7.19 shows the surface water CA of CBSs after different types of surface modifications at the time of 0, 1, 5, and 9 s from water dropping to stabilization [24]. For the CBSs without coating shown in Fig. 7.19a, they exhibited hydrophilic behavior with the initial CA of 79.2°. Subsequently, the water CA gradually decreased with the duration until to the smallest value of 70.1°. This implies that the water molecules could penetrate the CBSs due to the hydrophilic behaviors and porous structures of cement matrix. Consequently, the altered water content would be able to permanently affect the electrical and piezoresistive properties of CBSs, which indicates the necessity to coat the CBSs [25].

Fig. 7.19b–d shows the water CA of CBSs subjected to above-mentioned once coating, twice coating, and coating without isopropanol, respectively. It can be observed that all the treated CBSs exhibited hydrophobic behaviors with the water CA larger than 90°. For the CBSs after once coating without isopropanol, the water CA showed similar tendency to that of uncoated CBSs with the continual reduction of CA with increasing time. Differently, the once- and twice-coated CBSs showed a slight fluctuation rather than a continual decline. It may be due to the various porosity-filling efficiency of silane with or without isopropanol, which is discussed later in Section 3.1.4. Moreover, the twice-coated CBSs showed the largest CA of 163.4°, followed by the second-highest CA of 142.7° and the lowest CA of 127.0° for the CBSs after once coating and coating without isopropanol, respectively. Overall, regarding the water CA, the surface modification of CBSs by 4% silane with 96% isopropanol as solvent is better than those coated by 100% silane without isopropanol. Twice coating on CBSs is more efficient in improve the hydrophobicity of the sensor surfaces than once coating.

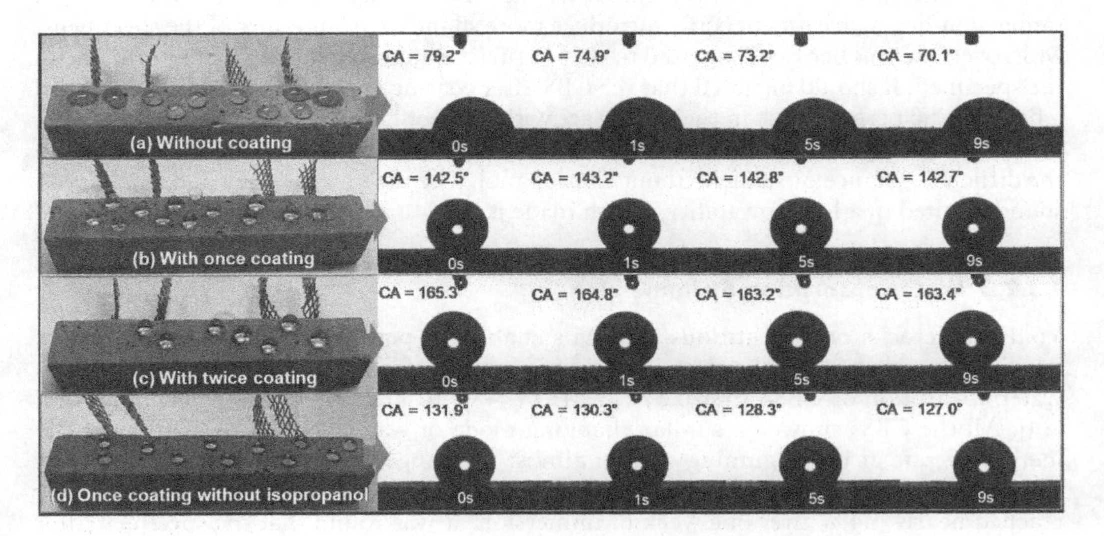

Figure 7.19 Water CA of CBSs after different types of surface modification [24].

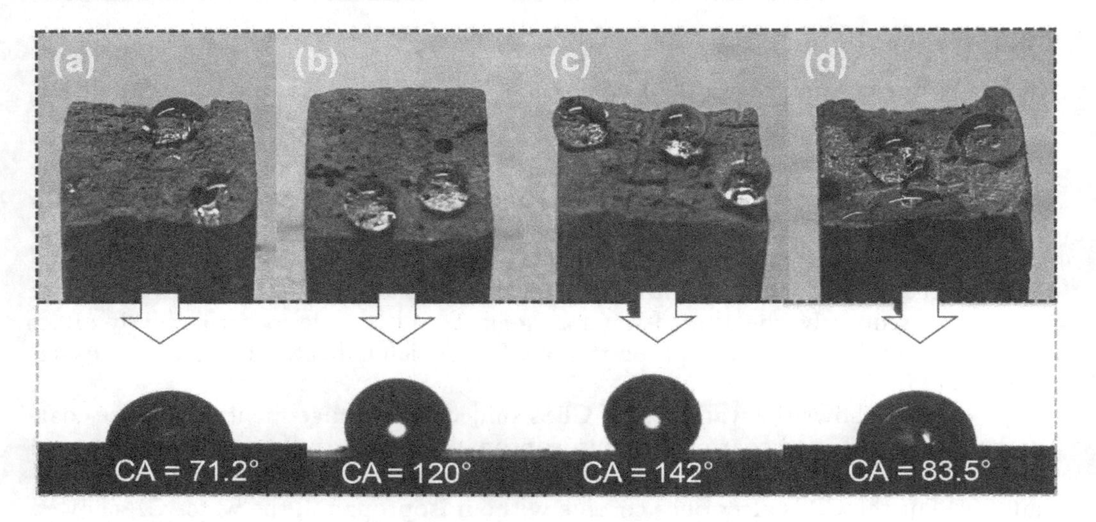

Figure 7.20 Water CA of the interiors of CBSs. (a) Without coating. (b) With once coating. (c) With twice coating. (d) Once coating without isopropanol [24].

7.3.2.2 Interior water CA

Fig. 7.20a–d shows the water CA of CBSs with or without multiple surface modifications regarding the cut cross-sectional area to display the hydrophobic or hydrophilic behaviors of the interior of the CBSs [24]. For the CBS without coating, it displayed hydrophilic behaviors and a similar CA of 71.2° to that of surface. On the contrary, the interior of the sensor became hydrophobic with the CA values of 120° and 142°, respectively, for the CBSs after the once and twice coating. It demonstrates that the silane–isopropanol solution could penetrate the core CBSs through micropores and microcracks, and result in the hydrophobicity of the cut cross section. Similarly, the CBSs subjected to the twice coating showed larger CA than that of the ones after the once coating, indicating that the second immersion/drying treatment could introduce more silane to the interiors of the specimens. Moreover, the smaller CA compared to that of surface indicated a less amount of silane in the specimen. It should be noted that the CBS after coating without isopropanol exhibited a hydrophilic performance in the inner part with a water CA reaching 83.5°. In comparison to the hydrophobic behavior on the surface, the transformation is most likely due to the difficult entrance of silane without isopropanol. The high consistency and viscosity of silane resulted in a low flowability, which made it hard to penetrate into CBSs [26].

7.3.2.3 Water absorption by immersion

Voids and cracks of cementitious materials enable the penetration of water into CBSs, which is bound to affect the electrical and piezoresistive behaviors [27]. Fig. 7.21 shows water absorption of CBSs after 0.5, 1, 2, 4, 24, 48, 72, and 168 h full water immersion [24]. All the CBSs showed a similar changing mode of water absorption with a rapidly increasing rate at the beginning and an almost unchanged stage at the end of the test. The CBS without coating exhibited the largest water absorption, and the final value reached nearly 4.4% after one week of immersion. It was found that the specimen after the twice coating had the minimum water absorption of approximately 2.3%, which

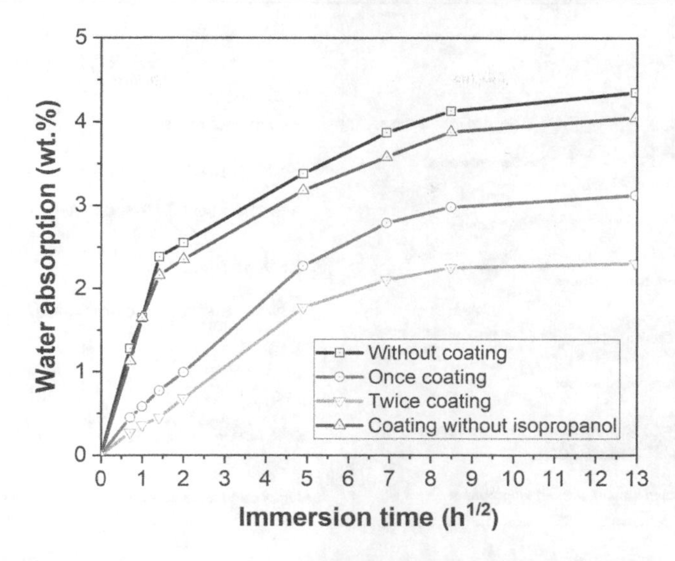

Figure 7.21 Water absorption of immersion for the CBSs subjected to multiple surface modifications [24].

was reduced almost by half compared to that of the sensor without coating. Based on the previous studies that developed the waterproofing CBSs with reduced capillary water absorption by more than ten times [25], this work further decreased water absorption by additional surface modification on the waterproofing sensors [28]. It demonstrated the excellent waterproofing efficiency of the twice coating for CBSs. For the CBS after once coating, the final water absorption reached 3.1%, indicating an improved waterproofing property that could be further enhanced by the second coating treatment. For the CBS coated with 100% silane without isopropanol, the increasing tendency of water absorption was very similar to that of the specimen without coating for the first several hours. Its final water absorption was only slightly lower than that of the sensor without coating. Therefore, the coating efficiency of 100% silane without isopropanol seemed worse than 4% silane with isopropanol on the CBSs. It is again likely due to the difficult entrance of silane into voids and cracks of cementitious materials due to its high viscosity. More explanations are further explored in the section of mechanism and microstructural discussions.

7.3.2.4 Mechanism discussion

The hydrophobic mechanism of CBSs after various surface modifications can be explained by the integrated physical intrusion and chemical reactions. Fig. 7.22 schematically displays the physical intrusion of silane into the CBS and the hydrolysis and polycondensation reactions in the transition zones of cement and silane [24]. For the CBS without coating, the hydrophilic behavior is mainly due to the microcracks and pores in the cement matrix that allows the ingress of water. For the sensors subjected to once and twice coating by silane/isopropanol solution, given the viscosity of silane can be reduced by mixing with isopropanol, the silane can penetrate as well as isopropanol by capillary suction to fill the cracks and pores [29]. In the process of drying, the isopropanol was removed because of thermal evaporation and only the silane was left in the cracks and pores. To provide

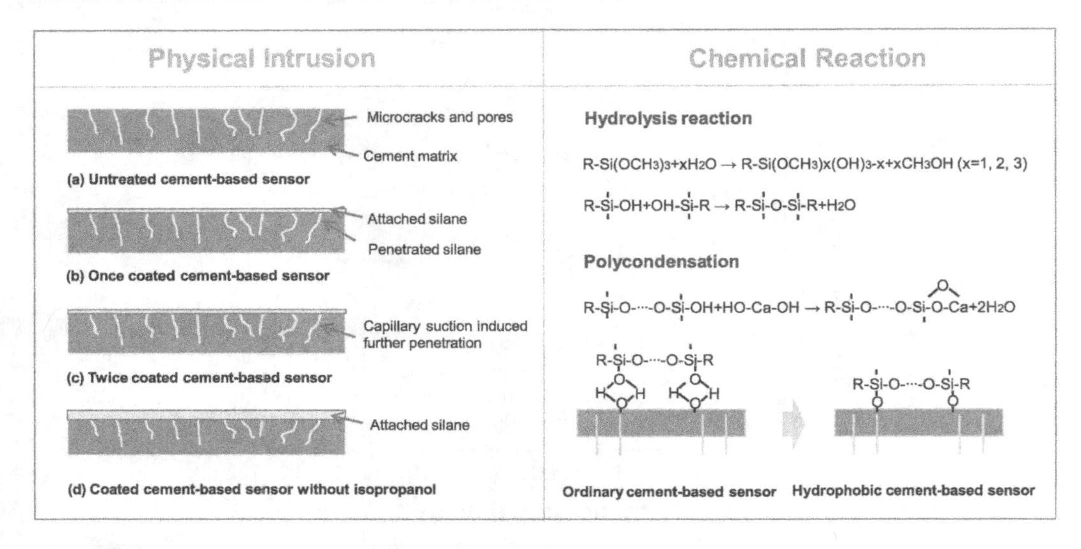

Figure 7.22 Schematic diagram of physical intrusion, hydrolysis, and polycondensation reactions in the transition zones of silane and cement [24].

the CBSs with hydrophobicity, the intruded silane had two chemical reactions with the cement matrix, including hydrolysis reaction and polycondensation. These two chemical reactions could result in the formation of covalent bond between the cement and silane, which is responsible for the transformation of hydrophilic cementitious materials into hydrophobic ones [30].

In addition, compared to the once-coated CBS, the specimen subjected to the twice coating enables more silane to attach to the surface and penetrate the inner sensor, as shown in Fig. 7.22b and c. The drying process of CBS after the first coating removes the isopropanol, as well as some pore solutions. It can enhance the capillary suction and provide more space for silane and isopropanol during the second immersion process. That is why the CA is higher and water absorption is lower for the CBSs after twice coating. Furthermore, because it is much easier for silane to be attached to the surface rather than inner cracks and pores, the water CA of the surface is always higher than that of the inner sensor. For the coated specimen by 100% silane without isopropanol, despite that the surface shows hydrophobic behavior, the performance is not the same as the specimens subjected to the once and twice coating. Moreover, the CA of interiors of specimen was significantly lower than that of the surface, indicating the poor penetration of 100% silane without the assistance from isopropanol. The difficulty of silane to attach and penetrate into the CBS is mainly due to the high viscosity and the different hydrophilic groups of silanes to cement matrix.

7.3.3 Self-sensing performance

The stress-sensing performances of CBSs without/with different types of surface modifications are displayed in Fig. 7.23a [24]. It could be found that all the FCR of CBSs exhibited an excellent relationship to the compressive stress, with the first decreased and then returned resistivity in the loading and unloading processes. It demonstrates that the silane-based surface modification would not eliminate the piezoresistivity of CBSs.

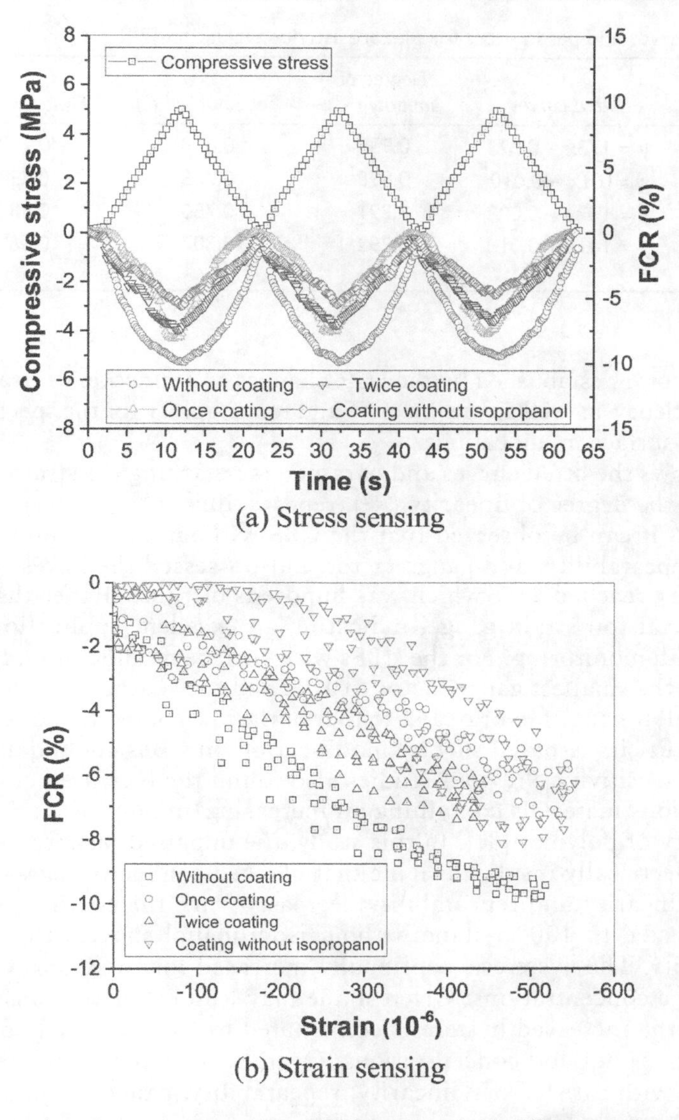

Figure 7.23 Stress- and strain-sensing capacities of CBSs without/with different surface modifications [24].

The graphene-filled CBS without coating showed the highest FCR value to 12.6%, followed by the average FCR of 7.7%, 7.2%, and 5.3% for the cement-based after the once coating, twice coating, and coating without isopropanol, respectively. It implies that the stress-sensing efficiency might be weakened by silane modification. In addition, small fluctuations can be found especially for the silane-treated CBSs. This was mainly due to the brittleness and heterogeneity of cementitious materials that led to the sudden changes of electrical resistivity. Second, the intruded silane aggravated the fluctuation because of its poor electrical conductivity. For the strain-sensing performance shown in Fig. 7.23b, the FCR possessed an excellent relationship with the compressive strain and showed a similar changing mode to the compressive stress. Despite that the CBSs showed unrepeatable

Table 7.2 Fitted curves and parameters for piezoresistivity evaluation [24]

Enhancements	*Fitted curves*	*Degree of linearity (R^2)*	*Repeatability (L)*	*Hysteresis (H)*	*Gauge factor (G)*
Without coating	$y = 173x - 0.023$	0.939	0.878	0.203	173
Once coating	$y = 101x - 0.010$	0.898	0.765	0.333	101
Twice coating	$y = 142x - 0.008$	0.891	0.750	0.382	142
Coating without isopropanol	$y = 141x + 0.010$	0.792	0.507	0.560	141

behaviors, the curves established by the FCR changes as a function to strain are to assess the sensing efficiency, as well as the linearity and repeatability for the specimens subjected to the different surface modifications.

Table 7.2 shows the fitted curves and parameters regarding the strain-sensing behaviors, including the degree of linearity (R^2), repeatability (L), hysteresis (H), and gauge factor (G) [24]. It can be observed that the CBS without coating presented the highest linearity, repeatability, and gauge factor, but possessed the lowest hysteresis. The strain sensitivity reached 216, which was hundreds of times higher than the sensitivity of commercial foil strain gauge, indicating an excellent application potential for structural health monitoring. For the CBSs with surface modification, the once coated sensor showed the smallest gauge factor to 101, and the sensor after the twice coating exhibited a similar gauge factor of 142 to that of the one coated without isopropanol. It indicates that the silane-based surface modification on CBSs could damage the piezoresistivity and sensitivity. Previous studies also found the increased electrical resistivity of cementitious materials containing an increasing amount of latex because of the nonconductivity of polymer [31]. In this study, the impaired piezoresistivity might be owing to the electrically insulated silane that blocked conductive passages. Moreover, the degree of linearity and repeatability decreased with the coating procedures, and the CBSs subjected to 100% silane without isopropanol showed the worst linearity and repeatability. The hysteresis continually increased for the CBSs with the coating times and silane concentrations. Given silane may lead to fluctuations on the electrical resistivity, the increased hysteresis contributed to the increased content of silane with the coating times and concentrations. Overall, the CBSs after the twice coating were provided with satisfactory linearity, repeatability, gauge factor, and acceptable hysteresis.

7.3.4 Microstructural characterization

Fig. 7.24 shows microstructures and EDX results of CBSs without coating and after once coating, twice coating, and coating without isopropanol, to distinguish the substances from each other [24]. For the GNP with platelike structures, it can be found that carbon occupied the highest atomic proportion among all elements in Fig. 7.24a. The other elements were sourced from the attached cement hydrations products. Previous chemical reactions claimed that silane could react with cement materials and water to form hydrophobic groups connected to the cement matrix. Therefore, the formed products should contain a high proportion of carbon and oxygen. For the once-coated CBS, the microsilane monomers can be observed where the elements of oxygen and carbon accounted

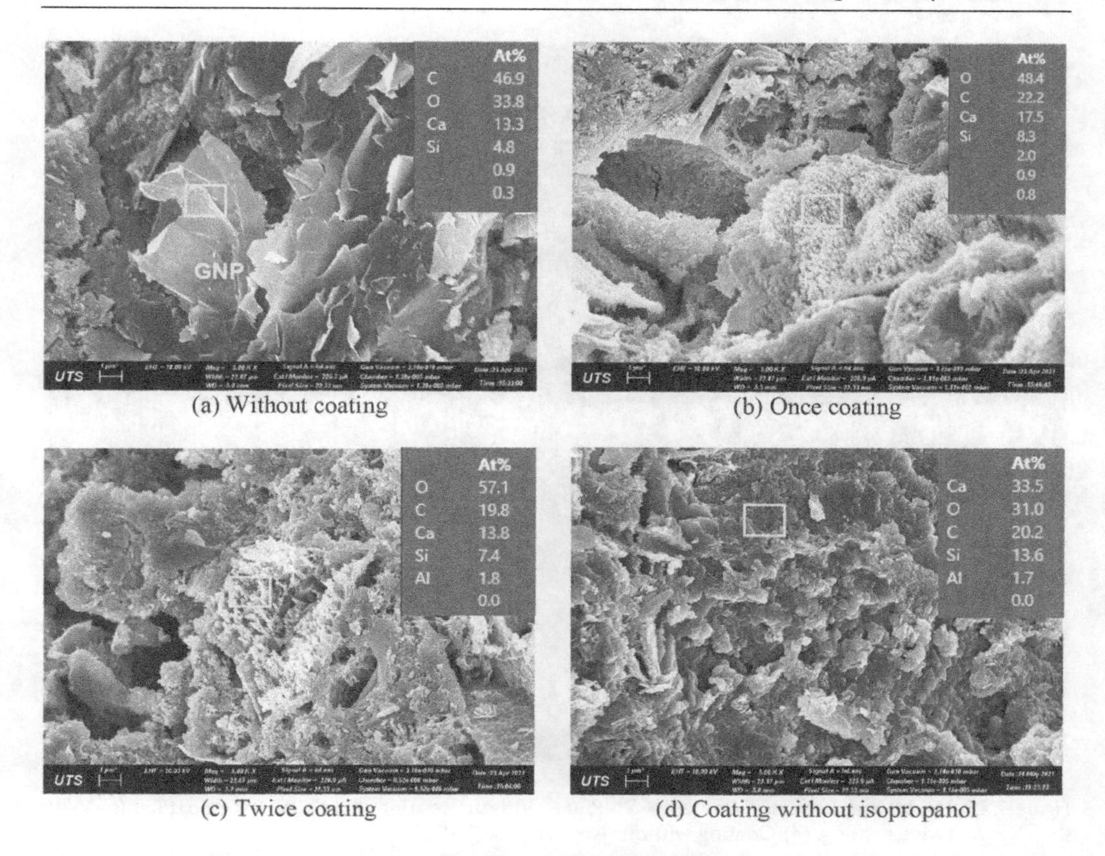

Figure 7.24 Microstructure features of CBSs after multiple surface enhancements [24].

for the largest portions, and meanwhile the calcium and silicon from the cement matrix took the relatively small percentages. More silane monomers could be observed for the CBSs after the twice coating. In particular, the silane monomers agglomerated to clusters with diameter less than 10 μm. The studies found micropapillae with a diameter of approximately 10 μm for the cementitious materials with dimethylsiloxane, and proposed that the micropapillae were responsible for the enhanced hydrophobicity [32]. Instead of micropapillae, the silane monomers found in this study could contribute to the improved waterproofing and hydrophobic behaviors. In terms of CBSs after coating without isopropanol, the calcium rather than oxygen and carbon played a dominant role, indicating that the cement matrix failed to be enclosed by silane.

The microstructural differences of CBSs after multiple surface modifications are compared in Fig. 7.25 [24]. It can be found that the compactness of CBSs without coating or coating without isopropanol was poorer than those of CBSs after the once or twice coating. This can be reflected by the number of microvoids in which the former sensors showed widespread microcracks and voids, while the latter displayed fewer pores in the specimens after the once and twice coating. Given the compactness of cementitious materials affects the durability and water absorption ability, the reduced voids by intruded silane can explain the reduced water absorption and improved hydrophobicity [33]. In addition, silane monomers with similar size and shapes to the previous SEM

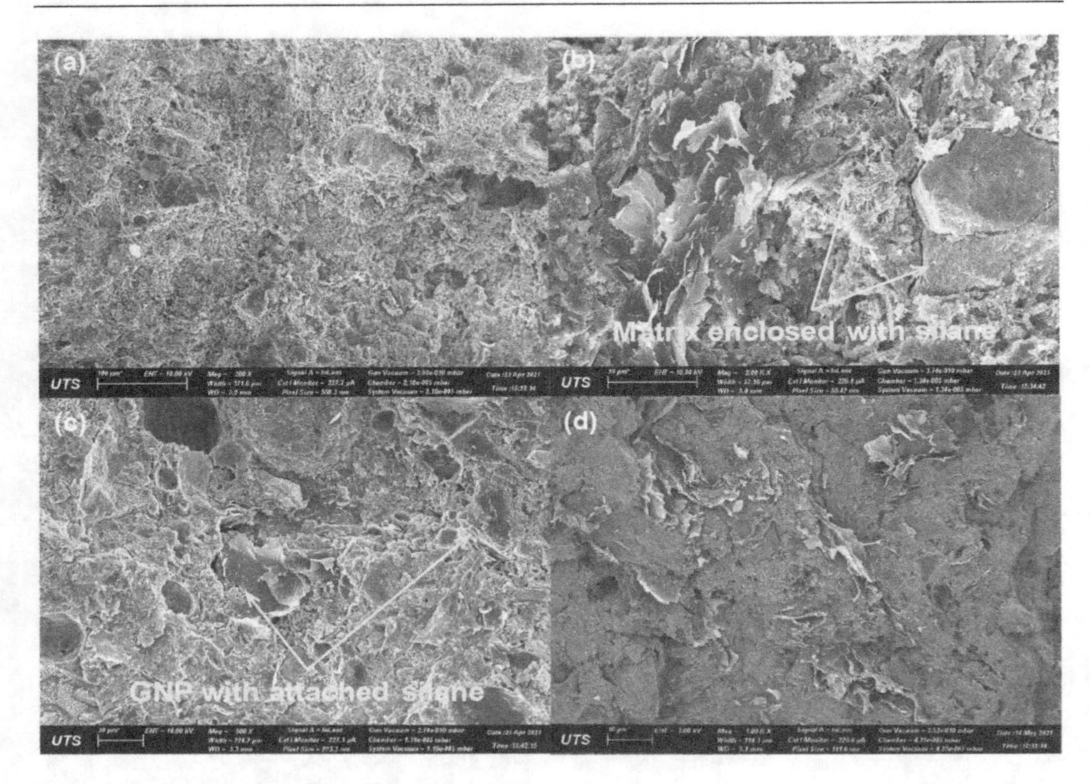

Figure 7.25 Microstructures of CBSs with GNP. (a) Without coating. (b) With once coating. (c) With twice coating. (d) Coating without isopropanol [24].

images with EDX attached to the cement matrix for the specimens after once coating, indicating an excellent cohesion between cement matrix and silane monomers. The cohesion between silane and GNP seemed to be worse than that between silane and cement matrix because less silane was found on the surface of GNP. However, for the CBSs after the twice coating, the GNP enclosed with silane clusters could be observed on the edges. Studies showed that GNP was able to increase the flow path of water into cementitious material because of its platelike structures. To further increase the hydrophobicity and waterproofing properties of cementitious materials, GNP may be suitable to interact with silane in CBSs.

To explain the considerable differences in hydrophilic and hydrophobic performances of interior and surface of CBSs after coating without isopropanol, Fig. 7.26a and b display the microstructural morphology of samples collected from the inside and on the surface of CBSs after coating without isopropanol, respectively [24]. Because the pure silane without isopropanol had high viscosity and low permeability, the morphology was similar to that of the sensor without coating for the internal CBS. However, on the surface of specimen, because of the high concentration of silane, it can be observed that cement hydration products of C—S—H and ettringite were completely enclosed by silane gels. The cracks and pores in the surface were also sealed by silane gels, which resulted in a harder entrance of more silane penetrating internal CBSs. Differently, for the CBSs coated by 4% silane solution isopropanol, the silane could continually penetrate after the

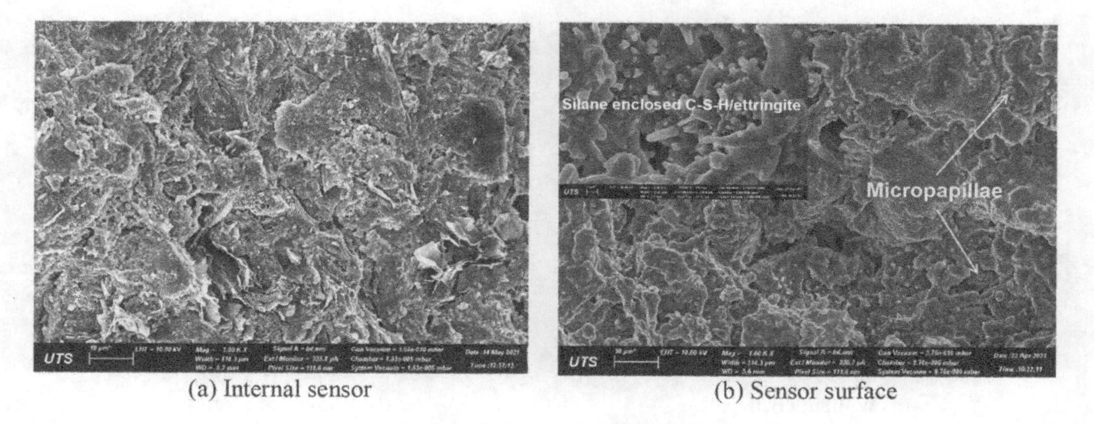

Figure 7.26 Microstructural morphology of the interior and surface of CBSs subjected to coating without silane [24].

volatilization of isopropanol. Some randomly distributed micropapillae could be found on the surface of CBSs but were hardly observed in the internal sensors after coating without isopropanol. Similar to the micropapillae on the surface of lotus leaf, it provides the CBSs with hydrophobicity and waterproofing ability. The microstructural discrepancy provides a reason for the hydrophilic and hydrophobic performances of CBSs regarding the interior and the surface.

7.3.5 Phase analysis

Fig. 7.27 shows TG and DSC results of CBSs with increasing temperature after different types of surface modifications [24]. Generally, the mass reduction during heating is due to the dehydration of C—S—H, while the intensified mass reductions in the ranges of 65–135°C, 420–450°C, and 660–730°C were, respectively, due to the evaporation of absorbed water from ettringite, portlandite, and the degradation of calcium carbonate [34]. Based on the aforementioned hydrolysis and polycondensation reactions, silane could react with calcium hydroxide to form a covalent bond between silane and the cement, thus creating a hydrophobic film on CBSs. It can be found that the CBSs without coating showed the largest mass reduction of portlandite (from 420°C to 450°C), reaching approximately 0.79%. Compared to the portlandite content from our previous studies which applied similar mix proportion without SHP [35], the less amount of portlandite in this study was likely due to the additional SHP that contained silane to react with calcium hydroxide in polycondensation reaction. Similarly, for the CBSs after surface modification, the content of portlandite decreased because more intruded silane reacted with calcium hydroxide, especially for the specimens after the twice coating. Moreover, the DSC results can further demonstrate the deduction, through providing the largest area for the CBS without surface modification and the smallest area for the twice-coated CBS [36]. For the CBS coated without isopropanol, the portlandite content was smaller than that of the sensor subjected to once coating, indicating that the high content of silane might be easier to react with calcium hydroxide, despite it would be harder for 100% silane to penetrate into the CBS.

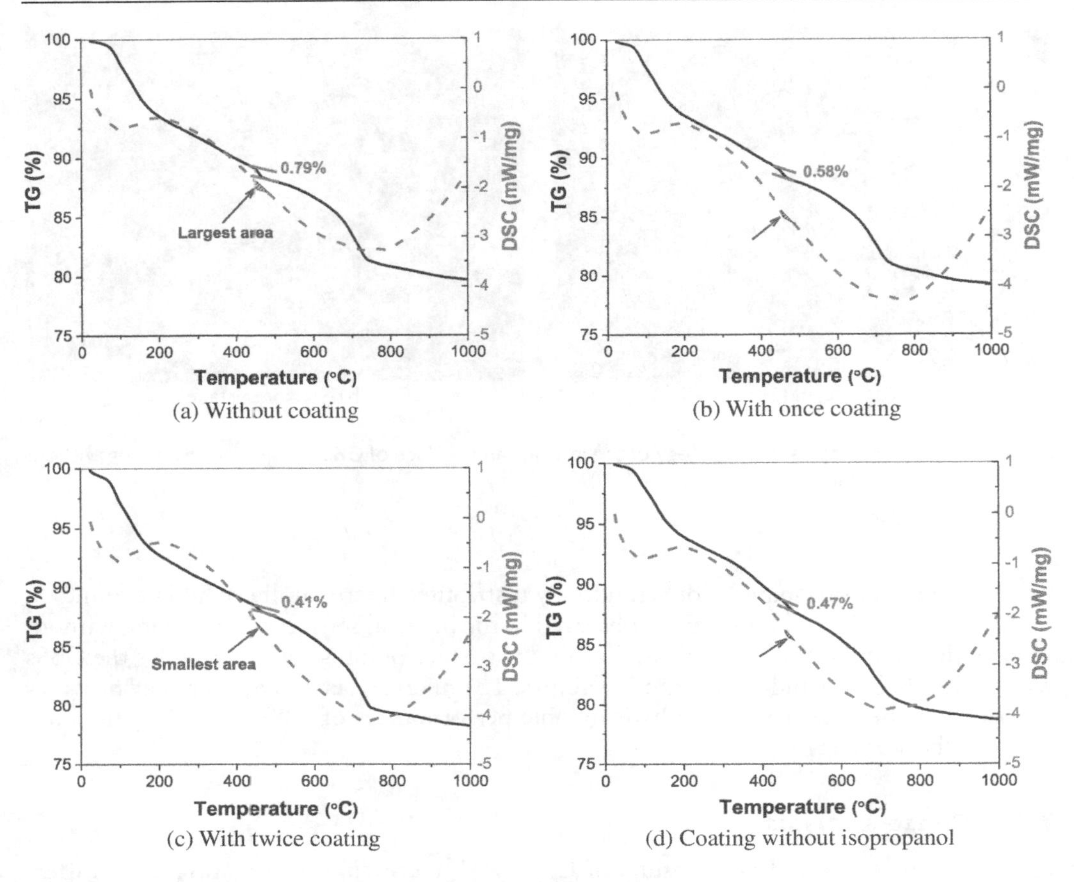

Figure 7.27 The TG and DSC results of CBSs with GNP subjected to (a) without coating, (b) with once coating, (c) with twice coating, and (d) coating without isopropanol [27].

Fig. 7.28 shows the FTIR spectra and changes of chemical groups of CBSs after different surface modifications [24]. For all CBSs, the peak intensity around 445 cm^{-1} was mainly due to the stretching vibration of O-Si-O tetrahedron [37]. The two peaks at approximately 874 cm^{-1} and 1,408 cm^{-1} represented the stretching vibration of O—C—O from CO_3^{2-} [38]. It can be found that the peaks were weakened for the CBSs after the once and twice coating and almost kept unchanged for the sensor after coating without isopropanol. It implies that the surface modifications by once/twice coating enable to enhance the resistance of carbonation of CBSs, through increasing above-mentioned waterproofing and hydrophobic behaviors [39]. However, the specimen after coating without isopropanol failed to reduce their carbonation and showed similar results to that of the one without coating. In addition, peaks at 3,394 cm^{-1} indicate the stretching vibration of hydroxyl in water. Given the powders for FTIR were dried at temperature of 50°C for three days, the maintained water was more likely to be constitutional water rather than free water. The smallest peak of CBSs after twice coating demonstrated that the hydroxyl groups of C—S—H were replaced by covalent bonding with silane. The last peak around 3,643 cm^{-1} was due to the —OH from calcium hydroxide [40]. It showed

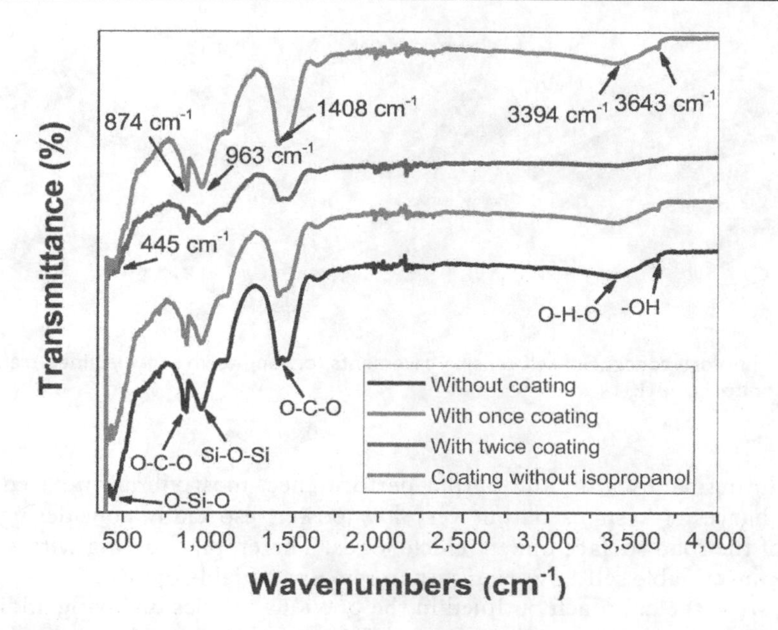

Figure 7.28 FTIR spectra of CBSs after multiple surface modifications [24].

the smallest peak for the CBSs after the twice coating, which was consistent with the aforementioned TG results. Conclusively, the FTIR results could further demonstrate the entrance of silane into CBSs that provided superhydrophobicity, especially for the CBSs after the twice coating.

7.4 INTEGRATED SELF-SENSING AND ENERGY HARVESTING PROPERTIES

CBSs typically require extensive wiring and external power sources to provide the necessary current for accurate and continuous measurement of resistance changes under different pressure signals, which hinders the further development of CBSs in engineering applications. Currently, research on self-powering or self-charging CBSs remains scarce. Researchers are attempting to propose novel structures and materials of CBSs, allowing electrical energy to be collected and stored within the sensor. This enables repeated charging and discharging, integration of multifunctional devices, and the realization of a self-powering operational mode for sensing systems.

As shown in Fig. 7.29, in the highway system, the road surface is composed of self-sensing materials such as cement or asphalt, while the power supply and data acquisition (DAQ) systems of intelligent road surface sensors are installed at the road edges to minimize impact on traffic [41]. In this study, the authors argue that the conventional solar rechargeable power sources may not meet the needs of intelligent cement-based or asphalt-based sensor road surfaces, especially during nighttime or overcast conditions. Therefore, they supplemented this with a piezoelectric energy harvester, which exploits the ability of piezoelectric materials to generate an electric field when subjected to external vibrations and mechanical stimuli. However, it is important to note that due to limitations in input

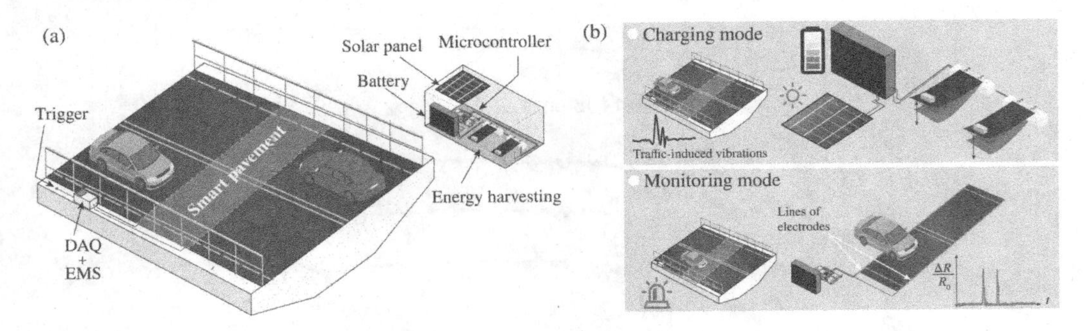

Figure 7.29 Technology concept of self-sensing pavements for long-term self-sustained traffic and WIM monitoring [41].

power, operating bandwidth, and output performance, most vibration-based piezoelectric energy harvester systems output very low power, especially considering the small vibrations of the road surface under vehicle loads. Therefore, coupling with solar charging or other sustainable self-powering system remains a viable option.

Differing from the approach outlined in the previous articles involving additional sustainable piezoelectric self-powering devices, Reference [42] introduces the development of a conductive CBS as a piezoelectric energy harvester. It converts external force loads applied into electrical energy to sustain its own sensor's electrical resistance measurement and DAQ system. Compared to the additional power source and DAQ systems set up on the roadside, this self-charging CBS exhibits greater autonomy and flexibility. This enhances its utility in locations where conventional power sources are unavailable or in applications where frequent battery replacement is to be avoided. The self-powering CBSs are produced by mixing well-dispersed rGO with cement, and two different ways of rGO dispersion by ultrasonic cleaner and tip are proposed. As shown in Fig. 7.30, the voltage generated by compression load reached nearly 30 mV for the CBSs produced by ultrasonic cleaner method. Based on the results, the electric power generated in the CBSs has been identified as a suitable magnitude for driving self-powering strain-sensing applications due to its strong correlation with mechanical strain.

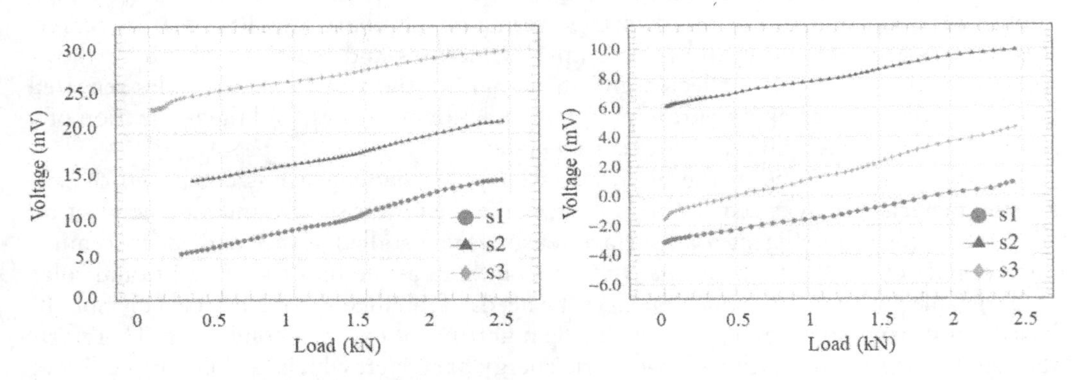

Figure 7.30 Piezoelectric voltage versus compression load for rGO-based cement composites with rGO dispersed by (a) ultrasonic cleaner (method 1) and (b) ultrasonic tip set (method 2) [42].

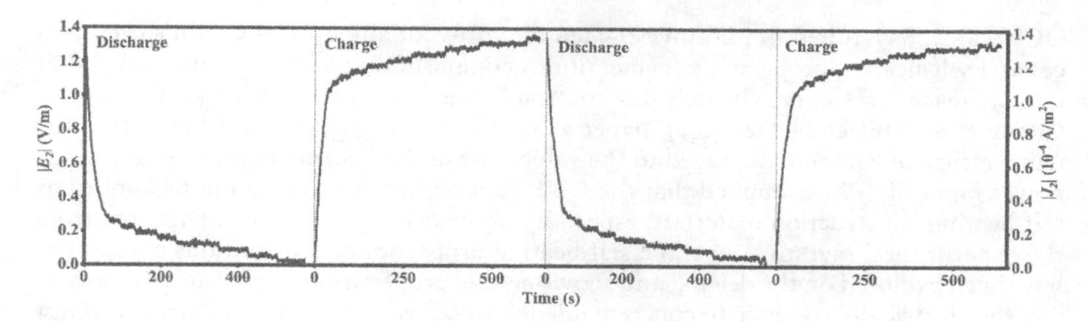

Figure 7.31 Discharge–charge characteristics for two cycles, showing electric field $|E_2|$ and current density $|J_2|$. The inter-electrode distance is 20 mm [43].

In contrast to self-powering sensors based on piezoelectric or Seebeck effects, Xi and Chung have proposed a self-powering material based on the polarization effect of ordinary cement concrete [43]. This novel charger does not require additional temperature differentials or vibrations, and it can generate electrical energy based on the movement of ions in the pore solution. Fig. 7.31 shows the discharge–charge characteristics of cement paste. It shows that the discharge–charge characteristics are similar for the various cycles, with the charging density of 7.12×10^{-2} C/m³. However, the energy output of the cement-based charger based on polarization is minimal, and its practical implementation is currently challenging. Future research directions for self-powering CBSs involve various aspects, including the integration of multiple energy sources (solar energy, piezoelectric energy harvester, etc.), enhancing energy conversion efficiency, environmental adaptability, cost considerations, etc.

7.5 INTEGRATED SELF-SENSING AND THERMOELECTRICAL PROPERTIES

Currently, there is no study that combines the self-sensing and self-heating properties of cementitious materials. Due to the good electrical conductivity of CBSs, it is feasible to achieve self-heating cement-based materials based on the Joule effect. There are many challenges in this, including the amount of electrical energy applied, electrode positions, environmental influences, and perhaps one of the biggest challenges is how to eliminate the influence on the self-sensing performance of CBSs during temperature changes. In this chapter, the self-heating efficiency of conductive cementitious materials and their potential applications are displayed. Based on the Joule effect, the electrically conductive cementitious composites under applied current would generate heat energy. The heat energy followed Eq. (7.1):

$$Q = I^2Rt \tag{7.1}$$

where Q is the heat energy; I is the current generated through the composite; R is the electrical resistivity, and t is the power-on duration.

Wang et al. [44] manufactured the cement mortar with excellent conductivity, which could raise its temperature from room temperature to 90°C within an hour under 8 V

DC voltage. Sassani et al. [45] investigated the different amounts of CF on the electrical and self-heating properties of cementitious composites, and they recommended that the CF dosage was approximately 1.0 vol.% to achieve the best heating performances. Chang et al. [46] embedded CNF paper as thermistor in concrete and found that the temperature of concrete increased to the values above the freezing point of water at the power input of 6 W. It implied that the CNF paper–filled concrete could be applied as self-heating construction materials, especially in the deicing pavements. Different from the experimental investigations, the self-heating properties of cementitious composites have been explored on the deicing and snow-melting properties of pavements. For example, the electrically conductive concrete filled with CF was applied in an airport to test the anti-icing properties [47]. Despite that the more reduced conductivity of concrete was detected in the broad-scale application than that in the laboratory, the concrete generated 300–350 W/m^2 power density that was sufficient to melt the ice and snow, as depicted in Fig. 7.32a. In addition, the concrete pavement with sandwiched conductive graphite–PET sheets also showed highly efficient snow melting in Fig. 7.32b [48]. Gomis et al. [49] proposed that the conductive cement paste could control the ice layers in pavements. Still, the cost was high because of the high-applied voltage. Therefore, the conductive cementitious composites should be further optimized to higher heating efficiency.

(a) Self-heating concrete slab at an airport in US [47]

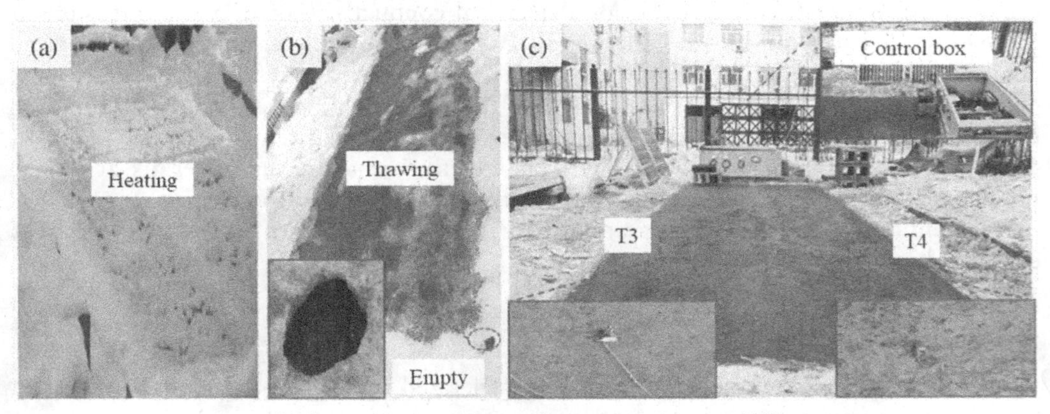

(b) Snow-thawing process of concrete pavement [48]

Figure 7.32 Practical application of self-heating concrete for pavement deicing.

7.6 SUMMARY

This chapter comprehensively summarizes the CBSs with integrated capacities for multifunctional applications. CBSs with integration of self-healing and self-sensing characteristics offer a more durable and autonomous solution for structural monitoring. The superhydrophobic CBS isolates the influence of external moisture, expanding the application scope of CBSs to underwater and even marine environments. CBSs incorporating self-powering capabilities can be applied more sustainably without power supply, eliminating the need for additional interventions and recharging. The self-heating CBS maximizes the utilization of electrical energy used for resistance measurement, significantly improving the energy efficiency. The following are some of the key conclusions that can be drawn:

1. The dispersion of CB nanoparticles was more stable in the superplasticizer–water solution, rather than the SL–superplasticizer–water solution. CB-filled CBS without SL displayed limited self-healing efficiency with partial small cracks being healed. However, with the added SL, the cracks were efficiently healed and only some of the cracks were without sufficient fillers because of the uneven distribution of SL. During the self-healing process, the water spray transferred calcium ions from dissolved SL to the cracks. The self-healing products mainly include the precipitated CH from saturated CH solution, cement hydration products from the unhydrated cement, and the calcium carbonate from the carbonation of calcium ions. The CBS showed satisfactory piezoresistivity, and the addition of SL increased the fractional changes of resistivity. Unfortunately, after the self-healing process, it showed variable piezoresistivity due to the induction of cracks and the self-healing products.
2. The superhydrophobic behaviors with the highest water CA of 163.4° and 142° were achieved on the surface and interior of CBSs after the twice coating by 4% silane with isopropanol. The once coated CBS exhibited slightly worse performance than that of the twice coated counterparts. The CBSs after 100% silane treatment without isopropanol presented hydrophobic behaviors only on the surface but hydrophilic behaviors interior. In terms of the waterproofing cementitious materials, the surface modification of the twice coating further greatly reduced the water absorption of CBSs by half. The additions of GNP and SHP can enhance the microstructures by reducing the porosity and increasing the compactness, and the silane-based surface modification could improve the hydrophobicity and block the contact between water and cementitious materials.
3. The self-sensing property/piezoresistivity of superhydrophobic CBSs seemed to decrease after surface modification, with poorer linearity and repeatability, lower gauge factor, and higher hysteresis. However, the CBSs after the twice coating exhibited acceptable linearity and repeatability, with even dozens of times higher gauge factor than that of commercially available strain gauge. The microstructures showed products from silane for the CBSs after twice coating. However, it was hard to observe silane for the sensors after coating without isopropanol. Moreover, CBSs after once/twice coating seemed more compact than the ones without coating or after coating without isopropanol.
4. Three types of energy harvesting approaches have been introduced in this chapter, including solar energy with integrated piezoelectric energy harvester, the

piezoelectric GO–cement composites as power generator, and the self-powering plain cementitious materials based on polarization. The first method requires additional piezoelectric equipment, while the second and third methods both are based on cement-based materials. Although current research is limited, it remains one of the crucial directions for future studies to achieve intelligent construction and smart transportation.

5. In this era that advocates low carbon emissions, the maximum efficiency in energy utilization is a crucial step for the future. The self-heating CBS presents a concept that utilizes the excellent electrical conductivity of CBSs to not only use them as sensors but also make use of the heat they generate during this process. It can be employed in applications such as deicing road surfaces, snow removal from roofs, indoor heating, etc., thereby serving a dual purpose of multifunctionality in building materials and maximizing energy utilization.

REFERENCES

1. W. Dong, W. Li, Z. Tao, K. Wang, Piezoresistive properties of cement-based sensors: Review and perspective, Construction and Building Materials 203 (2019) 146–163.
2. W. Li, F. Qu, W. Dong, G. Mishra, S.P. Shah, A comprehensive review on self-sensing graphene/cementitious composites: A pathway toward next-generation smart concrete, Construction and Building Materials 331 (2022) 127284.
3. W. Li, W. Dong, Y. Guo, K. Wang, S.P. Shah, Advances in multifunctional cementitious composites with conductive carbon nanomaterials for smart infrastructure, Cement and Concrete Composites 128 (2022) 104454.
4. Z. Deng, W. Li, W. Dong, Z. Sun, J. Kodikara, D. Sheng, Multifunctional asphalt concrete pavement toward smart transport infrastructure: Design, performance and perspective, Composites Part B: Engineering 265 (2023) 110937.
5. K. Gopalakrishnan, H. Ceylan, S. Kim, S. Yang, H. Abdualla, Electrically conductive mortar characterization for self-heating airfield concrete pavement mix design, International Journal of Pavement Research and Technology 8(5) (2015) 315.
6. T. Wang, S. Faßbender, W. Dong, C. Schulze, M. Oeser, P. Liu, Sensitive surface layer: A review on conductive and piezoresistive pavement materials with carbon-based additives, Construction and Building Materials 387 (2023) 131611.
7. E. Richalot, M. Bonilla, M.-F. Wong, V. Fouad-Hanna, H. Baudrand, J. Wiart, Electromagnetic propagation into reinforced-concrete walls, IEEE Transactions on Microwave Theory and Techniques 48(3) (2000) 357–366.
8. E.Q. Zhang, L. Tang, Rechargeable concrete battery, Buildings 11(3) (2021) 103.
9. X. Xi, D. Chung, Electret, piezoelectret and piezoresistivity discovered in steels, with application to structural self-sensing and structural self-powering, Smart Materials and Structures 28(7) (2019) 075028.
10. W. Dong, W. Li, K. Wang, S.P. Shah, D. Sheng, Multifunctional cementitious composites with integrated self-sensing and self-healing capacities using carbon black and slaked lime, Ceramics International 48(14) (2022) 19851–19863.
11. W. Dong, W. Li, K. Wang, Y. Guo, D. Sheng, S.P. Shah, Piezoresistivity enhancement of functional carbon black filled cement-based sensor using polypropylene fibre, Powder Technology 373 (2020) 184–194.
12. O. Mendoza, G. Sierra, J.I. Tobón, Influence of super plasticizer and $Ca(OH)_2$ on the stability of functionalized multi-walled carbon nanotubes dispersions for cement composites applications, Construction and Building Materials 47 (2013) 771–778.
13. L. Jiang, L. Gao, J. Sun, Production of aqueous colloidal dispersions of carbon nanotubes, Journal of Colloid and Interface Science 260(1) (2003) 89–94.
14. P. Mira, V.G. Papadakis, S. Tsimas, Effect of lime putty addition on structural and durability properties of concrete, Cement and Concrete Research 32(5) (2002) 683–689.

15. A.K. Das, C.K. Leung, D.K. Mishra, J. Yu, Smart self-healing and self-sensing cementitious composites: Recent developments, challenges, and prospects, Advances in Civil Engineering Materials 8(3) (2019) 554–578.
16. N. De Belie, E. Gruyaert, A. Al-Tabbaa, P. Antonaci, C. Baera, D. Bajare, A. Darquennes, R. Davies, L. Ferrara, T. Jefferson, C. Litina, B. Miljevic, A. Otlewska, J. Ranogajec, M. Roig-Flores, K. Paine, P. Lukowski, P. Serna, J.-M. Tulliani, S. Vucetic, J. Wang, H.M. Jonkers, A review of self-healing concrete for damage management of structures, Advanced Materials Interfaces 5(17) (2018) 1800074.
17. W. Dong, W. Li, L. Shen, Z. Sun, D. Sheng, Piezoresistivity of smart carbon nanotubes (CNTs) reinforced cementitious composite under integrated cyclic compression and impact, Composite Structures 241 (2020) 112106.
18. W. Dong, W. Li, N. Lu, F. Qu, K. Vessalas, D. Sheng, Piezoresistive behaviours of cement-based sensor with carbon black subjected to various temperature and water content, Composites Part B: Engineering 178 (2019) 107488.
19. B. Han, X. Yu, J. Ou, Effect of water content on the piezoresistivity of MWNT/cement composites, Journal of Materials Science 45(14) (2010) 3714–3719.
20. B. Lubelli, T. Nijland, R. Van Hees, Self-healing of lime based mortars: Microscopy observations on case studies, Heron 56(1/2) (2011) 75–91.
21. L. Alarcon-Ruiz, G. Platret, E. Massieu, A. Ehrlacher, The use of thermal analysis in assessing the effect of temperature on a cement paste, Cement and Concrete research 35(3) (2005) 609–613.
22. I. Odler, S. Abdul-Maula, Possibilities of quantitative determination of the AFt-(ettringite) and AFm-(monosulphate) phases in hydrated cement pastes, Cement and Concrete Research 14(1) (1984) 133–141.
23. H. Fares, S. Remond, A. Noumowe, A. Cousture, High temperature behaviour of self-consolidating concrete: Microstructure and physicochemical properties, Cement and Concrete Research 40(3) (2010) 488–496.
24. W. Dong, W. Li, Z. Sun, I. Ibrahim, D. Sheng, Intrinsic graphene/cement-based sensors with piezoresistivity and superhydrophobicity capacities for smart concrete infrastructure, Automation in Construction 133 (2022) 103983.
25. W. Dong, W. Li, X. Zhu, D. Sheng, S.P. Shah, Multifunctional cementitious composites with integrated self-sensing and hydrophobic capacities toward smart structural health monitoring, Cement and Concrete Composites 118 (2021) 103962.
26. E. Franzoni, B. Pigino, C. Pistolesi, Ethyl silicate for surface protection of concrete: Performance in comparison with other inorganic surface treatments, Cement and Concrete Composites 44 (2013) 69–76.
27. S. Li, W. Zhang, J. Liu, D. Hou, Y. Geng, X. Chen, Y. Gao, Z. Jin, B. Yin, Protective mechanism of silane on concrete upon marine exposure, Coatings 9(9) (2019) 558.
28. C. Christodoulou, C.I. Goodier, S.A. Austin, J. Webb, G.K. Glass, Long-term performance of surface impregnation of reinforced concrete structures with silane, Construction and Building Materials 48 (2013) 708–716.
29. L. Shen, H. Jiang, T. Wang, K. Chen, H. Zhang, Performance of silane-based surface treatments for protecting degraded historic concrete, Progress in Organic Coatings 129 (2019) 209–216.
30. Z. Ma, F. Zhu, T. Zhao, Effects of surface modification of silane coupling agent on the properties of concrete with freeze-thaw damage, KSCE Journal of Civil Engineering 22 (2018) 657–669.
31. X. Fu, D.D.L. Chung, Degree of dispersion of latex particles in cement paste, as assessed by electrical resistivity measurement, Cement and Concrete Research 26(7) (1996) 985–991.
32. P. Liu, Y. Gao, F. Wang, J. Yang, X. Yu, W. Zhang, L. Yang, Superhydrophobic and self-cleaning behavior of Portland cement with lotus-leaf-like microstructure, Journal of Cleaner Production 156 (2017) 775–785.
33. A. Azevedo, D. Cecchin, D. Carmo, F. Silva, C. Campos, T. Shtrucka, M. Marvila, S. Monteiro, Analysis of the compactness and properties of the hardened state of mortars with recycling of construction and demolition waste (CDW), Journal of Materials Research and Technology 9(3) (2020) 5942–5952.

34. W. Dong, W. Li, Y. Guo, X. He, D. Sheng, Effects of silica fume on physicochemical properties and piezoresistivity of intelligent carbon black-cementitious composites, Construction and Building Materials 259 (2020) 120399.
35. W. Dong, W. Li, K. Wang, S.P. Shah, Physicochemical and piezoresistive properties of smart cementitious composites with graphene nanoplates and graphite plates, Construction and Building Materials 286 (2021) 122943.
36. X. Chen, Y. Geng, S. Li, D. Hou, S. Meng, Y. Gao, P. Zhang, H. Ai, Preparation of modified silane composite emulsion and its effect on surface properties of cement-based materials, Coatings 11(3) (2021) 272.
37. X. Xue, Y.-L. Liu, J.-G. Dai, C.-S. Poon, W.-D. Zhang, P. Zhang, Inhibiting efflorescence formation on fly ash–based geopolymer via silane surface modification, Cement and Concrete Composites 94 (2018) 43–52.
38. B. Subeshan, A. Usta, R. Asmatulu, Deicing and self-cleaning of plasma-treated superhydrophobic coatings on the surface of aluminum alloy sheets, Surfaces and Interfaces 18 (2020) 100429.
39. Y.-G. Zhu, S.-C. Kou, C.-S. Poon, J.-G. Dai, Q.-Y. Li, Influence of silane-based water repellent on the durability properties of recycled aggregate concrete, Cement and Concrete Composites 35(1) (2013) 32–38.
40. E. Teomete, The effect of temperature and moisture on electrical resistance, strain sensitivity and crack sensitivity of steel fiber reinforced smart cement composite, Smart Materials and Structures 25(7) (2016) 075024.
41. H.B. Birgin, E. García-Macías, A. D'Alessandro, F. Ubertini, Self-powered weigh-in-motion system combining vibration energy harvesting and self-sensing composite pavements, Construction and Building Materials 369 (2023) 130538.
42. D.A. Triana-Camacho, J.H. Quintero-Orozco, E. Mejía-Ospino, G. Castillo-López, E. García-Macías, Piezoelectric composite cements: Towards the development of self-powered and self-diagnostic materials, Cement and Concrete Composites 139 (2023) 105063.
43. X. Xi, D. Chung, Deviceless cement-based structures as energy sources that enable structural self-powering, Applied Energy 280 (2020) 115916.
44. Z. Wang, Z. Wang, M. Ning, S. Tang, Y. He, Electro-thermal properties and Seebeck effect of conductive mortar and its use in self-heating and self-sensing system, Ceramics International 43(12) (2017) 8685–8693.
45. A. Sassani, A. Arabzadeh, H. Ceylan, S. Kim, S.M.S. Sadati, K. Gopalakrishnan, P.C. Taylor, H. Abdualla, Carbon fiber-based electrically conductive concrete for salt-free deicing of pavements, Journal of Cleaner Production 203 (2018) 799–809.
46. C. Chang, M. Ho, G. Song, Y.-L. Mo, H. Li, A feasibility study of self-heating concrete utilizing carbon nanofiber heating elements, Smart Materials and Structures 18(12) (2009) 127001.
47. A. Sassani, H. Ceylan, S. Kim, A. Arabzadeh, P.C. Taylor, K. Gopalakrishnan, Development of carbon fiber-modified electrically conductive concrete for implementation in Des Moines International Airport, Case Studies in Construction Materials 8 (2018) 277–291.
48. Q. Zhang, Y. Yu, W. Chen, T. Chen, Y. Zhou, H. Li, Outdoor experiment of flexible sandwiched graphite-PET sheets based self-snow-thawing pavement, Cold Regions Science and Technology 122 (2016) 10–17.
49. J. Gomis, O. Galao, V. Gomis, E. Zornoza, P. Garcés, Self-heating and deicing conductive cement: Experimental study and modeling, Construction and Building Materials 75 (2015) 442–449.

Chapter 8

Potential application of CBSs in various infrastructures

8.1 INTRODUCTION

The potential applications of cement-based sensors (CBSs) are diverse and can contribute to various fields. As shown in Fig. 8.1, the potential application of CBSs includes traffic detection in the pavement and structural health evaluation of bridges, buildings, high-speed railways, oil well, and tunnels. Vehicle and traffic detection has become important in traffic management. Many types of sensors have appeared in the past few decades like magnetic sensors and video detector sensors [1–4]. Most of those sensors needed a complex and time-consuming process of installation and calibration. However, the CBSs provided a brand-new method to address this problem because they present clear benefits of easy installation and maintenance, great compatibility, and cost-saving when compared to the traditional sensors [5]. In addition to vehicle and traffic detection of pavements. Cracks are one of the important reasons that influence the durability of pavements. Incorporating self-sensing and self-healing CBSs into a structural health monitoring system could substantially improve its efficiency and reduce maintenance costs. Moreover,

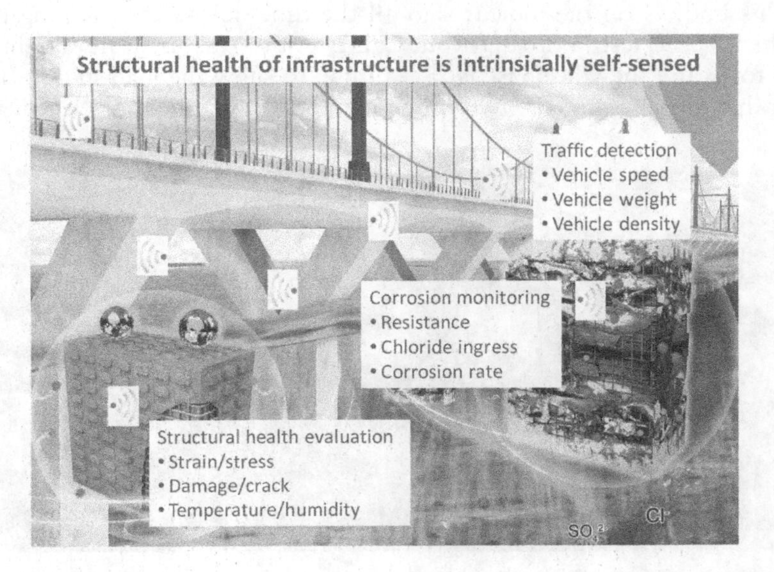

Figure 8.1 Potential application of functional/smart self-sensing concrete for monitoring of health of traffic, bridge, and corrosion [3].

DOI: 10.1201/9781032663685-8

researchers have also explored the application of CBSs in other infrastructures, such as in high-rise buildings, bridges, railways, oil walls, and tunnels, where they serve as sensors to monitor structural health conditions [6–8]. It should be noted that the practical application of CBSs in practical engineering projects is very limited. Apart from the complexity of its influencing factors, the key to whether CBSs can be practically applied in various engineering structures lies in how to enhance their reliability and serviceability in the future.

This chapter introduces the potential application of CBSs in pavement first for traffic speed and human motion detection, followed by a summary of the comparison between different sensing methods in pavement. Subsequently, the CBSs applied in various building structures of columns, beams, and brick walls are reviewed, followed by the application scenarios of bridges, high-speed railways, oil wells, and tunnels. This chapter can encourage investigators to address various challenges in the application of CBSs and promote their usage in future civil infrastructures.

8.2 PAVEMENT

8.2.1 Traffic speed and human motion detection

Fig. 8.2 displays three tests carried out for the evaluation of pavement application: (a) self-sensing performance of cement mortar slab under experimental compression machine; (b) self-sensing performance of mortar slab to detect human motions, and (c) vehicle speed [9]. Three duplicates of the cement mortar slab with embedded CBSs are applied for the tests. For the lab compression test, the experimental results can be found in Chapter 3. The human motion test is conducted by jumping and "up–down" foot movements on the mortar slab. The weight of the tester is approximately 90 kg, for the "up–down" feet motion test, his whole body is totally on/off the mortar slab with the foot movements. However, his body is on the mortar slab all the time during the jumping motion test. For the vehicle speed test, a 2013 Toyota Camry Altise was used. Its weight is approximately 3.0 tons, and the wheelbase is 2,775 mm. To show the load-monitoring capacity of CBSs in the mortar slabs, the two-probe method was applied for electrical resistivity

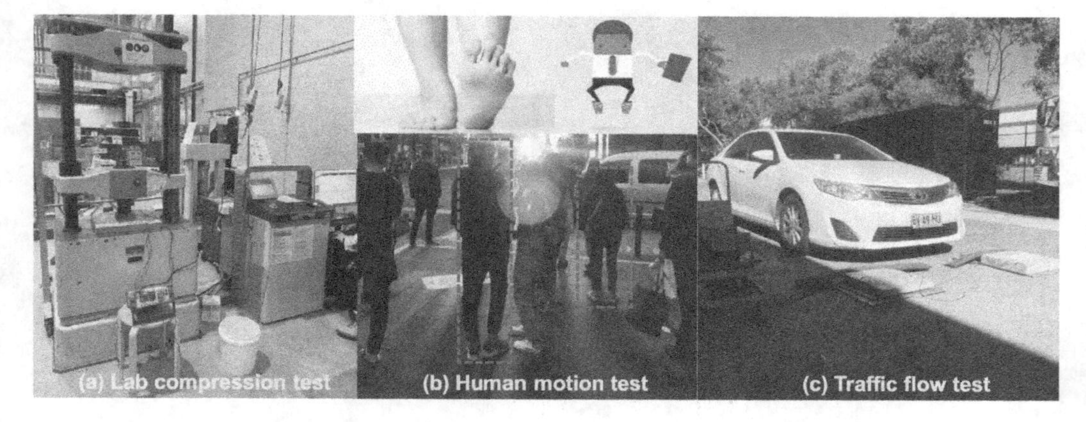

Figure 8.2 Mortar slab with embedded CBSs for compression, human motions, and vehicle speed detections for pavement application [9].

measurement. The multimeters are connected to the cement mortar slab to record the electrical resistivity changes of CBSs in all the tests.

8.2.1.1 Human motion detection

The application of CBSs for human motion detection is a key step for future smart infrastructures. It has wide application potential, including people counting, crowd flow monitoring, housing security, and even body weighing. Fig. 8.3 shows the electrical resistivity changes of cement mortar slab with the human motions of feet up and feet down [9]. It could be found that the FCR possessed a clear relationship with the human motion of feet up or feet down. The largest FCR was in the range of –1.3% to –1.0% in the first ten cycles when the feet were totally on the slab, and the smallest FCR was around –0.15% as the feet down the mortar slab. The average FCR was approximately –1.09% at the peaks. For the second ten cycles, the average FCR was gently decreased to –0.92%, indicating a released irreversibility after preloading [10, 11]. It should be noted that the FCR could almost return to its initial value after the feet up and feet down tests, indicating that the human weight will not greatly affect the inner structures of the CBSs and disturb the conductive passages.

A small irreversibility of FCR values during human motions of feet up and feet down test was observed. Two potential reasons were responsible for the irreversibility. First, the ten cycles of feet up and feet down were carried out in 22 s (2.2 s for each cycle). Different from the static compression test, the irreversible FCR could be enhanced in the dynamic loading because of the hysteresis behavior of CBSs [12, 13]. Second, the data collection rate in this study is five datums per second, increasing the rate of electrical resistance acquisition might reduce the FCR irreversibility during human motion detection. In addition, the noise of FCR especially in the step of feet down test could be observed. Since nine CBSs were connected in series and embedded in the mortar slab, the noise mainly depended on the exact location of feet landing on the cement mortar slab. To improve the accuracy and serviceability of CBSs, the relationships between FCR changes and weight location on the CBSs should be further investigated.

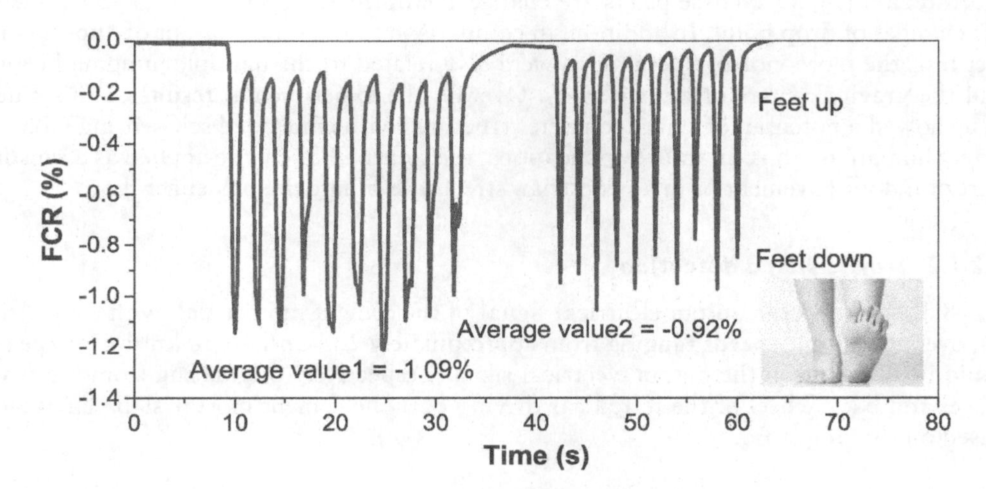

Figure 8.3 Self-sensing behaviors of cement mortar slab under human motion of feet up and feet down [9].

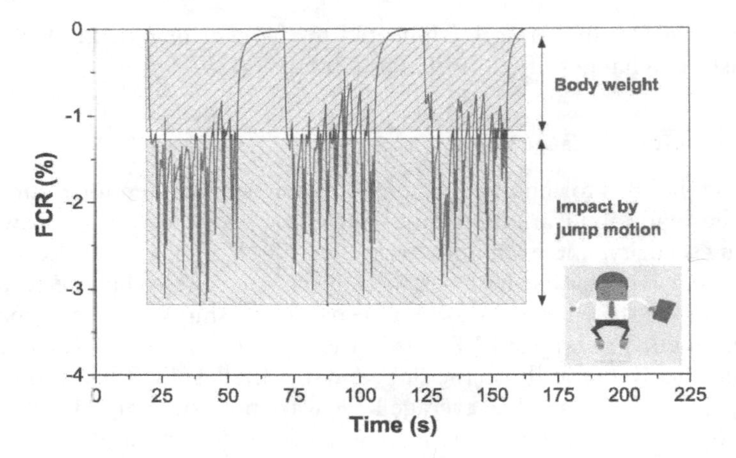

Figure 8.4 Self-sensing behaviors of cement mortar slab under human motion of jumping up and down [9].

Fig. 8.4 displays the electrical resistivity changes of embedded CBSs when the mortar slab is subjected to human jump load [9]. In total, three cycles with each cycle containing ten times of jumping were carried out. It should be noted that the FCR of CBSs did not return to the initial value during jumping, because of the pushing force at the beginning of jumping and the hysteresis behavior of CBSs. Therefore, different from the motion of "up–down" feet, the red and blue regions with an irreversible FCR and different FCR peaks could be observed, respectively. As aforementioned that the body weight was applied on the mortar slab all the time during jumping, it could be found that the FCR changes of CBSs out of body weight were around –1% in red region. This value is very close to the above-mentioned FCR changes of CBSs exposure to motion of "up–down" feet, when all the body weight was applied on the mortar slab. During jumping on the mortar slab, two dynamic loads by pushing off mortar slab and the following drop impact can be produced. The sudden impact load can generate larger FCR peaks, as shown in the green region [14, 15]. These peaks are changeable in the ranges of –2% to –3%, due to the changes of drop point. In addition, in comparison to the stable output of "up–down" feet test, the more violent fluctuations were also related to the multiple jumping heights and the gravity centers of human body. Overall, the experimental results of this study first showed a potential of smart concrete structures with embedded self-sensing CBSs to detect human motions. In future applications, the smart CBSs can be applied as a sensory part of indoor pavement or in a pedestrian street to evaluate the passenger flow.

8.2.1.2 Traffic speed detection

Fig. 8.5a–h shows the output electrical signal of the cement mortar slab with a car driving over at multiple speeds ranging from approximately 2.0 km/h to 40 km/h. Two peaks could be observed in the output electrical signal, respectively originating from the front wheel and back wheel of the tested car driving over the cement mortar slab. Therefore, based on the simple Eq. (8.1):

$$V = \frac{L}{t} \tag{8.1}$$

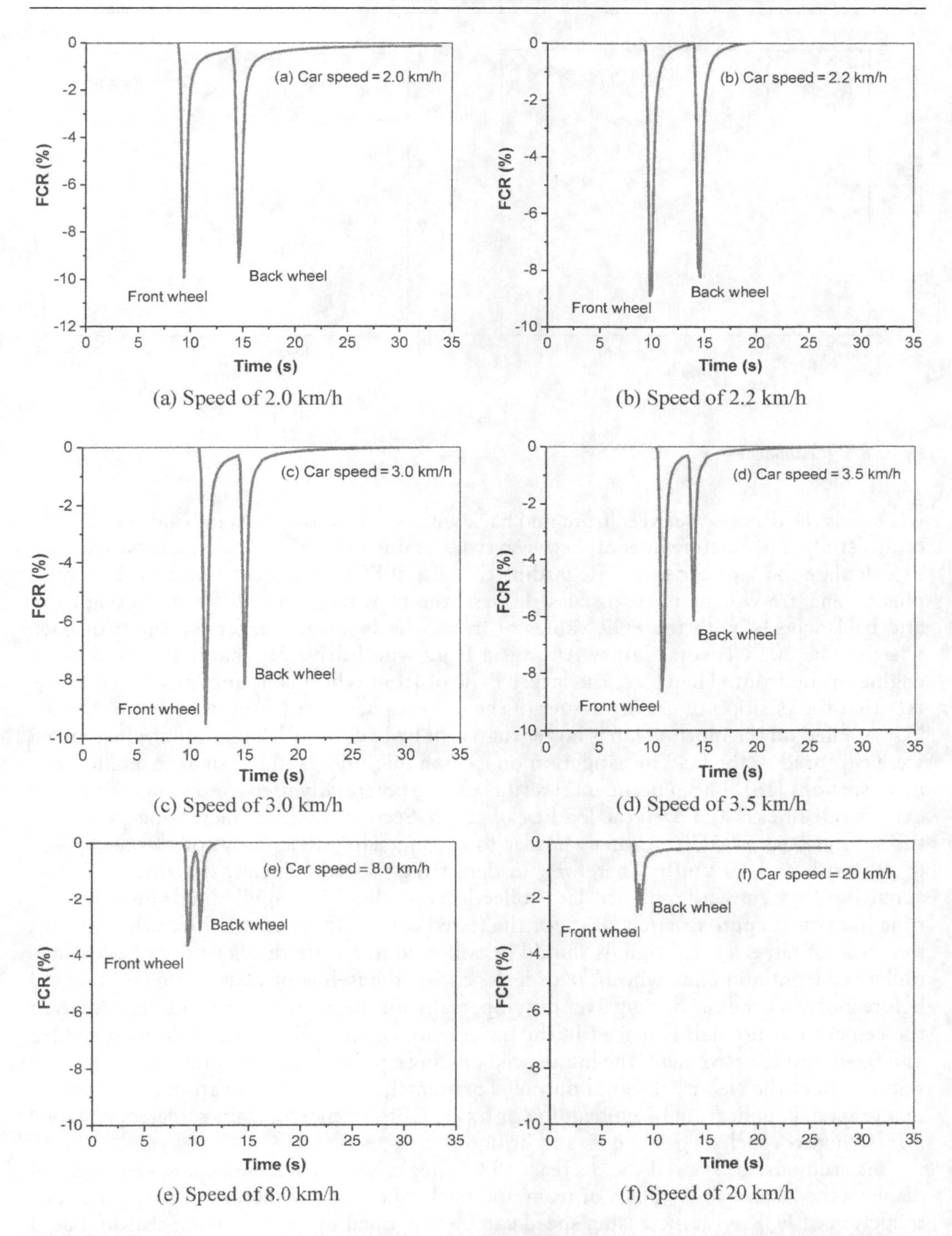

(a) Speed of 2.0 km/h

(b) Speed of 2.2 km/h

(c) Speed of 3.0 km/h

(d) Speed of 3.5 km/h

(e) Speed of 8.0 km/h

(f) Speed of 20 km/h

Figure 8.5 Self-sensing performance of cement mortar slab under traffic load with different speeds [9]. *(Continued)*

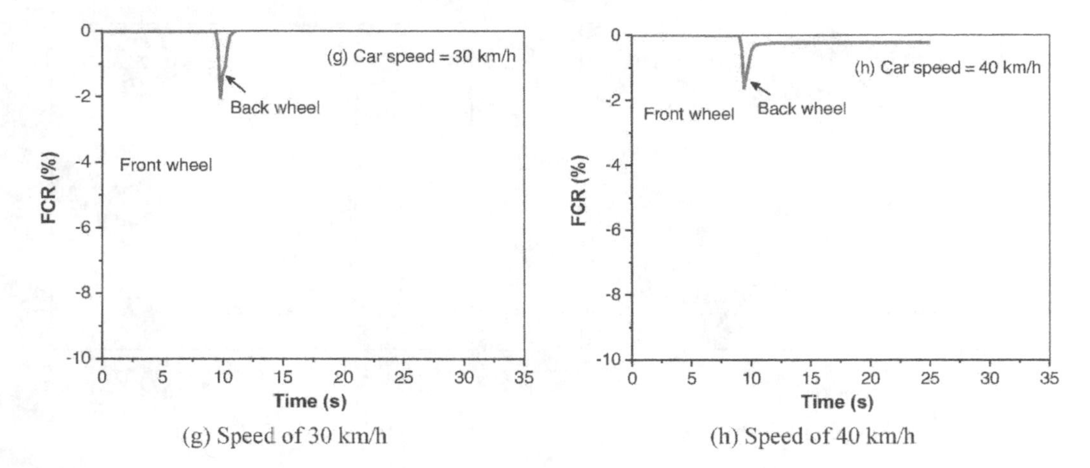

(g) Speed of 30 km/h

(h) Speed of 40 km/h

Figure 8.5 (Continued)

where L is the distance between front and back wheels, which is a constant value of 2.775 m in this study; t is the time interval between two obtained electrical peaks, and we can easily calculate the V of car speed. Regarding the output FCR of cement mortar slab, several phenomena are worthy to be discussed. First, the FCR peaks are different among front and back wheels, with the FCR values of front wheels always larger than that of back wheels. The 2013 Toyota Camry Altise is a front-wheel-drive car that is powered by an engine in the front. Therefore, the larger FCR of front wheel is mainly attributed to the fact that the position of gravity center of the car is closer to the front wheels, so the carload applied on the mortar slab is larger than that by back wheel. The similar phenomena were captured in the field investigation on carbon microfibers–filled smart asphalt pavement sections [16]. The experimental results can be potentially used for distinguishing the drive mode of cars and even the loading of cargo. Second, with the increasing car speed, the second peak of FCR becomes harder to distinguish, particularly for the car speeds of 30 km/h and 40 km/h. Therefore, to detect high-speed vehicles, the advanced data acquisition system with a higher data collection rate should be applied. For instance, the time interval is approximately 0.1 s for the tested car at the speed of 100 km/h. It means that at least ten electrical signals should be collected to ensure the capture of FCR peaks under the front and back wheels. Third, the changed baseline of FCR could be observed before and after the car driving over test, especially for the carload with high speed. Given the cement mortar slab is placed in the pavement and moveable, rather than embedded and fixed into the pavement, the high-speed passing car can drag and mobilize the mortar slab to affect the electrical signal output. Fortunately, this problem can be easily solved in a practical application by embedding or fixing CBSs or mortar slabs to detect concrete infrastructures such as pavements and bridges.

This study first observed the decreased FCR peaks with increasing car speed. Fig. 8.6 displays the largest FCR values of front and back wheels as a function of car speed. The reduction of FCR with increasing speed can be explained by the following reason, based on the Bernoulli equation:

$$p+\frac{1}{2}\rho v^2+\rho g h=C \tag{8.2}$$

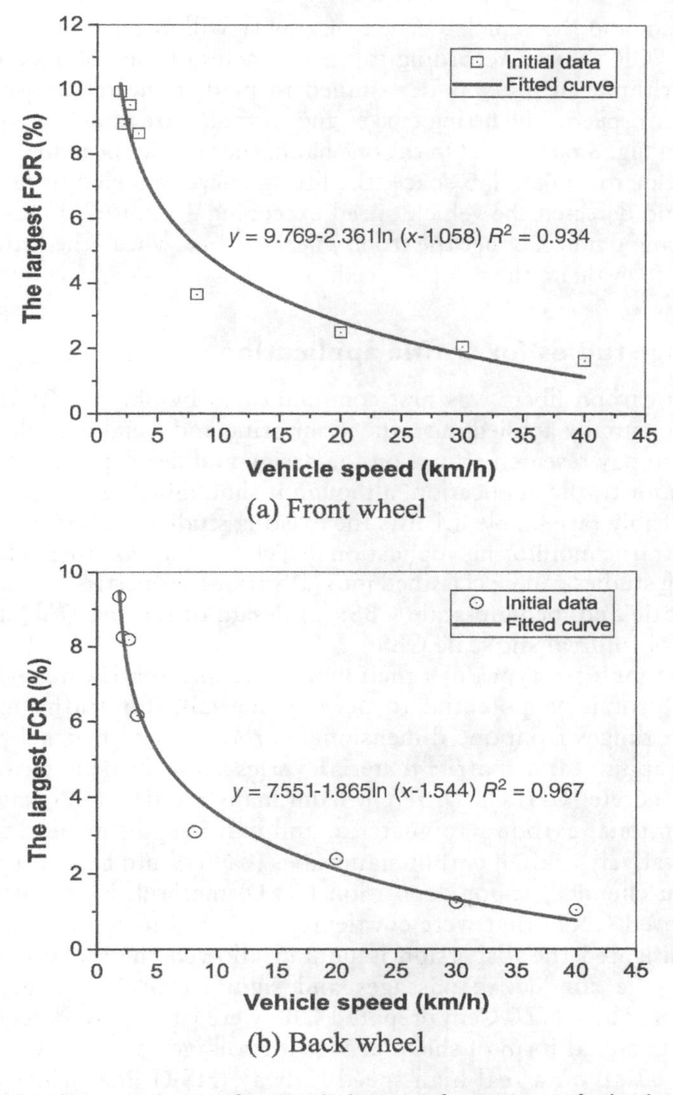

Figure 8.6 Relationship between largest fractional changes of resistivity of wheels and vehicle speed under logarithmic fitting [9].

where p is the intensity of pressure; ρ is the fluid density; v is the fluid speed; C is the constant; g is the acceleration of gravity, and h is the height. With increasing car speed, the air above car moves faster than the air under the car, which results in a larger air pressure under the car. Therefore, one of the reasons could be the reduced carload applied on the cement mortar slab, because of the generated lift force by the high-speed moved air. Also, the upward thrust has offset a portion of the load acting on the concrete slab, which might be the main reason for the reduced FCR. Anyways, the relationship between the largest FCR and the vehicle speed could be another index to predict the car speed, to adjust the obtained car speed based on the FCR peaks. Moreover, the properties of wheels and suspension system and the vehicle's center of gravity will also affect the grip to the slab to decrease the FCR values. According to the character of the obtained data, the three-parameter logarithmic fitting was determined to predict the relationship between the largest FCR to car speed. The fitting curves and formulas are attached with the obtained data as shown in Fig. 8.6a and b. On the one hand, the vehicle speed higher than 40 km/h was not tested due to limited lab space, the fitting curves enabled to predict the largest FCR of CBSs and detected the vehicle speed exceeding 40 km/h. On the other hand, in the case of missing data on either the front wheel or back wheel, the fitting curves could be another way to evaluate the vehicle speed.

8.2.2 Existing studies for traffic application

Since CBS with carbon fibers was first communicated by Shi and Chung [17] in 1999 that could be potentially applied in traffic monitoring and weighing vehicles in motions, scholars began to pay research efforts on the design and development of CBS-integrated sensing systems for traffic application, although it should be highlighted that the related works are noticeably rare. Table 8.1 lists the existing studies of the CBS-integrated sensing systems for traffic monitoring application in a chronological order. The main research focus of existing studies can be classified into (1) various properties and sensing effectiveness of small-scale and medium-scale CBS; (2) design of sensing DAQ systems; and (3) field validation of full/realistic scale CBS.

It is seen that multiple types of functional fillers and matrix materials possess the feasibility and promising potential to develop the CBS for traffic monitoring. The functional filler ranges from one-dimensional carbon black or nickel particle to two-dimensional graphite, and matrix material varies from cement paste to ultrahigh-performance concrete (UHPC). Different from many studies, CBS was prepared with usage of conventional carbon nanomaterial and multistep dispersion treatment, Ding et al. [18] innovatively grafted carbon nanotubes (CNTs) into cement particles (CNT@ Cem) by in situ chemical vapor deposition (CVD) method, the nest-like and hierarchically structured CNTs that were covalently anchored to cement particles not only applausively mitigated the dispersion issue and allowed the scalable production but also optimized the conductive passages and significantly enhance the self-sensing property of CBS. The CNT@Cem prepared CBS were capable to detect the number of passing bogies in signal form of sharp peak with voltage variation when embedded in a concrete track slab on a real high-speed railway (HSR) line, which was analogous to the vehicular axle counter characteristics for CBS applied in a pavement section. Another promising point is that UHPC-typed CBS with both deformation sensing and damage sensing capacities, respectively in elastic stage and post-crack stage, open a possibility for dual functionalities of traffic monitoring and structural health monitoring in large-scale traffic systems [19].

In view of previous studies, the common research strategy is that once the optimal material design of CBS was assured by small/medium-scaled laboratory validation, the design of CBS-integrated sensing system and its field investigation are of importance to implement. As shown in Table 8.1, the early works such as those by Han et al. [20] mainly focused on the deployment form of CBS integrated into pavement section and the feasibility exploration whether proposed CBS-integrated systems attained a basic characteristic sensing signal (i.e., peaks in voltage change) in response to passing vehicles, recent works were started to shift the eyes to other essential components composing

Table 8.1 Summary of CBSs-integrated sensing system for traffic monitoring application [9]

Work type	*Research target*	*Functional filler*	*Matrix type*	*Application form and dimensions*	*Electrical configuration*	*Remark*	*Contribution and limitation*	*Reference*
Laboratory work	Development of wireless DAQ system for CBS				The wireless DAQ hardware system composed of a transformation circuit, a microprocessing module, a wireless transceiver, and an outside terminal module	The Wireless DAQ system was developed based on CC1000 wireless transceiver radio module	Converting circuit in transformation circuit can eliminate noise signal due to the environmental impact by adopting bridge or differential circuit Lack of adequate experimental validation on small-scale CBS Lack of full-scale measurement and field validation	Yu et al. [24]
Field investigation	Field investigation of smart concrete pavement-integrated system	Nick particle	Cement paste	Small-scale cube: 50.8 mm^3 Eight small-scale CBS implemented as a sensor array embedded in the pavement 1.2 m long spanning half width of traffic lanes	Two mesh electrodes with a distance of 10 mm	The CBS pavement section was tested in the real-field environment	The study only presented a preliminary feasibility study that each passing axle can be identified as a sharp peak of voltage change, other traffic-monitoring characteristics such as train speed and WIM measurements were not addressed	Han et al. [20]

(Continued)

Table 8.1 Summary of CBSs-integrated sensing system for traffic monitoring application [9] *(Continued)*

Work type	*Research target*	*Functional filler*	*Matrix type*	*Application form and dimensions*	*Electrical configuration*	*Remark*	*Contribution and limitation*	*Reference*
Field investigation	Field investigation of smart concrete pavement-integrated system	Carbon nanotube	Mortar	$160 \times 23 \times 10$ cm^3 slab	Two mesh electrodes with a distance of 5 cm	The CBS pavement section was tested in the real-field environment Both cast-in-paste sensor and pre-cast sensor were tested	The CBS pavement section could synchronously sense the passing axles as diagnosed by companion strain gauge The study only presented a preliminary feasibility study that each passing axle can be identified as a sharp peak of voltage change, other traffic-monitoring characteristics such as train speed and WIM measurements were not addressed The maximum tested speed was 32 km/h^{-1}	Han et al. [27]
Laboratory work	Piezoresistivity of CBS	Graphite	Cement paste	Small-scale cube: 50 mm^3 Medium-scale slab: $30 \times 15 \times 5$ cm^3	Four embedded wire electrodes with central distance of 8 cm	Small-scale tests indicated 20% graphite-to-cement CBS demonstrated comparable mechanical properties to traditional pavement materials and best piezoresistive performance	The medium-scale CBS exhibited good FCR–load/displacement correlation and low noise-to-signal ratio under step compressive stress with amplitudes in accordance with the fatigue traffic loads defined by Eurocodes Lack of full-scale measurement and field validation	Birgin et al. [22]

(Continued)

Table 8.1 Summary of CBSs-integrated sensing system for traffic monitoring application [9] *(Continued)*

Work type	*Research target*	*Functional filler*	*Matrix type*	*Application form and dimensions*	*Electrical configuration*	*Remark*	*Contribution and limitation*	*Reference*
Numerical study	A proposed WIM characterization algorithm for piezoresistive signals			Full-scale slab embedded in pavement; 3 m long spanning the width of traffic lanes 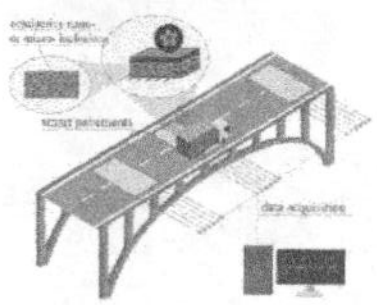	Six embedded wire electrodes with central distance of 60 cm 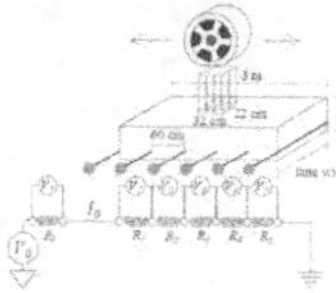	The proposed smart pavement-based WIM system was numerically simulated by FEM modeled approach bridge equipped with smart pavement sections Individual smart pavement layer with six wire electrodes spaced evenly was built in the FEM in terms of thermal-electrical user-defined components The piezoresistivity was simulated through subroutine of user-defined elements based on developed piezoresistive model	Encounter the challenge of correlating piezoresistive signals to WIM characteristics Essential information, including axle weights, axle numbers, and vehicle numbers were extracted The proposed algorithm was numerically validated by passing through three different types of trucks over a modeled bridge Two trucks driving in opposite directions across the same pavement sections can be properly characterized The proposed algorithm is only numerically validated and limited to certain scenarios such as constant speed and speed cases up to 54 km/h Lack of full-scale laboratory and reality-scale field verifications	Birgin et al. [26]

(Continued)

Table 8.1 Summary of CBSs-integrated sensing system for traffic monitoring application [9] *(Continued)*

Work type	*Research target*	*Functional filler*	*Matrix type*	*Application form and dimensions*	*Electrical configuration*	*Remark*	*Contribution and limitation*	*Reference*
Field investigation	Development and field investigation of wireless DAQ system for CBS	Carbon nanotube	Cement paste	Small-scale cube: 50 mm^3	Four plate electrodes with a central distance of 10 mm The wireless DAQ system composed of a transmitter module and a receiver module 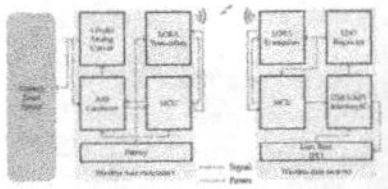	The reliability of wireless transmission system in terms of transmission distance was evaluated in a real railway environment 	CBS with wireless DAQ system exhibited similar piezoresistive performance to the counterpart with conventional DC wired DAQ system Effective transmission distance can be up to 200 m in open space and 100 m at railway sleeper environment, laying a data foundation of determining the sensor node topology of real-time wireless monitoring system for railway concrete infrastructures The local sensor network of proposed wireless DAQ system can be simply connected to national network because of LORA communication protocol The battery service life of wireless DAQ system was expected up to 3.5 years Self-sensing performance of CBS with wireless DAQ system was only tested under cyclic compression in lab condition All wireless transmission modules were integrated into a small box with dimensions of ca. 50 mm^3 Deployment and application form of CBS-integrated wireless system in real railway environment were not addressed	Lee et al. [21]

(Continued)

Table 8.1 Summary of CBSs-integrated sensing system for traffic monitoring application [9] *(Continued)*

Work type	*Research target*	*Functional filler*	*Matrix type*	*Application form and dimensions*	*Electrical configuration*	*Remark*	*Contribution and limitation*	*Reference*
Laboratory work	Piezoresistivity of CBS under impulsive traffic-like loadings	Carbon black	Mortar	$100 \times 100 \times 60$ mm³	Two mesh electrodes with a distance of 40 mm	Low amplitudes impulsive loading simulating traffic-like contact Pressures with three dynamic loading paths were assessed	The FCR signals were highly repetitive and synchronous to loadings with GF reached 40–50. The effects of temperature variation between 15°C and 45°C on CBS were examined in terms of resistance and stress sensitivity. Lack of full-scale measurement and field validation	Monteiro et al. [28]
laboratory work		Steel fibers; Fine steel slag aggregates; Carbon nanotubes	UHPC	Small-scale cube: 50 mm³ Small-scale dog-bone shape specimen: 475 mm in length; 25 mm in thickness	Cubic specimen: two mesh electrodes with a distance of 20 mm Dog-bone shape specimen: two perimetral electrodes with a distance 50 mm (single-channel measurement) and 30 mm (two-channel measurements)  The wireless DAQ hardware system composed of a BLE beacon-integrated microcontroller unit, an electrical resistance sensing circuit, and a powder management circuit.	Two matrix composition of UHPC were tested. CBS with wireless DAQ system were tested in compression and tension. The Wireless DAQ system was developed based on Bluetooth low energy (BLE) protocol.	The battery service life of wireless DAQ system was expected up to 3.6 years taking the advantage of a deployed power management circuit CBS with wireless DAQ system exhibited comparable stress and damage sensing capacities to the counterpart with conventional DC wired DAQ system Multiple channels measurement was feasible for the CBS with wireless DAQ system Lack of full-scale measurement and field validation	Le et al. [19]

(Continued)

Table 8.1 Summary of CBSs-integrated sensing system for traffic monitoring application [9] *(Continued)*

Work type	*Research target*	*Functional filler*	*Matrix type*	*Application form and dimensions*	*Electrical configuration*	*Remark*	*Contribution and limitation*	*Reference*
Field investigation	Field investigation of smart asphalt pavement-integrated system for WIM sensing	Carbon microfiber	Asphalt	Full-scale slab: 300 × 150 × 8 cm^3 	12 embedded wire electrodes forming 3 groups with central distance of 30 cm 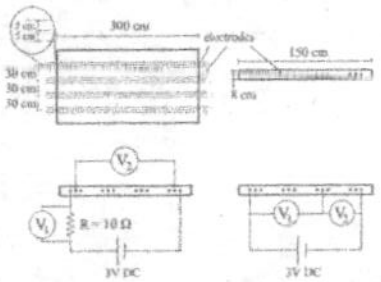	The WIM system consisted a sensing pavement section with tailed electrical configurations, a low-cost DAQ hardware system, and a software of implementing WIM algorithms for vehicle weight estimation WIM algorithms included signal reconstruction algorithm, extremum-seeking algorithm, and supervised artificial neural network Various properties	A homogeneous dispersion validated by linear growth model was achieved in bulk asphalt pavement section with realistic size Sensitive and repeatable responses proportional to vehicle weights were captured in extensive range of weight Common WIM characteristics can be attained on vehicles with both two and five axles Field-applicability of WIM characterization algorithm developed by Birgin et al. [16] was first verified Long-term autonomous sensing of the gross weight of passing trucks on proposed WIM smart pavement system in field condition for two months was achieved with an estimation error less than 20% Field deployment of proposed WIM smart pavement system was designed at realistic scale The prototyping cost for hardware and electronics of DAQ system is only 50USD Vehicle speed estimation and vehicle passing at higher speed were not addressed	Birgin et al. [16]

(Continued)

Table 8.1 Summary of CBSs-integrated sensing system for traffic monitoring application [9] *(Continued)*

Work type	*Research target*	*Functional filler*	*Matrix type*	*Application form and dimensions*	*Electrical configuration*	*Remark*	*Contribution and limitation*	*Reference*
laboratory work & Field investigation	Various properties of CBS	In situ synthesizing carbon nanotubes on cement (CNT@ Cem)	Cement paste	Small-scale cube 20 mm^3 connected with two shielded wires directly embedded into track slab	Two coper electrodes with a distance of 10 mm	The CBS-embedded smart track slab was tested on a real high-speed railway (HSR) line (Shanghai–Hangzhou HSR, K005, China) The chemical properties, microstructure, and synthesizing parameters of CNT@Cem were discussed	CNT@Cem eliminated the ultrasonication and functionalities steps that are unrealizable for the massive and onsite production of CBSs with traditional functional fillers such as CNT Nest-like CNT@Cem tailed the electrical structure, thereby enhancing the electrical conductivity and self-sensing sensitivity of the CBSs CNT@Cem is highly scalable in the production of CBSs The sensor was deployed in the shoulder of track slab aside from the rail, which compensated the conductivity and sensitivity of the sensor The study only presented a preliminary feasibility study that each passing bogie can be identified as a sharp peak of voltage change, other traffic monitoring characteristics such as train speed and WIM measurements were not addressed The tests were conducted at a low-speed maintenance window	Ding et al. [18]

the CBS-integrated sensing system such as hardware and software of data acquisition systems, and the development of characterization algorithms postprocessing the original signal [21, 22]. It is worthy to point out that the conventional wired DAQ system used for laboratory characterization has suffered from many shortcomings. In conventional DAQ system with complex installation requirement, CBSs are required to be wired to an external powder. The long-wired deployment can be easily damaged from harsh ambient conditions without proper protective measurements. In some particular locations of large-scale traffic infrastructures, the deployment of wired CBS might be unrealistic [23]. These disadvantaged features made conventional wired DAQ system with low adoptability, very costly, and inferior reliability, which fatally hinder the field-applicability and long-term operation of CBS-integrated sensing system into large-scale traffic systems [16]. One alternative strategy to effectively mitigate these drawbacks is to develop a proper wireless DAQ system replacing the conventional ones. Although there are few attempts specifically concentrated on the development of wireless DAQ systems with different wireless communication protocol such as LORA and BLE and had received some positive results, as detailed in Table 8.1. The research status on wireless DAQ system for CBS seems to be in a feasibility investigation stage since there were many essential considerations involved in wireless DAQ system. In general, a wireless DAQ system for CBS is composed of a transmitter model and a receiver model, which companioned circuits for specific purpose. For example, Yu et al. [24] deployed a converting circuit to decouple or denoise the environmental noise signal and Le et al. [19] integrated a powder management circuit to secure a long-term battery life longevity. In a mature and realistic CBS-integrated sensing system in a large-scale traffic system, multiple CBS should be deployed in an extensive range of zones to collect sufficient data required for the extraction of the detected parameters of traffic monitoring. In such a situation, the maximum effective transmission distance for an individually distributed CBS is crucial for the determination of sensor node topology.

On the other hand, it was proposed that weigh-vehicles-in-motion (WIM) is the most reliable and advanced detecting features among various available traffic load monitoring system [25]. The WIM feature can detect the passing vehicles and categorize the vehicular types without disturbing the traffic flow. However, the original signal acquired in all present studies is in a form of peaks in voltage variation that might only give the indication of passing counts of vehicular axles (i.e., peak point) and distance between two axles (i.e., distance between peaks). To enable WIM feature, the characterization/learning algorithms that extract more information by postprocessing the original signal are necessary. An innovative WIM characterization algorithm for the CBS-integrated sensing system was proposed by Birgin et al. [26] provoked by numerical verification, and later validated in another full-scaled field investigation by the same authors [16], while this is realized based on a specific electrode of CBS that horizontally divided the bulk pavement CBS slab into several equal-width segments. Therefore, there are many considerations, such as electrode design, hardware and software of DAQ system, CBS deployment, learning/characterization algorithms, life-cost analysis, techniques of multichannel measurement, etc., that should be set as equal or even high-level importance compared to the self-sensing performance of CBS. All these aspects should be properly considered and designed, in order to realize the field application of the robust, easy-to-deploy, low-cost, and highly autonomous CBS-integrated system with advanced functions and detecting features for traffic monitoring.

8.2.3 Comparison between different methods

Fig. 8.7 presents the comparative results with previous field investigations of CBS-integrated sensing system for traffic-monitoring features in terms of detecting features [9]. It is seen that the present study is the first attempt that simultaneously explore the detecting features of vehicle speed and human motion, vis-à-vis the majority of studies concentrated on axle/bogie counter and WIM. Another innovative merit of present study is that the CBS deployment (i.e., eight $60 \times 10 \times 10$ mm^3 small-scale CBS connected in series) embedded greatly save the cost, when compared to the full-scale CBS as bulk conductive pavement section produced with conductive cementitious material adopted in many previous studies [29]. However, it is worthy to point out that the present study has some shortcomings such as the wired DAQ system and lack of hardware/circuit integrated with characterization/learning algorithms for advanced signal extraction. As a result, future analogous works might emphasize on the development of more advanced DAQ systems such as portable wireless DAQ system and the characterization/learning algorithms that can enable more advanced and accurate detections. In addition, when referred to previous discussion, it can be seen that almost all studies ignored impacts from ambient environments. Although Monteiro et al. [28] examined the effects of temperature variation between 15°C and 45°C on the resistance and sensing sensitivity of small-scale CBS through applying simulated traffic-like loadings in lab condition, there was a lack of full-scale field investigation. Previous studies show that the moisture and temperature variations have significant effects on self-sensing stability of CBS [30, 31], while CBS inevitably suffer from climatic cycles when being integrated into traffic infrastructures systems. Studies also proposed that CBS-integrated pavement would become warmer because of the repeatedly rubbing by the vehicle wheel at a heavy traffic load scenario [17]. Therefore, future field investigations of CBS-integrated

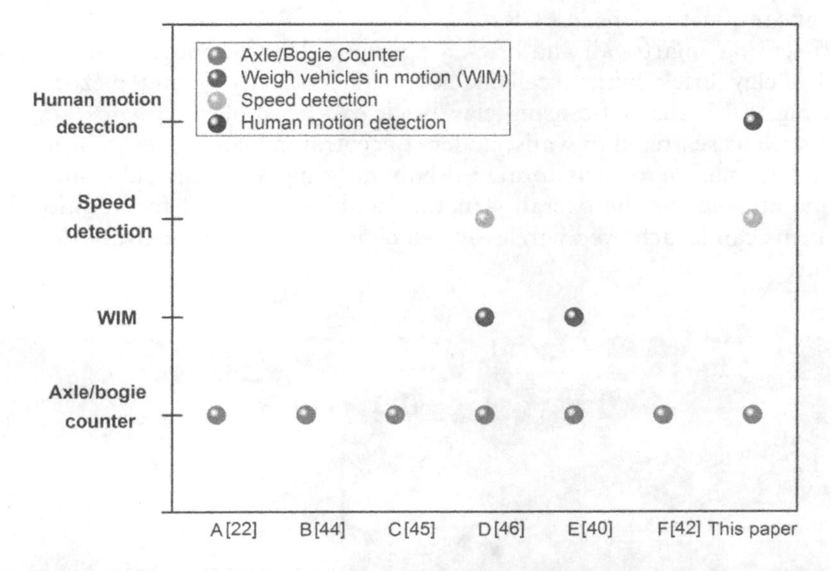

Figure 8.7 Comparative analysis with existing investigations of CBS-integrated sensing system for traffic monitoringapplicationintermsofdetectingfeatures(A[37];B[20];C[27];D[26];E[16];F[18];G[9]).

sensing system should also take into account the ambient impacts. In view of previous studies, some alternative solutions might be the development of hydrophobic CBS [32–34], temperature compensation circuit [35], or other noise signal extraction/separation techniques [36].

8.3 BUILDING

Concrete beams are capable of bearing and dispersing the loads imposed by the superstructure, playing a crucial role in providing support and connection in the overall building structure. Health monitoring of concrete beams contributes to maintaining the structural well-being of buildings, enhancing their long-term performance, reducing maintenance costs, and ensuring the safety of the structure. Rao and Sasmal [38] proposed a CBS for vibration monitoring of reinforced concrete (RC) beams, as shown in Fig. 8.8. The CBSs are fixed to the upper surface of the RC beam with the help of a mechanical anchor device. The vibration testing is employed by an impact hammer to randomly strike various locations on a RC beam, and the DAQ system is to collect the electrical signal. The vibration responses of the CBSs and commercially available accelerometer is displayed in Fig. 8.9. Each signal peak represents one hammer strike. It can be observed that the sensitivity of the CBS is comparable to commercial accelerometer. Additionally, its noise levels are within the controlled range. Wang and Aslani [39] investigated the structural and piezoresistive performance of 3D printed CBSs embedded in RC beams. They found that the 3D printed CBS can contribute to the load-carrying capacity of the RC beam and meanwhile work as embedded sensors for strain and crack monitoring. In a state of advanced damage to RC beams, the effectiveness of 3D printed CBSs is heightened. The occurrence and enlargement of cracks lead to increased gaps between embedded sensor layers, consequently disrupting the conductive pathways among each layer and resulting in the significant fluctuations in FCR.

The self-sensing smart wall and brick are proposed by Filippo Ubertini's group, who developed a clay brick with excellent electrical conductivity and piezoresistivity. As shown in Fig. 8.10, the self-sensing clay brick can be applied in some key regions of structures such as scattered in walls, under concentrated loading point, or in an arch, to track local deformation for structural health monitoring [40]. This individual smart brick serves a monitoring for the overall structure without the need for additional sensors. Measurements can be achieved wirelessly, which will not affect the overall appearance of the structures.

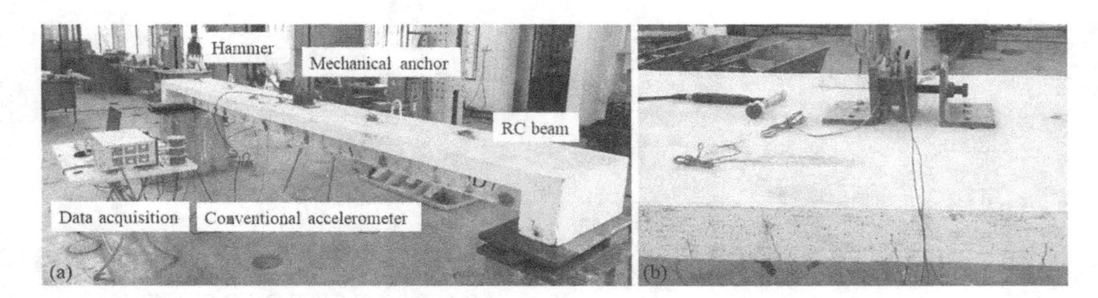

Figure 8.8 Experimental setup (a) for vibration sensing and (b) detail view of CBS [38].

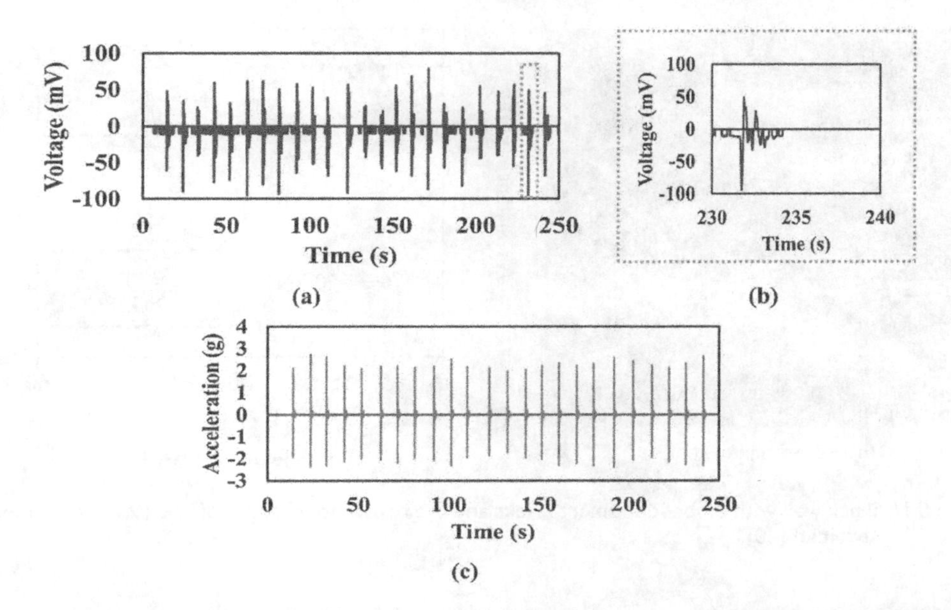

Figure 8.9 Electrical signature of (a) CBS, (b) magnified output, and (c) commercial accelerometer [38].

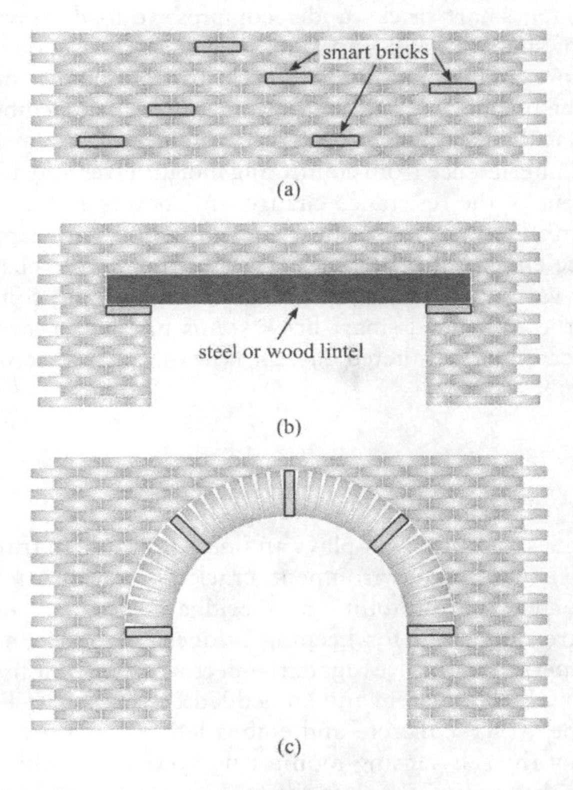

Figure 8.10 Potential application of self-sensing bricks: (a) scattered in walls; (b) under concentrated loading point; (c) at key locations in an arch [40].

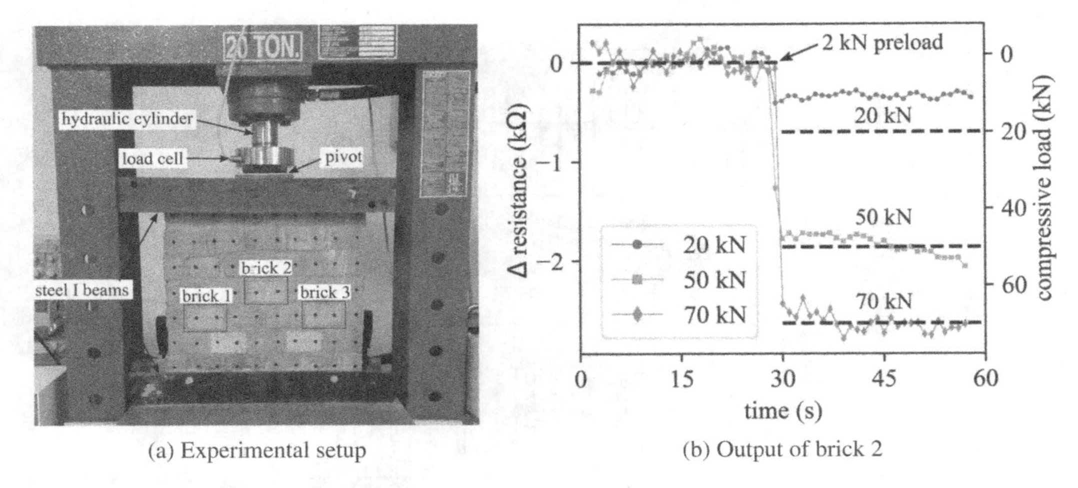

(a) Experimental setup

(b) Output of brick 2

Figure 8.11 Brick wall with embedded smart bricks and the resistance changes of brick 2 under compressive load [40].

Fig. 8.11 shows the experimental small-scale brick wall consisting of 35 bricks embedded with 3 self-sensing smart bricks under compressive load, as well as the electrical resistance changes of smart brick 2 under multiple compressive loads of 20 kN, 50 kN, and 70 kN [40]. Before conducting the test, the electrical resistance of the sensor needs to be corrected from the unembedded to the resistance after being embedded into the wall. Moreover, the electrical measurement should be conducted separately to ensure there is no significant signal interference from connecting mortar layer and adjacent smart bricks. The perfect alignment of the resistance change time points and amplitude corresponds precisely to the magnitude and timing of the applied stress. As the applied load increases, the rate of resistance change also becomes larger. Although the changes are not linear, a certain pattern of variation can still be observed. In addition, compared to the bricks compressed solely, the embedded smart brick seems had better strain-sensing capacity in masonry wall because of the altered current flow out of connecting mortar layer and normal bricks.

8.4 BRIDGE

Bridges are one of the structures that plays an important role in transportation system. However, due to their working environment, cracks, vibrations, fatigue damages, and corrosion will influence the durability of a bridge structure. Thus, self-monitoring and maintenance are important for keeping bridges in good condition. Wang et al. [41] designed an experiment to investigate the performance of a bridge superstructure which was made by GFRP-concrete and embedded FBG sensors. Their results demonstrated that with the GFRP-concrete and embedded FBG sensors, the internal strain could be detected by the self-sensing monitoring system. Another study made use of cementitious composites with CF to test the fatigue damage of bridge decks. Fig. 8.12

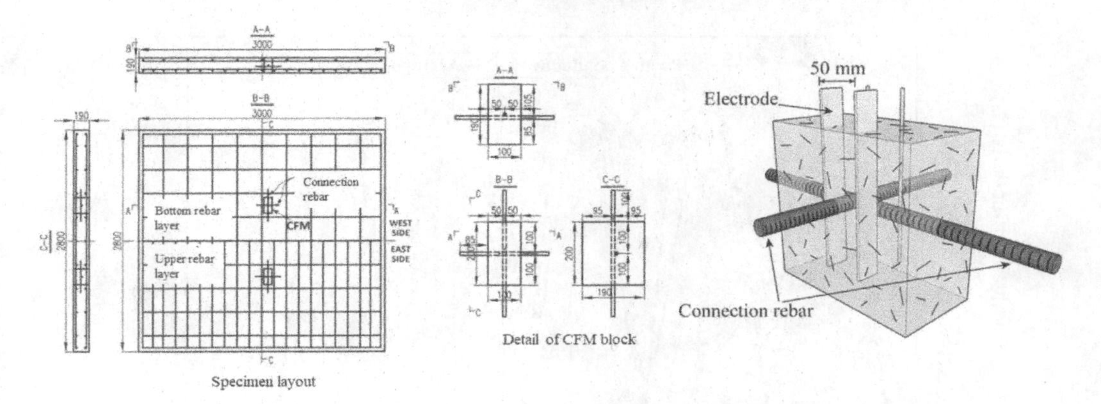

Figure 8.12 Concrete deck slab specimen layout and electrode configuration of CBS [42].

shows the concrete deck slab specimen layout and the electrodes configuration [42]. The experiment showed that a thermal effect on the piezoresistivity property of the system was more apparent than humidity effects. The system could detect the fatigue damage of the decks. The reason why some dispersed points was observed is the relationship between the conductivity of materials and fatigue cracks. The results also showed that the conductivity of CCFC and the steel strain embedded in CCFC blocks has a nonlinear regression relationship. It is obvious that self-sensing cementitious composites had a better performance and a wider application area in the SHM system than FBG sensors. In the experiment on FBG sensors, the self-sensing system could detect the change in the internal strain when the load changed with the installation of FBG sensors [41]. However, it could not exactly monitor the health condition of the bridge, and the installation process of sensors were complex. The test by Xu et al. [42] used CBSs to indicate that cementitious CF composites could detect not only the strain but also the change in the conductivity of the specimen, which is as shown in Fig. 8.13. As the sensors were made of cement, their compatibility with the bridge decks was much better compared with other kinds of sensors. It was observed that the CBSs have shown better compatibility with their mother structure, greater sensing performance, and lower fabrication cost.

Some studies have made efforts to achieve a better control of vibration using self-sensing capacity. An analytical model was developed to test the performance of a self-sensing piezoelectric (PZT) actuator. The experiment made use of capacitors to improve the stability of the control system. This study demonstrated that the suitable temperature for the PZT material is 22°C and the PZT material could enhance the performance of the self-sensing actuation system [43]. Because the collocation of sensors and actuator could provide a better performance in feedback controls, the self-sensing actuators were introduced. A smart patch was designed in the control system, and it suppressed the vibration under free decay condition. By using this self-sensing sensor, the structural vibration was controlled. To identify the signals produced by sensors, a novel method using the artificial neural networks was introduced which improved the ability of the actuation system to control the vibration [44].

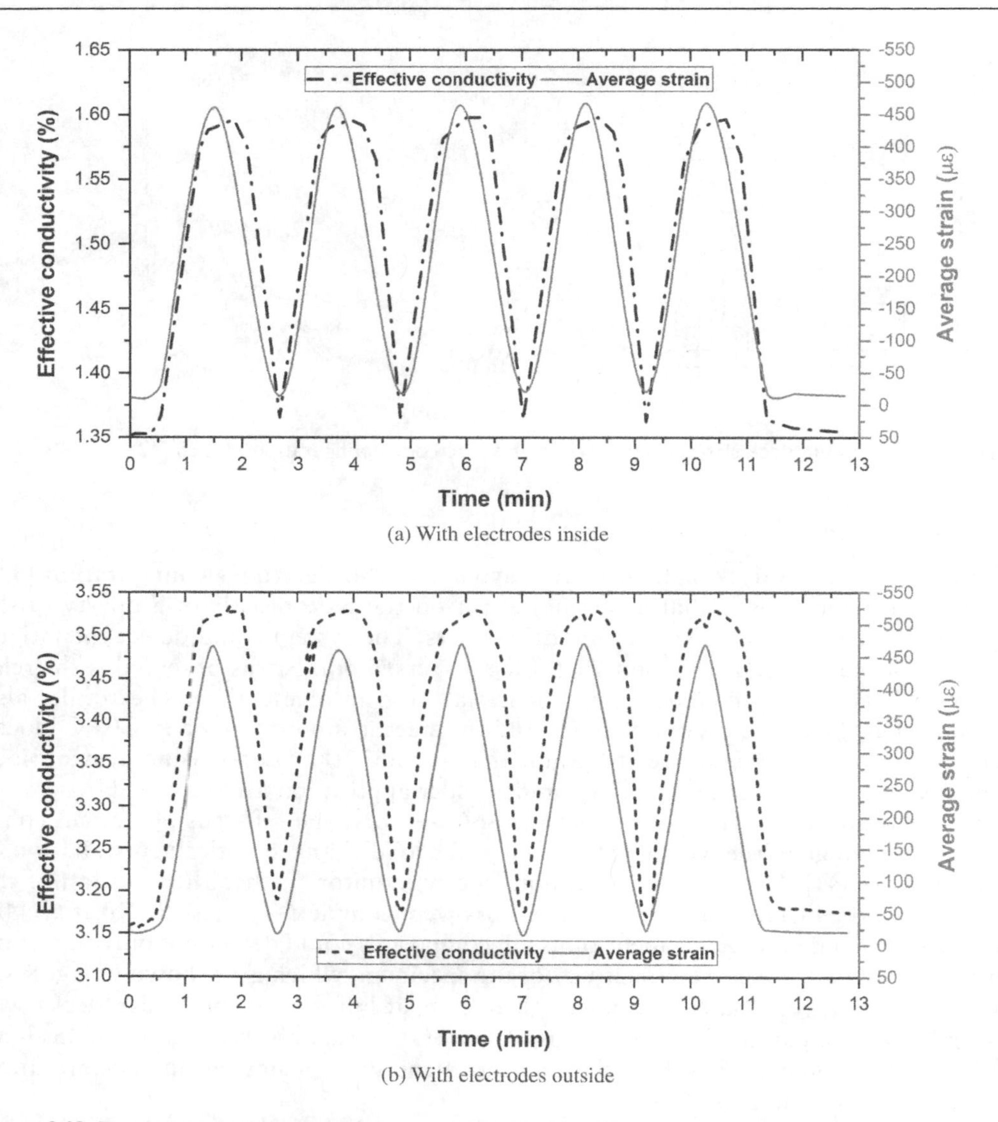

Figure 8.13 Piezoresistivity of two self-sensing bridge decks [42].

8.5 HIGH-SPEED RAILWAY

Nowadays, the railway transportation system has been widely used around the world. The railway system is used to transport passengers and goods due to its convenience and cost-saving characteristics. Concrete slab tracks have been used instead of the traditional ballasted tracks. In order to create a safe and high-efficiency transportation system, efficient safety detection and maintenance sensing system are needed to monitor the health condition of the railway [45]. Components of railways include rails, sleepers, fasteners, rail joints, and ballast or concrete slabs. Railway sleepers are an important part of the railway system, it is designed to transfer loads from the rail to the ground. As the train speed limit is continuously being raised, the railway system also requires sleepers to be insulated to avoid shunting the circuit of the track [46]. The structure response like the strain, deformation, and stress will influence the lifespan and the performance of sleepers. Therefore, it is essential to detect the service condition and to extend the service life of concrete sleepers and concrete slabs.

Butler et al. [47] used the FBG strain and temperature sensor arrays to estimate the load of a rail seat, monitor cracks, and detect sleepers' early-age performance. The results of this experiment demonstrated that FBG sensors could detect the change of the performance because of the process of manufacturing. The sensing system also indicated that within the design limits, the performance of the preservice sleeper was good. However, this experiment also had a drawback that there was a loss in prestress after casting which led to the loss of FGB measurements. Ruiz et al. [48] designed an instrumented concrete sleeper and found a linear relationship between the temperature gradient and the curling of the concrete sleeper. There were some challenges to design smart sleepers, including the cost of smart sleepers, the limited test of long-term performance, and the difficulties of maintenance. Keeping sleepers in good condition is essential for the whole system as well as developing a self-sensing system to support monitoring. However, as most of previous sensors were made of metal or polymer composites which means the compatibility with cementitious materials was not ideal, designing a more compatible sensor became the main objective [49]. In this regard, the self-sensing cementitious materials become a choice to replace previous sensors. Compared with traditional sensors, CBSs have the piezoresistivity property and the electrical resistance ability will change due to the load changes from outside. However, the cement-based sensing system needed to be wired to the sleepers, and the installation of the conventional self-sensing cement system was expensive and time-costing due to the large railway infrastructure. Due to these reasons, Lee et al. [21] came up with a wireless sensing system. The system used a wireless transmission module to transmit the data which was about the electrical resistance change of the sensor to the reception module, as shown in Fig. 8.14. The benefit of the wireless sensing system can be

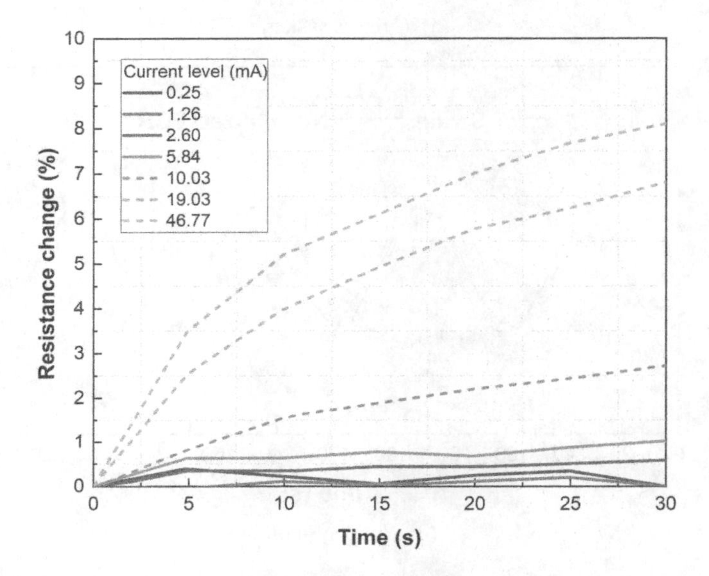

Figure 8.14 Change of the resistance under different current levels [21].

mounted and dismounted easily. The accuracy of the wireless system was quite similar to a wired system, and the lifetime of the battery is up to 3.5 years. This system solved the problem that self-sensing CBSs need wires to transfer the electrical signal, which reduces the influence of the wiring system. At the same time, with MWCNTs, the CBSs showed a good response to the change of the strain, as shown in Fig. 8.15. The curve that represents the normalized FCR follows closely the curve that represents the change of the strain.

However, this was a short-term experiment which means that the influences caused by the change in the temperature, the frequency of the train passing through sensors, and the change in the humidity of the environment were not considered. The experiment lacked a multi-sensor test and there was a distance limitation for the signal transportation of the current wireless sensing system. This wireless system lacked an algorithm which could collect all data from each sensor and analyze them. The design of this algorithm is an essential part for future studies.

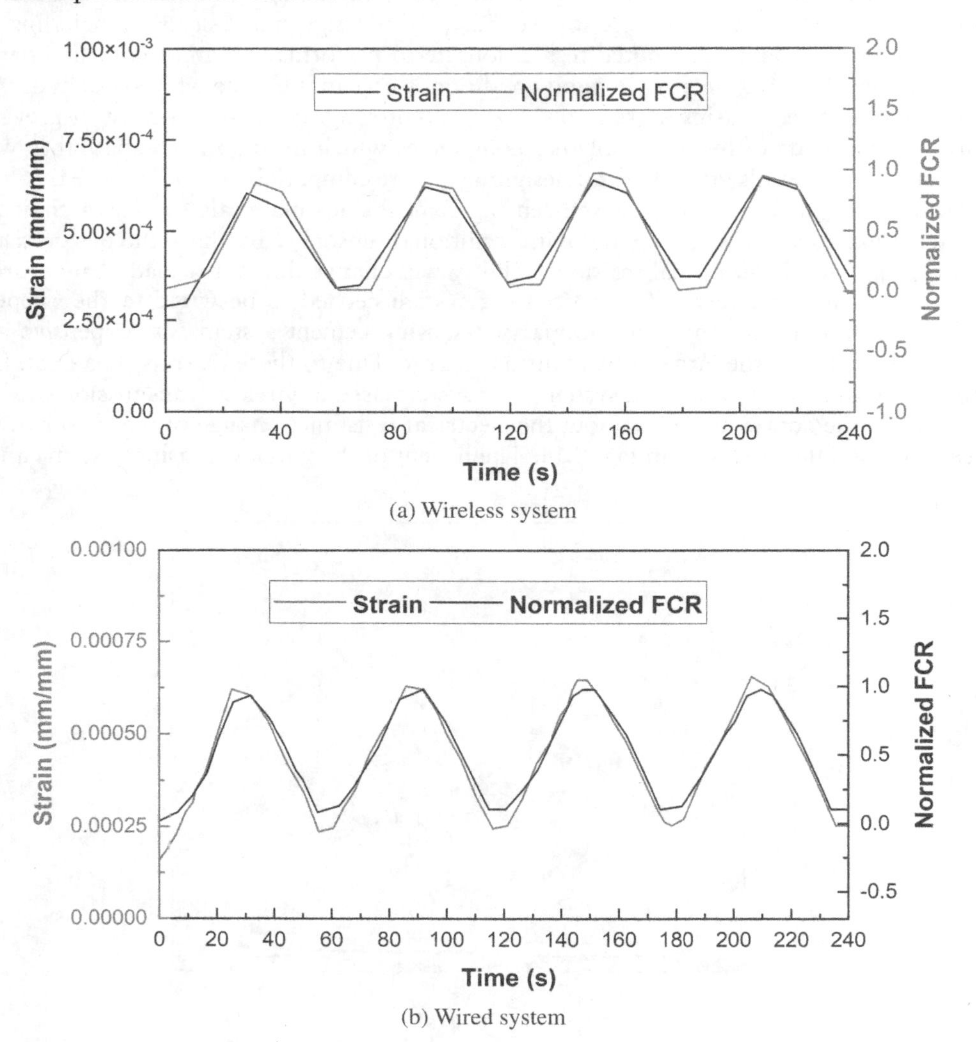

Figure 8.15 FCR of the CBSs with wired or wireless system [21].

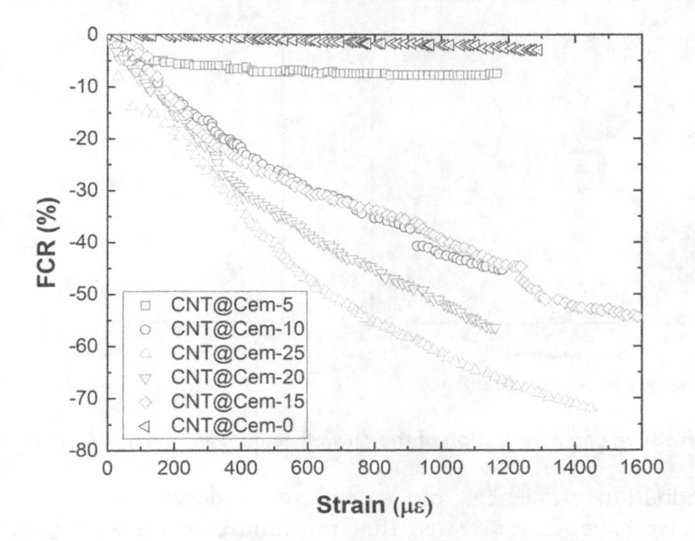

Figure 8.16 The relationship between FCR and strain for self-sensing cement-based composites with CNT@CEM under different stresses [18].

Ding et al. [18] applied a new method for the generation of CNTs on the cement, and the CNT–cement composite was structured by the cement which CNT could directly grow on. The CNT on cement presented great stability, fast recovery, and adaptability to different conditions which showed a great potential for long-term monitoring application work. This study also indicated that the CNT@CEM had a great ability to function as a self-sensing system. This method can save a lot of time in the construction process, and it also reduces the cost of nanomaterials. According to Fig. 8.16, under different compression stresses, the change of the FRC shows different trends. Some innovative methods about self-sensing and wireless technologies have been reviewed. Compared to a wired control system, the wireless system can provide a more efficient way of data collection and has less influence on the infrastructure. However, since there is a distance limitation on the current wireless system, more improvements are needed to solve this problem. Experiments in the real infrastructure systems should be made to analyze whether this wireless technology can make some differences in the construction process. For CNT@ CEM materials, it can achieve economical sustainability in a construction project, and this new technology can reduce the time-cost of nanomaterials producing process. Future studies should be designed to prove that this material can be used for a long time. As the aim of this research is to achieve multifunctional transport, the incorporation of a wireless control system with CNT@CEM materials can also be studied.

8.6 OIL WELL AND TUNNEL

Oil well cement is a specialized type of cement used in the construction process of oil wells. It is employed to isolate the spaces between different geological formations in the well, preventing the movement of underground water, gases, or oil from one formation to another. It can also be utilized to reinforce the structure of the wellbore, preventing collapse or failure, ensuring the stability of the oil well, preventing contamination,

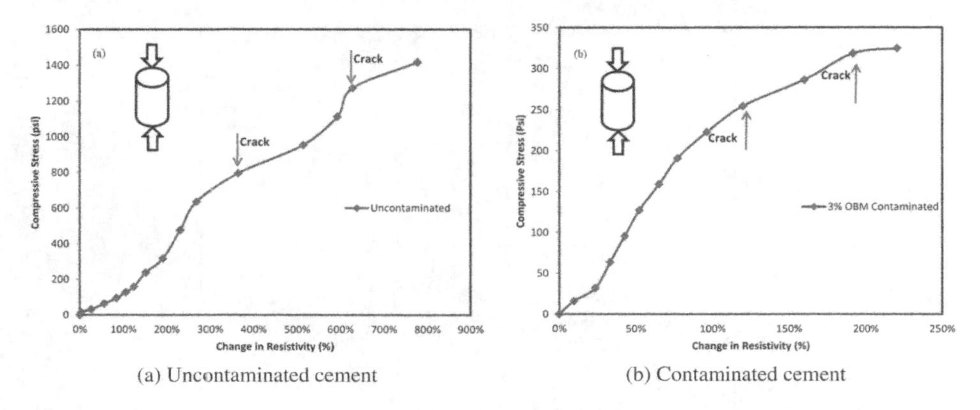

(a) Uncontaminated cement (b) Contaminated cement

Figure 8.17 Piezoresistive characterization of the oil well cement at 28 days of curing [52].

enhancing production efficiency, and safeguarding downhole equipment. Extensive reports or articles have demonstrated that the failure of oil well cement is the main reason for the well blowing and gas explosion [50, 51]. Therefore, it is crucial to appropriately monitor and track the performance of oil well cement throughout the entire lifespan to ensure the integrity of the oil well. Vipulanandan et al. [52] investigated the behavior of piezoresistive smart cement contaminated with oil-based drilling mud and concluded that even piezoresistivity shows that at the smart cement with/without contamination, the oil-based drilling mud reduces the piezoresistivity at failure at all curing ages. As shown in Fig. 8.17, the electrical resistivity increased with increasing compressive load and especially a sharp increase occurred with the formation of crack. Moreover, similar studies have been conducted by Vipulanandan et al. [53] on the oil well cement with conductive CF and the piezoresistive behaviors are plotted in Fig. 8.18.

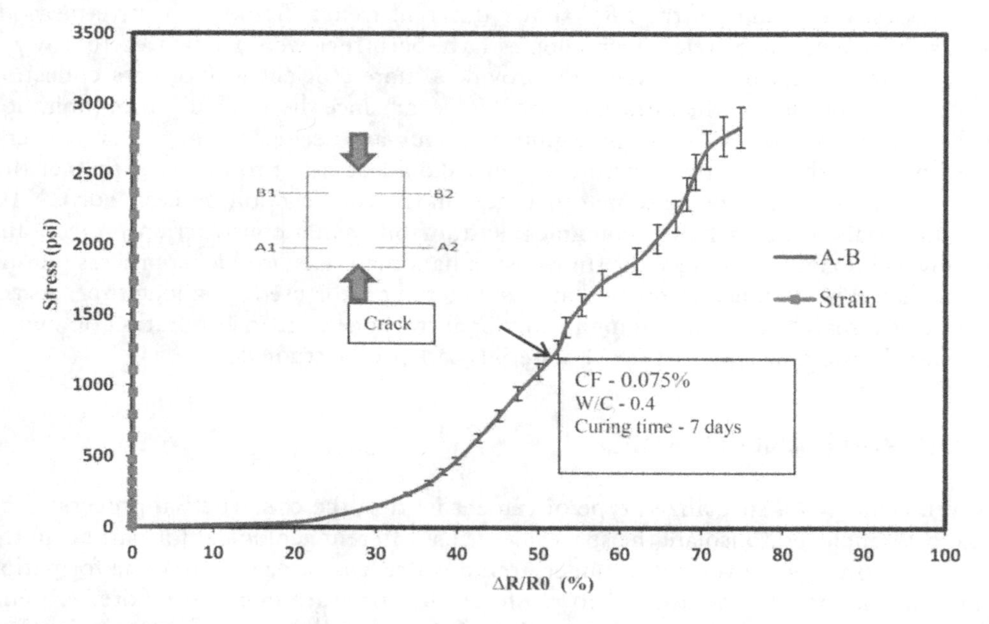

Figure 8.18 Piezoresistive behaviors of oil well cement with 0.075% CF [53].

It seems that the addition of 0.075% CF could significantly reduce the electrical resistivity of oil well cement and provide the piezoresistivity, without considerable interference on the rheological properties. Compared to the changes in compressive strain, the changes in electrical resistivity was nearly 400× higher.

Previous studies have explored the application of smart materials to improve the durability of tunnels. The fire protection of a tunnel plays an important role in improving the safety level of the tunnel. There are various fire protection systems, such as ventilation system, water-based fixed system, and passive protection system [54]. Application of geopolymer-based linings in concrete tunnels has been explored for the passive fire protection. Sakkas et al. [55] conducted the experiment to use the FeNi slag which was doped with the metallurgical alumina. The alkaline potassium hydroxide solution was used in this study. Based on experiment results, the hardness of the geopolymer could only be identified as medium hard and could not be used as a loadbearing element while mechanical properties of the geopolymer fitted well to the requirements of the fire protection material. Although the fire resistance of this geopolymer showed great performance in small scale, more experiments should be designed to test the performance of this material in large scales to prove its applicability. Another study conducted by Duan et al. [56] tested the performance of the fire resistance coating by using a scale test method to analyze the fire-resistive coating performance. According to the experiment, the concrete temperature was not high because of the fire resistance coating, and the wire mesh in this experiment helped improve the performance of the coating. The result also demonstrated that the cracking of the immersed tunnel should be considered during the fire resistance design of the tunnel. These studies have taken into account the impact of material performance on the fire resistance of tunnels, but they have not considered using the tunnel itself as a sensor to monitor internal temperature, humidity, and crack detection. Future studies can be more focused on the field by utilizing tunnel construction materials to produce the sensors itself for the achievement of environmental, liquid, and gas sensing potentials.

8.7 SUMMARY

This chapter summarizes the potential application of CBSs in various civil infrastructures, ranging from pavement, building, bridge, high-speed railway to oil well and tunnel. The CBSs can be applied not only for structural health monitoring but also for traffic speed and human motion detection. It can also be applied in railway sleeper to monitor the railway condition and the speed of high-speed carriage, and isolate the different geological formations when applied in the oil well. Some key conclusions that can be drawn are as follows:

1. For the traffic speed and human motion detection in pavement, the exact vehicle speed can be calculated based on the two peaks induced by the front and back wheels. With the increase of vehicle speed, the FCR gradually decreased because the slope and the lift force on the bottom of vehicle increased to decrease the applied vehicle load. In the "up–down" feet test, the FCR showed a peak at the feet down, and then returned to the initial values at the feet up. The small irreversible FCR could be due to the low data collection rate of multimeters. This demonstrated that the smart mortar slab with the CBSs can be used to detect human traffic. In the jump movement test, in addition to the human weight–induced FCR, the dynamic

loads by pushing off mortar slab at the beginning of jumping and the following drop impact can generate another FCR peaks. It can be an index for the smart mortar slab to detect several specific human movements.
2. The CBSs can be well embedded into various building structures, such as columns, beams, and arches. It can be produced in the form of smart brick with conductive fillers to provide the satisfactory piezoresistivity for masonry structures. The smart brick can measure internal strain changes within the masonry and can detect alterations in strain after the formation of compressive cracks, particularly nearing the limit state conditions.
3. The application of CBSs on bridges includes usage on the bridge concrete deck and piers. When applied on the concrete deck, they can serve for health monitoring and traffic flow monitoring. When applied on the bridge piers, they can provide real-time monitoring of the safety conditions of the bridge piers.
4. In terms of high-speed railway, the CBSs can be utilized in concrete slab track and sleeper for deformation and crack monitoring. The wireless DAQ and control system are encouraged to be developed for more convenient application. The application of CBSs in oil wells and tunnels can monitor the stability of structures to prevent sudden collapses. For timely monitoring, they can be positioned on the exterior of structures and in contact with the geological layers.

REFERENCES

1. L. Lin, X. Han, R. Ding, G. Li, S.C. Lu, Q. Hong, A new rechargeable intelligent vehicle detection sensor, Journal of Physics: Conference Series 13 (2005) 102.
2. F. Azhari, N. Banthia, Cement-based sensors with carbon fibers and carbon nanotubes for piezoresistive sensing, Cement and Concrete Composites 34(7) (2012) 866–873.
3. W. Li, W. Dong, Y. Guo, K. Wang, S.P. Shah, Advances in multifunctional cementitious composites with conductive carbon nanomaterials for smart infrastructure, Cement and Concrete Composites 128 (2022) 104454.
4. W. Li, F. Qu, W. Dong, G. Mishra, S.P. Shah, A comprehensive review on self-sensing graphene/cementitious composites: A pathway toward next-generation smart concrete, Construction and Building Materials 331 (2022) 127284.
5. W. Dong, W. Li, Z. Tao, K. Wang, Piezoresistive properties of cement-based sensors: Review and perspective, Construction and Building Materials 203 (2019) 146–163.
6. P. Cassese, C. Rainieri, A. Occhiuzzi, Applications of cement-based smart composites to civil structural health monitoring: A review, Applied Sciences 11(18) (2021) 8530.
7. D. Chung, A critical review of electrical-resistance-based self-sensing in conductive cement-based materials, Carbon 203 (2023) 311–325.
8. G.H. Nalon, J.C.L. Ribeiro, E.N.D. de Araújo, R.M. da Silva, L.G. Pedroti, G.E.S. de Lima, Concrete units for strain-monitoring in civil structures: Installation of cement-based sensors using different approaches, Construction and Building Materials 394 (2023) 132169.
9. W. Dong, W. Li, Y. Guo, Z. Sun, F. Qu, R. Liang, S.P. Shah, Application of intrinsic cement-based sensor for traffic detections of human motion and vehicle speed, Construction and Building Materials 355 (2022) 129130.
10. W. Dong, W. Li, L. Shen, Z. Sun, D. Sheng, Piezoresistivity of smart carbon nanotubes (CNTs) reinforced cementitious composite under integrated cyclic compression and impact, Composite Structures 241 (2020) 112106.
11. W. Dong, W. Li, K. Vessalas, X. He, Z. Sun, D. Sheng, Piezoresistivity deterioration of smart graphene nanoplate/cement-based sensors subjected to sulphuric acid attack, Composites Communications 23 (2021) 100563.

12. A. Meoni, A. D'Alessandro, A. Downey, E. García-Macías, M. Rallini, A.L. Materazzi, L. Torre, S. Laflamme, R. Castro-Triguero, F. Ubertini, An experimental study on static and dynamic strain sensitivity of embeddable smart concrete sensors doped with carbon nanotubes for SHM of large structures, Sensors 18(3) (2018) 831.
13. D. Chung, Self-sensing concrete: From resistance-based sensing to capacitance-based sensing, International Journal of Smart and Nano Materials 12(1) (2021) 1–19.
14. S. Sasmal, N. Ravivarman, B. Sindu, K. Vignesh, Electrical conductivity and piezo-resistive characteristics of CNT and CNF incorporated cementitious nanocomposites under static and dynamic loading, Composites Part A: Applied Science and Manufacturing 100 (2017) 227–243.
15. F. Ubertini, A.L. Materazzi, A. D'Alessandro, S. Laflamme, Natural frequencies identification of a reinforced concrete beam using carbon nanotube cement-based sensors, Engineering Structures 60 (2014) 265–275.
16. H.B. Birgin, A. D'Alessandro, M. Favaro, C. Sangiorgi, S. Laflamme, F. Ubertini, Field investigation of novel self-sensing asphalt pavement for weigh-in-motion sensing, Smart Materials and Structures 31(8) (2022) 085004.
17. Z.-Q. Shi, D. Chung, Carbon fiber-reinforced concrete for traffic monitoring and weighing in motion, Cement and Concrete Research 29(3) (1999) 435–439.
18. S. Ding, Y. Xiang, Y.-Q. Ni, V.K. Thakur, X. Wang, B. Han, J. Ou, In-situ synthesizing carbon nanotubes on cement to develop self-sensing cementitious composites for smart high-speed rail infrastructures, Nano Today 43 (2022) 101438.
19. H.-V. Le, T.-U. Kim, S. Khan, J.-Y. Park, J.-W. Park, S.-E. Kim, Y. Jang, D.-J. Kim, Development of low-cost wireless sensing system for smart ultra-high performance concrete, Sensors 21(19) (2021) 6386.
20. B. Han, K. Zhang, X. Yu, E. Kwon, J. Ou, Nickel particle-based self-sensing pavement for vehicle detection, Measurement 44(9) (2011) 1645–1650.
21. S.-J. Lee, D. Ahn, I. You, D.-Y. Yoo, Y.-S. Kang, Wireless cement-based sensor for self-monitoring of railway concrete infrastructures, Automation in Construction 119 (2020) 103323.
22. H.B. Birgin, A. D'Alessandro, S. Laflamme, F. Ubertini, Smart graphite–cement composite for roadway-integrated weigh-in-motion sensing, Sensors 20(16) (2020) 4518.
23. B. Han, Y. Yu, B. Han, J. Ou, Development of a wireless stress/strain measurement system integrated with pressure-sensitive nickel powder-filled cement-based sensors, Sensors and Actuators A: Physical 147(2) (2008) 536–543.
24. Y. Yu, B. Han, J. Ou, Wireless acquisition and transmission system of smart cement-based sensors, Sensors and Smart Structures Technologies for Civil, Mechanical, and Aerospace Systems 2009, SPIE, 2009, pp. 104–111.
25. M. Sujon, F. Dai, Application of weigh-in-motion technologies for pavement and bridge response monitoring: State-of-the-art review, Automation in Construction 130 (2021) 103844.
26. H.B. Birgin, S. Laflamme, A. D'Alessandro, E. Garcia-Macias, F. Ubertini, A weigh-in-motion characterization algorithm for smart pavements based on conductive cementitious materials, Sensors 20(3) (2020) 659.
27. B. Han, K. Zhang, T. Burnham, E. Kwon, X. Yu, Integration and road tests of a self-sensing CNT concrete pavement system for traffic detection, Smart Materials and Structures 22(1) (2012) 015020.
28. A. Monteiro, P. Costa, M. Oeser, P. Cachim, Dynamic sensing properties of a multifunctional cement composite with carbon black for traffic monitoring, Smart Materials and Structures 29(2) (2020) 025023.
29. S. Gupta, Y.-A. Lin, H.-J. Lee, J. Buscheck, R. Wu, J.P. Lynch, N. Garg, K.J. Loh, In situ crack mapping of large-scale self-sensing concrete pavements using electrical resistance tomography, Cement and Concrete Composites 122 (2021) 104154.
30. W. Dong, W. Li, N. Lu, F. Qu, K. Vessalas, D. Sheng, Piezoresistive behaviours of cement-based sensor with carbon black subjected to various temperature and water content, Composites Part B: Engineering 178 (2019) 107488.
31. M. Adresi, F. Pakhirehzan, Evaluating the performance of self-sensing concrete sensors under temperature and moisture variations-a review, Construction and Building Materials 404 (2023) 132923.

32. W. Dong, W. Li, Y. Guo, F. Qu, K. Wang, D. Sheng, Piezoresistive performance of hydrophobic cement-based sensors under moisture and chloride-rich environments, Cement and Concrete Composites 126 (2022) 104379.
33. W. Dong, W. Li, X. Zhu, D. Sheng, S.P. Shah, Multifunctional cementitious composites with integrated self-sensing and hydrophobic capacities toward smart structural health monitoring, Cement and Concrete Composites 118 (2021) 103962.
34. Y. Guo, W. Li, W. Dong, K. Wang, X. He, K. Vessalas, D. Sheng, Self-sensing cement-based sensors with superhydrophobic and self-cleaning capacities after silane-based surficial treatments, Case Studies in Construction Materials 17 (2022) e01311.
35. J. Ou, B. Han, Piezoresistive cement-based strain sensors and self-sensing concrete components, Journal of Intelligent Material Systems and Structures 20(3) (2009) 329–336.
36. S. Ding, C. Xu, Y.-Q. Ni, B. Han, Extracting piezoresistive response of self-sensing cementitious composites under temperature effect via Bayesian blind source separation, Smart Materials and Structures 30(6) (2021) 065010.
37. B. Han, X. Yu, E. Kwon, A self-sensing carbon nanotube/cement composite for traffic monitoring, Nanotechnology 20(44) (2009) 445501.
38. R.K. Rao, S. Sasmal, Smart nano-engineered cementitious composite sensors for vibration-based health monitoring of large structures, Sensors and Actuators A: Physical 311 (2020) 112088.
39. L. Wang, F. Aslani, Structural performance of reinforced concrete beams with 3D printed cement-based sensor embedded and self-sensing cementitious composites, Engineering Structures 275 (2023) 115266.
40. A. Downey, A. D'Alessandro, S. Laflamme, F. Ubertini, Smart bricks for strain sensing and crack detection in masonry structures, Smart Materials and Structures 27(1) (2017) 015009.
41. Y. Wang, Y. Li, J. Ran, M. Cao, Experimental investigation of a self-sensing hybrid GFRP-concrete bridge superstructure with embedded FBG sensors, International Journal of Distributed Sensor Networks 8(10) (2012) 902613.
42. C. Xu, J. Fu, L. Sun, H. Masuya, L. Zhang, Fatigue damage self-sensing of bridge deck component with built-in giant piezoresistive cementitious carbon fiber composites, Composite Structures 276 (2021) 114459.
43. G.E. Simmers Jr, J.R. Hodgkins, D.D. Mascarenas, G. Park, H. Sohn, Improved piezoelectric self-sensing actuation, Journal of Intelligent Material Systems and Structures 15(12) (2004) 941–953.
44. H. Ji, J. Qiu, Y. Wu, J. Cheng, M. Ichchou, Novel approach of self-sensing actuation for active vibration control, Journal of Intelligent Material Systems and Structures 22(5) (2011) 449–459.
45. G. Zi, S.-J. Lee, S.Y. Jang, S.C. Yang, S.-S. Kim, Investigation of a concrete railway sleeper failed by ice expansion, Engineering Failure Analysis 26 (2012) 151–163.
46. Y. Pang, S.N. Lingamanaik, B.K. Chen, S.F. Yu, Measurement of deformation of the concrete sleepers under different support conditions using non-contact laser speckle imaging sensor, Engineering Structures 205 (2020) 110054.
47. L.J. Butler, J. Xu, P. He, N. Gibbons, S. Dirar, C.R. Middleton, M.Z. Elshafie, Robust fibre optic sensor arrays for monitoring early-age performance of mass-produced concrete sleepers, Structural Health Monitoring 17(3) (2018) 635–653.
48. A.E.C. Ruiz, Y. Qian, J.R. Edwards, M.S. Dersch, Analysis of the temperature effect on concrete crosstie flexural behavior, Construction and Building Materials 196 (2019) 362–374.
49. G. Song, Y.L. Mo, K. Otero, H. Gu, Health monitoring and rehabilitation of a concrete structure using intelligent materials, Smart Materials and Structures 15(2) (2006) 309.
50. D. Izon, E. Danenberger, M. Mayes, Absence of fatalities in blowouts encouraging in MMS study of OCS incidents 1992–2006, Drilling contractor 63(4) (2007) 84–89.
51. K.M. Carter, E. van Oort, A. Barendrecht, Improved regulatory oversight using real-time data monitoring technologies in the wake of Macondo, SPE Deepwater Drilling and Completions Conference, SPE, 2014, p. D011S007R001.
52. C. Vipulanandan, M. Heidari, Q. Qu, H. Farzam, J.M. Pappas, Behavior of piezoresistive smart cement contaminated with oil based drilling mud, Offshore Technology Conference, OTC, 2014, p. D031S030R007.

53. C. Vipulanandan, R. Krishnamoorti, R. Saravanan, J. Liu, Q. Qu, G. Narvaez, B. Hughes, D. Richardson, J. Pappas, Development and characterization of smart cement for real time monitoring of ultra-deepwater oil well cementing applications, Offshore Technology Conference, OTC, 2014, p. D031S030R003.
54. M.S. Tomar, S. Khurana, Impact of passive fire protection on heat release rates in road tunnel fire: A review, Tunnelling and Underground Space Technology 85 (2019) 149–159.
55. K. Sakkas, D. Panias, P.P. Nomikos, A.I. Sofianos, Potassium based geopolymer for passive fire protection of concrete tunnels linings, Tunnelling and Underground Space Technology 43 (2014) 148–156.
56. J. Duan, Y. Dong, J. Xiao, D. Zhang, W. Zheng, S. Zhang, A large-scale fire test of an immersed tunnel under the protection of fire resistive coating, Tunnelling and Underground Space Technology 111 (2021) 103844.

Chapter 9

Challenges and future directions

9.1 INTRODUCTION

It shall be mentioned that the durability, thermal, and mechanical properties of cementitious composites have been widely explored and the results are encouraging. The electrical property of cementitious composites has been proven as one of the most promising characterizations for multifunctional cementitious composites. Cement-based sensors (CBSs) have gained attention for their potential applications in monitoring the structural health of concrete infrastructure. However, there are several challenges associated with these sensors, and researchers are exploring various directions to address these challenges and improve the effectiveness of CBSs. First, the CBSs need to withstand harsh environmental conditions, including exposure to moisture, chemicals, and temperature variations, to ensure long-term performance [1, 2]. Second, achieving high sensitivity and accuracy in detecting structural changes or damage is crucial for effective monitoring [3, 4]. Third, CBSs should be seamlessly integrated into concrete structures without compromising the structural integrity. The cost of manufacturing and implementing CBSs can be a barrier to widespread adoption because the price of nanomaterials is still costly. Moreover, the CBSs require wired connections for data transmission, which can be impractical for large structures or remote locations. As a consequence, further studies are needed on the effective control of critical factors such as application environments, data monitoring systems, dispersion, high production cost, degraded workability, and health-related issues [5, 6].

This chapter reviews the limitations and challenges of using CBSs for civil infrastructure health monitoring, from environmental interference, DAQ system, dispersion of nanomaterials, and degraded workability to sustainability of sensors. In terms of future development, the long-term performance and durability of CBSs must be explored to enhance the sensor's serviceability. The self-powered CBSs and wireless sensing can be developed for automatic and energy-saving self-sensing capacity. For future smart city applications, the CBSs can be integrated with Internet of Things (IoT) technology and building information modeling (BIM). For the wide application of CBSs, the standardization of raw materials, dispersion, manufacturing, curing, installation, and maintenance processes should be determined, and the certification of CBSs is the last step for the commercial application of CBSs.

DOI: 10.1201/9781032663685-9

9.2 LIMITATIONS AND CHALLENGES

9.2.1 Consistency and uniformity

Ensuring consistent and uniform properties of the CBSs throughout the production process is crucial for reliable and accurate measurements. Due to the anisotropy of the cement matrix and variations in the distribution of conductive fillers within the same batch of cementitious materials, it is challenging to achieve fully consistent electrical conductivity and pressure-sensitive characteristics in the produced CBSs. Therefore, achieving a homogeneous mixture of cementitious materials and nanomaterials is critical to ensure the sensors' performance. In addition, there is no single universal standard specifically dedicated to the production of CBSs. Once there are specifications regarding the proportions of cement-based materials and conductive fillers, as well as guidelines for dispersion, mixing, casting, electrode configuration, curing, drying, testing, and other manufacturing processes, it is believed that the individual variability of CBSs can be significantly reduced. Some special standards and guidelines depending on the specific application and industry should also be proposed.

9.2.2 Environmental interference

Variations in the temperature and humidity affect the electrical resistivity of cementitious composites; thus, the correlation to the targeted index could be affected. Taking piezoresistivity for example, it has been demonstrated that the temperature from –20°C to 100°C almost made no difference in the electrical resistivity changes under compression [7]. However, the temperature altered the initial resistivity before compression which might cause misunderstanding on the monitoring index, let alone the unknown performance of conductive cementitious composites in extreme environments outside the range of above temperatures. The piezoresistivity also considerably fluctuated for the cementitious composites with different water contents from a completely dry state to a saturated state. Dong et al. [8] developed hydrophobic CBSs to eliminate water interference. Some other researchers attempted to use temperature and humidity compensation circuits to eliminate the effect of environmental factors, while the exact circuit configuration in conductive cementitious composites is very rare [9]. Moreover, the durability and severability of conductive cementitious composites under freeze-thaw cycles should be further investigated because the combined effects of temperature and humidity were applied to the cementitious composites.

9.2.3 DAQ system

The data collection system relates to the accuracy of evaluating the efficiency of multifunctional cementitious composites with capacities of self-sensing, self-heating, self-healing, and EMI shielding. Currently, the methods to measure the electrical resistivity of cementitious composites are multiple from the two-point method to the four-point method by using DC to AC. The two-point method has similar effectiveness to the four-point method in evaluating the resistivity as long as the electrical resistivity of electrodes and conductive wires themselves could be neglected. For the dried cementitious composites where the polarization is not significant, an easier configuration of DC can be applied to measure the electrical resistivity. However, for saturated composites or those with high water contents, the power supply of AC is highly recommended since the electrical signals under DC will take a long time to reach the electrostatic equilibrium in an electric field. In addition to the resistivity measurement methods, the layout of conductive cementitious composites and the data collection in the concrete structures should be paid more attention to sufficiently realize the functionality. For example, the conductive cementitious composites could be entirely used in the surface layer of self-deicing pavement, or it could only be needed in several key elements of concrete structures such as a tensile beam. Moreover, the exploration of CBSs for defective area mapping could be carried out in future investigations.

9.2.4 Dispersion and toxicity of nanomaterials

Uniform dispersion of carbon materials in cementitious composites has a firm relationship to electrical conductivity. The field operation of carbon materials dispersion is not only complicated to the labor force but also consumes additional energy for ultrasonication which increases both the project duration and expenditure. The commonly used carbon materials to improve the electrical conductivity of cementitious composites were high-priced such as the CNT or GNP. Hence, it dramatically increases the related cost based on the optimal mix proportion. Therefore, a better mix design of conductive cementitious composites should be explored by adopting less content of conductors, such as the utilization of electromagnetic fields or cheap fibers with a high aspect ratio [10]. Although rarely mentioned, according to the report from IARC of the World Health Organization, CNT has been regarded as a class 2B carcinogen, which might cause genetic disease to operators. Therefore, the sustainable alternative of conductive fillers such as recycled carbon fiber, recycled conductive rubber, or recycled ceramics could be applied for conductive cementitious composites. Moreover, even for the same mixing proportion of functional cementitious materials, the deviation of electrical conductivity, piezoresistivity, and other functions is high, especially under the specific environments. Furthermore, because of the improved electrical conductivity and EMI shielding properties, the traditional nondestructive testing techniques for concrete structures might be influenced, which should be calibrated based on the altered electrical conductivity [11].

9.2.5 Adverse effects on workability

Most researchers have mentioned that the workability of GBNs-based cementitious composites could be decreased with the increasing addition of GBNs, which could have an adverse influence on practical construction. The decrease in the workability of the GBCCs could cause some trouble for the builders to appropriately situate, and condense the cementitious composites with steel bars on the building stations. In consequence, the higher porosity in the uncondensed cementitious composites could remarkably decrease the durability and mechanical properties. It should also be noted that some researchers have adopted different water-reducing agents to enhance the workability, but some researchers have also mentioned that the excessive water-reducing agents could cause a reduction in the adhesiveness of cementitious composites, leading to an adverse effect in the durability and mechanical performances of GBCCs [12–15]. Therefore, it is essential to conduct many more experiments to optimize the contents of water-reducing agents in GBCCs.

9.2.6 Sustainability of nanoengineered cementitious composites

Although the conductive nanoparticles in the cementitious composites could promote the properties and provoke the smart functions of the cementitious composites, there is still a large scale of fundamental research about the sustainability effect on the modified cementitious composites. A life cycle assessment (LCA) for the substantial effect of GBCCs has recently been studied by Papanikolaou et al. [16]. They illustrated that the manufacture of 1.0 kg GNPs coincided with 0.17 kg CO_2 comparable, lower than the production of 1.0 kg of OPC, which coincides to 0.86 kg CO_2. They also demonstrated that GNPs have little effect on the environment and could be recognized as beneficial nanomaterials in cementitious composites to decrease the emission of CO_2 by decreasing the relative cement volume while simultaneously keeping the similar performance of the mechanical performance and durability. Nevertheless, they have not taken the decrease of CO_2 emission in the field of preservation of graphene-based cementitious composites into consideration. Therefore, it is necessary to conduct more experiments on the sustainability of GBCCs. In addition, Papanikolaou et al. [16] also stated that the applications of GBNs can cause some damage to human health in all procedures dealing with those materials. It is difficult to remove all nanomaterials from the human body, which could increase the possibility of affecting the lung system [17]. Thus, more related experiments about the toxicology of GBNs are urgent to be conducted. It also displays the schematic diagram of the sustainability of smart 2D GBNs-based civil infrastructure. In order to promote the sustainable application of 2D GBNs in practical civil projects, the following four aspects are supposed to be primarily considered: for the materials design, the 2D GBNs-based concrete structure based on GBNs should have the characteristics of good dispersion, low price, low carbon, and superior performance; for structure design, the properties of 2D GBCCs in actual structure should be appropriately adjusted according to the test objects where they will be used; for data collection, the self-sensing properties of 2D GBNs-based civil infrastructure are supposed to be combined with the wireless technology to realize the data cloud for further analysis; for data process, finite element analysis software with open-source code editing techniques are expected to be used to analyze data for structural forensics.

9.3 FUTURE DEVELOPMENTS

9.3.1 Durability and long-term performance

The durability and long-term performance of CBSs are critical considerations for their effective deployment in structural health monitoring and other applications. A large number of CBSs utilizing different types of cement, geopolymers, and conductive fillers have been developed. They need to be compatible with the concrete structures or construction material in which they are embedded. Currently, there is limited research on the compatibility of CBSs with structures. However, the compatibility issues may arise over time, potentially leading to a decline in performance and thereby affecting the accuracy and reliability of sensor data [18, 19]. In addition, the design of CBSs should be able to withstand the environmental conditions correspondingly, including resistance to humidity, temperature fluctuations, exposure to chemicals, and other external factors that may influence sensor's performance [20]. In particular, corrosion is a significant concern for sensors embedded in concrete, especially in structures exposed to harsh environments or corrosive substances [21]. Therefore, appropriate sealing and encapsulation methods are indispensable to protect the sensor from water ingress and other environmental contaminants. This involves enhancing its internal pore structure and hydrophobicity and adding external coatings [22].

9.3.2 Self-powered CBSs and wireless sensing system

Self-powered sensors are an important trend in the future development of sensors, and the CBSs are no exception. Developing a self-powered CBSs involves integrating energy harvesting technologies into the sensor design, and choosing an appropriate energy harvesting technology based on the sensor's environment and requirements is critical for the sensors' long-term application. Currently, there are four main energy harvesting technologies: thermoelectric generator (TEG), piezoelectric material, solar cell, and vibration harvester. In particular, a TEG is a device that directly converts heat energy into electrical power based on the Seebeck effect. This effect occurs when a temperature gradient across a material leads to the generation of an electric voltage. The piezoelectric material exhibits the piezoelectric effect, which is the ability to generate an electric charge in response to mechanical stress or deformation. The solar cell or photovoltaic cell is a device that directly transforms sunlight into electricity through the photovoltaic effect. This mechanism entails the creation of an electric current when specific materials are illuminated. The vibration harvester is a device designed to capture and convert mechanical vibrations or oscillations into electrical energy, which can harness ambient vibrations from the surrounding environment and convert them into usable electrical power. The development of self-powered CBSs involves incorporating the specific energy harvesting technology into the design of the CBS. Previous studies have demonstrated that the CBSs with conductive nanoparticles possess the piezoelectric property that can be applied for the development of self-powered and self-diagnostic materials [23–25]. Another key component is power storage equipment such as a rechargeable battery or a supercapacitor to store the harvested energy for later use. It is also vital to design an energy management system that efficiently controls the harvested energy, charging and discharging the storage component as needed. This circuitry should be designed to optimize power usage and extend the CBS's operational life.

A wireless CBS is a sensor embedded in concrete structures, engineered to function without the necessity of physical wires for data transmission. These CBSs utilize wireless communication technologies to send data to a central monitoring system or other devices. The incorporation of wireless technology enhances the ease of deployment and reduces the complexity associated with traditional wired sensor systems. The CBSs can incorporate a wireless communication module, such as Bluetooth, Wi-Fi, or other wireless protocols, enabling seamless communication with external devices.

9.3.3 Integration of CBSs to IoT technology

The IoT technology can be widely applied to civil infrastructure monitoring, such as intelligent traffic systems, environmental monitoring, and waste management, to improve the efficiency of city operations. Integrating CBSs into the IoT involves combining CBSs technology with communication and data processing capabilities. As mentioned, the CBSs are commonly utilized in construction for SHM and providing real-time data on the condition of concrete structures. To combine them together, the first preparation is to collect the specific CBSs for stress/strain/temperature/humidity monitoring. These sensors should be located in the key position of concrete structures for perfect monitoring. Before connecting to IoT, a communication protocol suitable for the application should be selected (e.g., CoAP and HTTP), ensuring compatibility with both the network infrastructure and IoT platform intended to employ. Afterward, an IoT platform that aligns with the goals of project should be selected, which offers the concrete structures for data storage, analysis, and visualization [26].

9.3.4 Integration of CBSs to BIM

The role of CBSs in building is to monitor and record key parameters related to concrete structures. BIM is a digital representation of the physical and functional characteristics of a building. Incorporating CBSs into BIM entails merging sensor data with the digital depiction of a structure's physical and functional features, which facilitates the real-time acquisition of insights into the SHM and performance of a building [27, 28]. After the selection of CBSs with specific capacity, these sensors should be placed within the key concrete structure, such as load-bearing elements, and areas prone to stress. Then, a DAQ system designed to gather information from CBSs should be established. This system must possess the capability to process and transmit the collected sensor data to a central repository for further analysis and standardization. The proper BIM software that facilitates the integration of sensor data can be selected, since some BIM platforms come equipped with dedicated functionalities or plugins specifically tailored for the management of sensor information. The integration of CBSs to BIM means the establishment of BIM objects or components that depict the CBSs in the digital model. These objects should encompass pertinent parameters and properties linked to the sensor data. Once the connection between the sensor data and the corresponding elements in the BIM model is achieved, this connection enables the visualization of real-time data within the digital representation of the building. The last step will be the visualization and analysis tools within the BIM software to interpret and analyze the sensor data. This process aids in identifying potential issues, monitoring structural health, and making informed decisions regarding maintenance or interventions.

9.3.5 Standardization and certification

With the increasing development of CBSs, there may be a push for standardization and certification processes to ensure the reliability and accuracy of these sensors in diverse applications. Standardization of CBSs involves establishing guidelines and specifications to ensure consistent quality, performance, and reliability of these sensors. The specific content of standard can vary based on factors such as the type of sensor, intended application, and regional requirements. In addition, the standard should include the specific materials used in the construction of the CBSs and outline manufacturing requirements to ensure uniformity and quality control. The crucial performance of CBSs is expected to demonstrate accuracy, sensitivity, response time, and durability in diverse environmental conditions. The standard should provide guidance for the correct installation of CBSs to guarantee precise and dependable measurements. This might involve suggestions for CBSs placement, substrate preparation, and installation procedures. The calibration and maintenance of CBSs should also be summarized, with the detailed procedures for calibrating the CBSs to maintain accuracy and routine maintenance over time. Other standards include the testing and evaluation methods, data output, safety and environmental consideration, etc.

The certification process for CBSs involves several stages to ensure that the sensors meet specific standards and criteria, involving pre-assessment, application submission, testing and evaluation, documentation review, etc. Manufacturers should opt for a certification body with industry recognition and specialized knowledge in the relevant standards for CBSs. The certification procedure plays a vital role in increasing confidence in the sensors' quality and reliability, thereby increasing their marketability and trustworthiness among potential users.

9.4 SUMMARY

CBSs show potential for monitoring the structural health of concrete infrastructure, but they face several challenges that researchers are actively addressing for future advancements. This chapter presents the challenges and future developments of CBSs; addressing these challenges and advancing research in these directions can contribute to the broader adoption and effectiveness of CBSs in ensuring the structural integrity and safety of concrete infrastructure. Some key conclusions are as follows:

1. Environmental interference can vary based on the purpose of CBSs, so that the CBSs can also work as temperature or humidity sensor for environmental sensing. The DAQ system for CBSs can vary based on the application, the type of sensors used, and the monitoring requirements of the concrete structure. These systems are crucial for ensuring the reliability and effectiveness of structural health monitoring in various engineering applications.
2. Dispersion of nanomaterials relates to the piezoresistivity and sensitivity of CBSs, the nanoparticle agglomeration should be removed in the CBSs as well as the adverse effects on workability of cement slurry. Considering the toxicity of some nanomaterials such as CNT or GNP, the health concerns must be solved for researchers, workforces, producers, and users. The sustainability evaluation of CBSs such as LCA should also be featured for sustainable construction.
3. Future development should figure out the sensitivity, reliability, sustainability, and serviceability of CBSs. Therefore, it includes the durability and long-term performance enhancement, the development of self-powered CBSs, and wireless sensing technologies. To combine the CBSs with smart city application, the integration of CBSs to IoT and BIM technology should be the next step. In addition, the standardization and certification are vital for the commercialization of CBSs, which will be indispensable in future studies.

REFERENCES

1. D. Jang, J. Bang, H. Jeon, Impact of silica aerogel addition on the electrical and piezo-resistive sensing stability of CNT-embedded cement-based sensors exposed to varied environments, Journal of Building Engineering 78 (2023) 107700.
2. B. del Moral, F.J. Baeza, R. Navarro, O. Galao, E. Zornoza, J. Vera, C. Farcas, P. Garcés, Temperature and humidity influence on the strain sensing performance of hybrid carbon nanotubes and graphite cement composites, Construction and Building Materials 284 (2021) 122786.

3. W. Dong, W. Li, K. Wang, Z. Luo, D. Sheng, Self-sensing capabilities of cement-based sensor with layer-distributed conductive rubber fibres, Sensors and Actuators A: Physical 301 (2020) 111763.
4. A. Monteiro, A. Loredo, P. Costa, M. Oeser, P. Cachim, A pressure-sensitive carbon black cement composite for traffic monitoring, Construction and Building Materials 154 (2017) 1079–1086.
5. Z. Deng, W. Li, W. Dong, Z. Sun, J. Kodikara, D. Sheng, Multifunctional asphalt concrete pavement toward smart transport infrastructure: Design, performance and perspective, Composites Part B: Engineering 265 (2023) 110937.
6. T. Wang, S. Faßbender, W. Dong, C. Schulze, M. Oeser, P. Liu, Sensitive surface layer: A review on conductive and piezoresistive pavement materials with carbon-based additives, Construction and Building Materials 387 (2023) 131611.
7. W. Dong, W. Li, N. Lu, F. Qu, K. Vessalas, D. Sheng, Piezoresistive behaviours of cement-based sensor with carbon black subjected to various temperature and water content, Composites Part B: Engineering 178 (2019) 107488.
8. W. Dong, W. Li, Z. Sun, I. Ibrahim, D. Sheng, Intrinsic graphene/cement-based sensors with piezoresistivity and superhydrophobicity capacities for smart concrete infrastructure, Automation in Construction 133 (2022) 103983.
9. J. Ou, B. Han, Piezoresistive cement-based strain sensors and self-sensing concrete components, Journal of Intelligent Material Systems and Structures 20(3) (2009) 329–336.
10. R. Mu, H. Li, L. Qing, J. Lin, Q. Zhao, Aligning steel fibers in cement mortar using electromagnetic field, Construction and Building Materials 131 (2017) 309–316.
11. W. Dong, Y. Huang, B. Lehane, G. Ma, XGBoost algorithm-based prediction of concrete electrical resistivity for structural health monitoring, Automation in Construction 114 (2020) 103155.
12. G. Jing, K. Xu, H. Feng, J. Wu, S. Wang, Q. Li, X. Cheng, Z. Ye, The non-uniform spatial dispersion of graphene oxide: A step forward to understand the inconsistent properties of cement composites, Construction and Building Materials 264 (2020) 120729.
13. J. Lin, E. Shamsaei, F.B. de Souza, K. Sagoe-Crentsil, W.H. Duan, Dispersion of graphene oxide–silica nanohybrids in alkaline environment for improving ordinary Portland cement composites, Cement and Concrete Composites 106 (2020) 103488.
14. J. Liu, J. Fu, Y. Yang, C. Gu, Study on dispersion, mechanical and microstructure properties of cement paste incorporating graphene sheets, Construction and Building Materials 199 (2019) 1–11.
15. W. Qin, Q. Guodong, Z. Dafu, W. Yue, Z. Haiyu, Influence of the molecular structure of a polycarboxylate superplasticiser on the dispersion of graphene oxide in cement pore solutions and cement-based composites, Construction and Building Materials 272 (2021) 121969.
16. I. Papanikolaou, N. Arena, A. Al-Tabbaa, Graphene nanoplatelet reinforced concrete for self-sensing structures: A lifecycle assessment perspective, Journal of Cleaner Production 240 (2019) 118202.
17. W. Dong, W. Li, Z. Tao, K. Wang, Piezoresistive properties of cement-based sensors: Review and perspective, Construction and Building Materials 203 (2019) 146–163.
18. W. Li, W. Dong, Y. Guo, K. Wang, S.P. Shah, Advances in multifunctional cementitious composites with conductive carbon nanomaterials for smart infrastructure, Cement and Concrete Composites 128 (2022) 104454.
19. W. Li, F. Qu, W. Dong, G. Mishra, S.P. Shah, A comprehensive review on self-sensing graphene/cementitious composites: A pathway toward next-generation smart concrete, Construction and Building Materials 331 (2022) 127284.
20. W. Dong, W. Li, Y. Guo, F. Qu, K. Wang, D. Sheng, Piezoresistive performance of hydrophobic cement-based sensors under moisture and chloride-rich environments, Cement and Concrete Composites 126 (2022) 104379.
21. W. Dong, W. Li, K. Vessalas, X. He, Z. Sun, D. Sheng, Piezoresistivity deterioration of smart graphene nanoplate/cement-based sensors subjected to sulphuric acid attack, Composites Communications 23 (2021) 100563.
22. W. Dong, W. Li, X. Zhu, D. Sheng, S.P. Shah, Multifunctional cementitious composites with integrated self-sensing and hydrophobic capacities toward smart structural health monitoring, Cement and Concrete Composites 118 (2021) 103962.

23. D.A. Triana-Camacho, J.H. Quintero-Orozco, E. Mejía-Ospino, G. Castillo-López, E. García-Macías, Piezoelectric composite cements: Towards the development of self-powered and self-diagnostic materials, Cement and Concrete Composites 139 (2023) 105063.
24. D.A. Triana-Camacho, R. Ospina-Ospina, J.H. Quintero-Orozco, Method for fabricating self-powered cement sensors based on gold nanoparticles, MethodsX 11 (2023) 102280.
25. Y.-F. Su, R.R. Kotian, N. Lu, Energy harvesting potential of bendable concrete using polymer based piezoelectric generator, Composites Part B: Engineering 153 (2018) 124–129.
26. V.M. Reddy, S. Hamsalekha, V.S. Reddy, An IoT enabled real-time monitoring system for automatic curing and early age strength monitoring of concrete, AIP Conference Proceedings, AIP Publishing. 2021.
27. R. Volk, J. Stengel, F. Schultmann, Building Information Modeling (BIM) for existing buildings: Literature review and future needs, Automation in Construction 38 (2014) 109–127.
28. Y. Zou, A. Kiviniemi, S.W. Jones, A review of risk management through BIM and BIM-related technologies, Safety Science 97 (2017) 88–98.

Index

Note: Locators in *italics* represent figures and **bold** indicate tables in the text.

D

E

F

J

L

M

N

O

P

R

S

Y

Z